21世纪高等学校规划教材 | 计算机科学与技术

Struts 2 框架应用开发教程

王建国　编著

清华大学出版社
北京

内 容 简 介

Struts 2 框架是 MVC 设计模式的具体实现,是创建企业级 Java Web 应用的优雅的、可扩展的框架。借助于 Struts 2 框架可以减少在运用 MVC 设计模式开发 Web 应用时的时间。

本书共分 13 章,内容包括概述、Struts 2 框架入门、Struts 2 框架的配置、Struts 2 框架进阶、Struts 2 框架的 OGNL、Struts 2 框架的标签、Struts 2 框架的国际化、Struts 2 框架的类型转换、Struts 2 框架的拦截器、Struts 2 框架的输入校验、基于 Struts 2 框架的文件上传和下载、Struts 2 框架中使用数据库以及在线图片管理综合实例。本书结合具体实例介绍各个知识点,所使用的开发环境是 JDK 1.6＋MyEclipse 5＋Tomcat 6＋MySQL 5,逐步引领读者从基础到各个知识点的学习,并提供了大量的实例说明。

本书可以作为高等院校计算机及相关专业 Struts 2 框架课程学习的教材,也可以作为 Struts 2 框架培训班的培训教材,并且也适合网站开发人员参考使用。使用本书需要具有 Java Web 及数据库(MySQL、SQL Server 或 Oracle)开发的基础。

本书封面贴有清华大学出版社防伪标签,无标签者不得销售。
版权所有,侵权必究。 举报:010-62782989,beiqinquan@tup.tsinghua.edu.cn。

图书在版编目(CIP)数据

Struts 2 框架应用开发教程/王建国编著. —北京:清华大学出版社,2012.7(2024.12重印)
(21 世纪高等学校规划教材·计算机科学与技术)
ISBN 978-7-302-28469-7

Ⅰ. ①S… Ⅱ. ①王… Ⅲ. ①软件工具－程序设计－高等学校－教材 Ⅳ. ①TP311.56

中国版本图书馆 CIP 数据核字(2012)第 064405 号

责任编辑:高买花　薛　阳
封面设计:傅瑞学
责任校对:李建庄
责任印制:曹婉颖

出版发行:清华大学出版社
　　　　网　　址:https://www.tup.com.cn,https://www.wqxuetang.com
　　　　地　　址:北京清华大学学研大厦 A 座　　邮　编:100084
　　　　社 总 机:010-83470000　　邮　购:010-62786544
　　　　投稿与读者服务:010-62776969,c-service@tup.tsinghua.edu.cn
　　　　质量反馈:010-62772015,zhiliang@tup.tsinghua.edu.cn
　　　　课件下载:https://www.tup.com.cn,010-62795954
印 装 者:涿州市般润文化传播有限公司
经　　销:全国新华书店
开　　本:185mm×260mm　　印　张:23.5　　字　数:583 千字
版　　次:2012 年 7 月第 1 版　　印　次:2024 年 12 月第11次印刷
印　　数:7901~8200
定　　价:59.00 元

产品编号:043833-02

编审委员会成员

（按地区排序）

清华大学	周立柱	教授
	覃 征	教授
	王建民	教授
	冯建华	教授
	刘 强	副教授
北京大学	杨冬青	教授
	陈 钟	教授
	陈立军	副教授
北京航空航天大学	马殿富	教授
	吴超英	副教授
	姚淑珍	教授
中国人民大学	王 珊	教授
	孟小峰	教授
	陈 红	教授
北京师范大学	周明全	教授
北京交通大学	阮秋琦	教授
	赵 宏	副教授
北京信息工程学院	孟庆昌	教授
北京科技大学	杨炳儒	教授
石油大学	陈 明	教授
天津大学	艾德才	教授
复旦大学	吴立德	教授
	吴百锋	教授
	杨卫东	副教授
同济大学	苗夺谦	教授
	徐 安	教授
华东理工大学	邵志清	
华东师范大学	杨宗源	教授
	应吉康	教授
东华大学	乐嘉锦	教授
	孙 莉	副教授

浙江大学	吴朝晖	教授
	李善平	教授
扬州大学	李　云	教授
南京大学	骆　斌	教授
	黄　强	副教授
南京航空航天大学	黄志球	教授
	秦小麟	教授
南京理工大学	张功萱	教授
南京邮电学院	朱秀昌	教授
苏州大学	王宜怀	教授
	陈建明	副教授
江苏大学	鲍可进	教授
中国矿业大学	张　艳	教授
武汉大学	何炎祥	教授
华中科技大学	刘乐善	教授
中南财经政法大学	刘腾红	教授
华中师范大学	叶俊民	教授
	郑世珏	教授
	陈　利	教授
江汉大学	颜　彬	教授
国防科技大学	赵克佳	教授
	邹北骥	教授
中南大学	刘卫国	教授
湖南大学	林亚平	教授
西安交通大学	沈钧毅	教授
	齐　勇	教授
长安大学	巨永锋	教授
哈尔滨工业大学	郭茂祖	教授
吉林大学	徐一平	教授
	毕　强	教授
山东大学	孟祥旭	教授
	郝兴伟	教授
厦门大学	冯少荣	教授
厦门大学嘉庚学院	张思民	教授
云南大学	刘惟一	教授
电子科技大学	刘乃琦	教授
	罗　蕾	教授
成都理工大学	蔡　淮	教授
	于　春	副教授
西南交通大学	曾华燊	教授

出版说明

随着我国改革开放的进一步深化,高等教育也得到了快速发展,各地高校紧密结合地方经济建设发展需要,科学运用市场调节机制,加大了使用信息科学等现代科学技术提升、改造传统学科专业的投入力度,通过教育改革合理调整和配置了教育资源,优化了传统学科专业,积极为地方经济建设输送人才,为我国经济社会的快速、健康和可持续发展以及高等教育自身的改革发展做出了巨大贡献。但是,高等教育质量还需要进一步提高以适应经济社会发展的需要,不少高校的专业设置和结构不尽合理,教师队伍整体素质亟待提高,人才培养模式、教学内容和方法需要进一步转变,学生的实践能力和创新精神亟待加强。

教育部一直十分重视高等教育质量工作。2007年1月,教育部下发了《关于实施高等学校本科教学质量与教学改革工程的意见》,计划实施"高等学校本科教学质量与教学改革工程"(简称"质量工程"),通过专业结构调整、课程教材建设、实践教学改革、教学团队建设等多项内容,进一步深化高等学校教学改革,提高人才培养的能力和水平,更好地满足经济社会发展对高素质人才的需要。在贯彻和落实教育部"质量工程"的过程中,各地高校发挥师资力量强、办学经验丰富、教学资源充裕等优势,对其特色专业及特色课程(群)加以规划、整理和总结,更新教学内容、改革课程体系,建设了一大批内容新、体系新、方法新、手段新的特色课程。在此基础上,经教育部相关教学指导委员会专家的指导和建议,清华大学出版社在多个领域精选各高校的特色课程,分别规划出版系列教材,以配合"质量工程"的实施,满足各高校教学质量和教学改革的需要。

为了深入贯彻落实教育部《关于加强高等学校本科教学工作,提高教学质量的若干意见》精神,紧密配合教育部已经启动的"高等学校教学质量与教学改革工程精品课程建设工作",在有关专家、教授的倡议和有关部门的大力支持下,我们组织并成立了"清华大学出版社教材编审委员会"(以下简称"编委会"),旨在配合教育部制定精品课程教材的出版规划,讨论并实施精品课程教材的编写与出版工作。"编委会"成员皆来自全国各类高等学校教学与科研第一线的骨干教师,其中许多教师为各校相关院、系主管教学的院长或系主任。

按照教育部的要求,"编委会"一致认为,精品课程的建设工作从开始就要坚持高标准、严要求,处于一个比较高的起点上。精品课程教材应该能够反映各高校教学改革与课程建设的需要,要有特色风格、有创新性(新体系、新内容、新手段、新思路,教材的内容体系有较高的科学创新、技术创新和理念创新的含量)、先进性(对原有的学科体系有实质性的改革和发展,顺应并符合21世纪教学发展的规律,代表并引领课程发展的趋势和方向)、示范性(教材所体现的课程体系具有较广泛的辐射性和示范性)和一定的前瞻性。教材由个人申报或各校推荐(通过所在高校的"编委会"成员推荐),经"编委会"认真评审,最后由清华大学出版社审定出版。

目前，针对计算机类和电子信息类相关专业成立了两个"编委会"，即"清华大学出版社计算机教材编审委员会"和"清华大学出版社电子信息教材编审委员会"。推出的特色精品教材包括：

（1）21世纪高等学校规划教材·计算机应用——高等学校各类专业，特别是非计算机专业的计算机应用类教材。

（2）21世纪高等学校规划教材·计算机科学与技术——高等学校计算机相关专业的教材。

（3）21世纪高等学校规划教材·电子信息——高等学校电子信息相关专业的教材。

（4）21世纪高等学校规划教材·软件工程——高等学校软件工程相关专业的教材。

（5）21世纪高等学校规划教材·信息管理与信息系统。

（6）21世纪高等学校规划教材·财经管理与应用。

（7）21世纪高等学校规划教材·电子商务。

（8）21世纪高等学校规划教材·物联网。

清华大学出版社经过三十多年的努力，在教材尤其是计算机和电子信息类专业教材出版方面树立了权威品牌，为我国的高等教育事业做出了重要贡献。清华版教材形成了技术准确、内容严谨的独特风格，这种风格将延续并反映在特色精品教材的建设中。

清华大学出版社教材编审委员会
联系人：魏江江
E-mail：weijj@tup.tsinghua.edu.cn

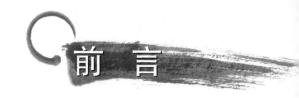

  本书针对 Struts 2 框架的编程进行了详细的介绍，以简单通俗易懂的示例，逐步引领读者从基础到各个知识点的学习。

  本书共分为 13 章。第 1 章为概述。主要介绍了 MVC 设计模式、Struts 2 框架的概念以及学习本书内容所需的基本开发环境。第 2 章为 Struts 2 框架技术入门。详细讲解了 Struts 2 框架的下载、安装以及基于 Struts 2 框架的开发环境的搭建，并通过一个示例体验了基于 Struts 2 框架的 Web 应用开发流程。第 3 章为 Struts 2 框架的配置。首先讲解了 Struts 2 框架的工作原理，然后详细讲解了 web.xml、struts.xml、struts.properties 等文件的配置。第 4 章为 Struts 2 框架进阶。主要讲解了 struts.xml 配置文件中 result 和 action 的配置，并讲解了 Struts 2 框架中的 Action 访问 Servlet API 的方法。第 5 章为 Struts 2 框架的 OGNL。讲解了 OGNL 的语法、OGNL 表达式和 OGNL 中的集合操作，最后重点讲解了 Struts 2 框架中 OGNL 的应用，并给出了具体示例。第 6 章为 Struts 2 框架的标签。介绍了模板和主题的概念，讲解了各个标签的使用。第 7 章为 Struts 2 框架的国际化。讲解了资源文件的格式、资源文件的分类及资源文件的加载顺序，重点介绍了如何在资源文件中使用参数及访问资源文件中的消息，并给出了基于 Struts 2 框架的 Web 应用的国际化体验示例。第 8 章为 Struts 2 框架的类型转换。介绍了 Struts 2 框架对类型转换的支持，重点强调了如何处理集合类型转换，并详细介绍了自定义类型转换器的开发步骤。第 9 章为 Struts 2 框架的拦截器。讲解了拦截器的工作过程和使用方法，通过一个示例讲解了自定义拦截器的开发步骤，详细讲解了 Struts 2 框架的内置拦截器的使用方法。第 10 章为 Struts 2 框架的输入校验。介绍了输入校验的分类，通过一个示例讲解了自定义校验器的开发步骤，详细讲解了 Struts 2 框架的内置校验器的使用方法，并介绍了输入校验国际化的方法。第 11 章为基于 Struts 2 框架的文件上传和下载。介绍了文件上传组件，重点介绍了文件上传的开发步骤，并详细介绍了基于 Struts 2 框架的文件下载应用开发。第 12 章为 Struts 2 框架中使用数据库。这是基于 Struts 2 框架的 Web 应用开发中非常重要的内容之一，讲解了连接 MySQL、Oracle 和 SQL Server 数据库的方式，并详细介绍了使用数据库的开发过程。第 13 章是一个综合示例：在线图片管理，按照实际 Web 应用的开发步骤，采用面向接口编程，讲解了基于 Struts 2 框架的 Web 应用开发方法。

  参加本书编写的人员有王建国、王建英和李小红。由于作者的水平有限，书中的错误和不妥之处在所难免，敬请读者批评指正。

王建国
2012 年 5 月

目 录

第1章 概述 ·· 1

1.1 JSP 模型 ·· 1
   1.1.1 JSP 模型 1 ·· 1
   1.1.2 JSP 模型 2 ·· 1
1.2 MVC 设计模式 ·· 2
1.3 Struts 2 框架的定义 ··· 2
   1.3.1 Servlet Filters ··· 3
   1.3.2 Struts Core ··· 4
   1.3.3 Interceptors ·· 5
   1.3.4 User Created ·· 5
1.4 基本开发环境 ·· 5
   1.4.1 JDK 的下载与安装 ·· 5
   1.4.2 Tomcat 服务器的下载与安装 ··· 7
   1.4.3 MyEclipse 的下载与安装 ··· 10
习题 ·· 16

第2章 Struts 2 框架技术入门 ·· 17

2.1 Struts 2 框架的下载与安装 ·· 17
   2.1.1 Struts 2 框架的下载 ·· 17
   2.1.2 Struts 2 框架的安装 ·· 19
2.2 搭建基于 Struts 2 框架的 Web 应用开发环境 ·· 21
2.3 基于 Struts 2 框架的 Web 应用示例 ··· 22
   2.3.1 创建视图 ·· 23
   2.3.2 创建业务控制器类 ··· 24
   2.3.3 创建 struts.xml 文件 ··· 25
   2.3.4 编辑 web.xml 文件 ··· 25
   2.3.5 Web 项目的发布与测试 ·· 26
2.4 基于 Struts 2 框架的 Web 应用开发流程总结 ·· 28
习题 ·· 28

第3章 Struts 2 框架的配置 ··· 29

3.1 Struts 2 框架的体系结构与工作原理 ·· 29

3.1.1 Struts 2 框架的体系结构 ……………………………… 29
3.1.2 Struts 2 框架的工作原理 ……………………………… 30
3.2 配置 web.xml ……………………………… 30
3.2.1 配置 Struts 2 框架的核心控制器 ……………………………… 30
3.2.2 配置第三方过滤器框架 ……………………………… 31
3.2.3 配置初始化参数 ……………………………… 32
3.3 配置 struts.xml ……………………………… 35
3.3.1 struts.xml 文件的基本框架 ……………………………… 35
3.3.2 package 及其包含的子元素 ……………………………… 37
3.3.3 include 子元素 ……………………………… 45
3.3.4 bean 子元素 ……………………………… 46
3.3.5 constant 子元素 ……………………………… 47
3.4 配置 struts.properties ……………………………… 49
3.4.1 开发模式属性 ……………………………… 49
3.4.2 国际化属性 ……………………………… 51
3.4.3 文件上传属性 ……………………………… 51
3.4.4 模板和主题属性 ……………………………… 52
3.4.5 url 属性 ……………………………… 53
3.4.6 freemarker 属性 ……………………………… 54
3.4.7 velocity 属性 ……………………………… 54
3.4.8 ognl 属性 ……………………………… 54
3.4.9 其他属性 ……………………………… 55
3.5 配置 struts-default.xml ……………………………… 59
3.6 配置 struts-plugin.xml ……………………………… 60
习题 ……………………………… 61

第 4 章 Struts 2 框架进阶 ……………………………… 62

4.1 result 配置 ……………………………… 63
4.1.1 Struts 2 框架中内置的 result 类型 ……………………………… 63
4.1.2 缺省配置 ……………………………… 70
4.1.3 "其他"result 配置 ……………………………… 71
4.1.4 动态 result 配置 ……………………………… 71
4.1.5 局部和全局 result 配置 ……………………………… 73
4.2 action 配置 ……………………………… 73
4.2.1 默认类 ……………………………… 73
4.2.2 method 属性 ……………………………… 74
4.2.3 动态方法调用 ……………………………… 77
4.2.4 默认的 action ……………………………… 78
4.2.5 默认的通配符 ……………………………… 79

		4.2.6 使用 param 子元素为 action 传递参数 ………………………………… 80

- 4.3 Struts 2 框架中的 Action ………………………………………………………… 81
  - 4.3.1 ActionSupport 类 ………………………………………………………… 81
  - 4.3.2 Action 访问 Servlet API ………………………………………………… 82
- 习题 ………………………………………………………………………………………… 85

第 5 章 Struts 2 框架的 OGNL ……………………………………………………… 86

- 5.1 OGNL 简介 ……………………………………………………………………… 86
- 5.2 OGNL 语法 ……………………………………………………………………… 86
- 5.3 OGNL 表达式 …………………………………………………………………… 87
  - 5.3.1 常量 ……………………………………………………………………… 87
  - 5.3.2 属性访问 ………………………………………………………………… 88
  - 5.3.3 操作符 …………………………………………………………………… 88
  - 5.3.4 设置值和检索值 ………………………………………………………… 89
  - 5.3.5 访问静态方法和字段 …………………………………………………… 89
  - 5.3.6 索引 ……………………………………………………………………… 90
  - 5.3.7 括号表达式 ……………………………………………………………… 91
  - 5.3.8 链接子表达式 …………………………………………………………… 91
  - 5.3.9 变量访问 ………………………………………………………………… 91
  - 5.3.10 表达式计算 ……………………………………………………………… 92
  - 5.3.11 Lambda 表达式 ………………………………………………………… 92
- 5.4 OGNL 的集合操作 ……………………………………………………………… 92
  - 5.4.1 创建集合 ………………………………………………………………… 93
  - 5.4.2 投影 ……………………………………………………………………… 94
  - 5.4.3 选择 ……………………………………………………………………… 94
- 5.5 Struts 2 中的 OGNL …………………………………………………………… 95
  - 5.5.1 值栈 ……………………………………………………………………… 95
  - 5.5.2 索引 ……………………………………………………………………… 96
  - 5.5.3 使用 top 访问栈顶对象 ………………………………………………… 96
  - 5.5.4 访问静态属性 …………………………………………………………… 96
  - 5.5.5 Struts 2 框架的命名对象 ……………………………………………… 97
  - 5.5.6 访问 Action 属性 ……………………………………………………… 97
  - 5.5.7 集合 ……………………………………………………………………… 98
- 5.6 OGNL 应用示例 ………………………………………………………………… 98
- 习题 ……………………………………………………………………………………… 101

第 6 章 Struts 2 框架的标签 ………………………………………………………… 102

- 6.1 普通标签 ………………………………………………………………………… 103
  - 6.1.1 控制标签 ………………………………………………………………… 103

		6.1.2	数据标签	109
	6.2	模板和主题		122
		6.2.1	模板	122
		6.2.2	主题	123
	6.3	UI 标签		125
		6.3.1	表单标签的公共属性	125
		6.3.2	表单标签	128
		6.3.3	非表单标签	138
		6.3.4	Ajax 标签	142
	习题			165

第 7 章 Struts 2 框架的国际化 167

7.1	资源文件		167
	7.1.1	名称格式	167
	7.1.2	资源文件的内容	168
7.2	基于 Struts 2 框架的 Web 应用的国际化体验		168
7.3	如何在资源文件中使用参数		173
7.4	访问资源文件中消息的方式		175
	7.4.1	在 Action 中访问资源文件中的消息	175
	7.4.2	在 JSP 页面中访问资源文件中的消息	177
	7.4.3	在表单标签中访问资源文件中的消息	178
	7.4.4	在资源文件中访问资源文件中的消息	178
7.5	资源文件的分类		179
	7.5.1	包资源文件	179
	7.5.2	Action 资源文件	180
7.6	资源文件的加载顺序		181
习题			183

第 8 章 Struts 2 框架的类型转换 184

8.1	Struts 2 框架对类型转换的支持		184
8.2	Struts 2 框架内置的类型转换器		185
8.3	类型转换体验		185
8.4	处理 List 类型转换		189
8.5	处理 Map 类型转换		192
8.6	自定义类型转换器		194
	8.6.1	创建基于 OGNL 的类型转换器	194
	8.6.2	创建基于 Struts 2 框架的类型转换器	196
习题			199

第 9 章　Struts 2 框架的拦截器 …… 200

9.1　拦截器概述 …… 200
9.1.1　AOP …… 200
9.1.2　拦截器的作用 …… 201
9.2　拦截器的工作过程 …… 202
9.3　拦截器的使用方法 …… 202
9.3.1　创建拦截器类 …… 203
9.3.2　注册拦截器 …… 204
9.3.3　使用拦截器 …… 204
9.4　自定义拦截器示例 …… 205
9.4.1　拦截器工作过程示例 …… 205
9.4.2　登录示例 …… 209
9.5　Struts 2 框架的内置拦截器 …… 213
9.5.1　内置的拦截器 …… 213
9.5.2　内置的拦截器栈 …… 230
9.5.3　内置拦截器的配置 …… 234
习题 …… 235

第 10 章　Struts 2 框架的输入校验 …… 236

10.1　输入校验概述 …… 236
10.2　服务器端输入校验 …… 237
10.2.1　使用编码进行输入校验 …… 237
10.2.2　使用配置文件进行输入校验 …… 240
10.3　客户端输入校验 …… 246
10.4　Ajax 校验 …… 249
10.5　输入校验的国际化 …… 252
10.6　Struts 2 框架的内置校验器 …… 254
10.6.1　类型转换校验器 …… 255
10.6.2　日期校验器 …… 255
10.6.3　双精度浮点数校验器 …… 256
10.6.4　电子邮件校验器 …… 257
10.6.5　表达式校验器 …… 257
10.6.6　字段表达式校验器 …… 257
10.6.7　整型校验器 …… 258
10.6.8　正则表达式校验器 …… 259
10.6.9　必填校验器 …… 259
10.6.10　必填字符串校验器 …… 260
10.6.11　字符串长度校验器 …… 260

10.6.12 网址校验器 261
10.6.13 visitor 校验器 261
10.6.14 conditionalvisitor 校验器 262
10.7 自定义校验器 263
10.7.1 创建自定义校验器类 263
10.7.2 注册自定义校验器类 264
10.7.3 使用自定义校验器 265
习题 267

第 11 章 基于 Struts 2 框架的文件上传和下载 268

11.1 文件上传概述 268
11.1.1 文件上传组件 268
11.1.2 基于 Struts 2 框架的文件上传开发体验 269
11.2 上传单个文件 271
11.2.1 不对保存上传文件的目录进行硬编码 271
11.2.2 使用新名称保存上传的文件 272
11.2.3 对上传文件的大小及类型进行限制 272
11.2.4 上传文件属性的配置 273
11.2.5 对上传文件错误消息进行国际化 273
11.2.6 上传单个文件示例 274
11.3 上传多个文件 277
11.3.1 使用数组方式实现多文件上传 277
11.3.2 使用 List 方式实现多文件上传 280
11.4 文件下载概述 284
11.5 基于 Struts 2 框架的文件下载 286
11.6 任意内容类型的文件下载 288
11.7 统计文件下载的次数 290
习题 293

第 12 章 Struts 2 框架中使用数据库 294

12.1 连接数据库 294
12.1.1 连接 MySQL 数据库 294
12.1.2 连接 Oracle 数据库 295
12.1.3 连接 SQL Server 数据库 297
12.2 MySQL 数据库的下载与安装 298
12.3 连接测试 300
12.4 使用数据库的示例 304
12.4.1 创建国际化资源文件 304
12.4.2 创建数据库操作的辅助类 305

	12.4.3 创建 Action 类	309
	12.4.4 创建输入校验的配置文件	312
	12.4.5 编辑配置文件	313
	12.4.6 创建 JSP 文件	314
	12.4.7 测试	316
习题		318

第 13 章 在线图片管理 ... 319

13.1 概述 ... 319
13.1.1 功能简介 ... 319
13.1.2 总体设计 ... 320

13.2 准备工作 ... 320
13.2.1 创建数据库和表 ... 320
13.2.2 使用 Log4j 输出信息 ... 321
13.2.3 国际化 ... 321
13.2.4 异步交互 ... 322
13.2.5 数据库配置 ... 322
13.2.6 Web 应用的目录结构 ... 322

13.3 辅助类 ... 323
13.3.1 封装数据库操作的辅助类 ... 323
13.3.2 数据分页的辅助类 ... 328

13.4 实现数据模型 ... 329
13.4.1 用户数据模型 ... 329
13.4.2 图片数据模型 ... 330

13.5 实现 DAO 层 ... 331
13.5.1 定义 DAO 层接口 ... 331
13.5.2 实现 DAO 层接口 ... 332

13.6 实现业务逻辑层 ... 337
13.6.1 定义 Service 层接口 ... 337
13.6.2 实现 Service 层接口 ... 338

13.7 实现控制器 Action ... 340
13.7.1 管理用户的控制器 ... 340
13.7.2 管理图片的控制器 ... 342

13.8 编写配置文件 ... 347
13.8.1 struts.xml ... 347
13.8.2 输入校验文件 ... 348

13.9 编写 JSP 文件 ... 349
13.9.1 操作入口界面 ... 349
13.9.2 用户注册界面 ... 350

13.9.3 用户登录界面 …… 351
13.9.4 用户列表界面 …… 352
13.9.5 图片上传与编辑界面 …… 352
13.9.6 图片列表界面 …… 353
13.9.7 图片查找界面 …… 354
13.10 测试 …… 355
13.10.1 上传图片 …… 355
13.10.2 显示图片 …… 355
13.10.3 查找图片 …… 356

参考文献 …… 357

第 1 章 概述

Struts 2 框架是 MVC 设计模式的一个具体实现。本章首先介绍了 JSP 开发的模型,然后介绍了 MVC 设计模式以及什么是 Struts 2 框架,最后介绍了基本的开发环境。

1.1 JSP 模型

在使用 JSP 技术进行 Web 应用开发时,主要采用 JSP 模型 1 和 JSP 模型 2 两种模型。具体采用哪种模型则根据具体的 Web 应用而定。

1.1.1 JSP 模型 1

JSP 模型 1 采用 JSP+JavaBean 的开发模式,实现了数据显示(视图层)和处理(业务逻辑)的分离,如图 1-1 所示。在 JSP+JavaBean 这种模式中,JavaBean 既要提供设置和获取数据的方法,又要提供数据的处理方法。因此,这种方式不适合复杂的 Web 应用开发。

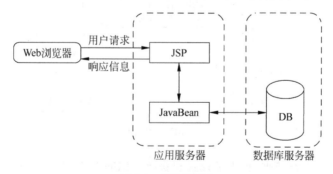

图 1-1 JSP 模型 1

1.1.2 JSP 模型 2

JSP 模型 2 采用 JSP+Servlet+JavaBean 的开发模式,如图 1-2 所示。在这种开发模式中,JavaBean 只负责提供设置和获取数据的方法,而对于数据的处理则由控制器(Servlet)来负责,即控制器负责处理数据,并将处理结果保存到 JavaBean 中,实现了数据处理与存储的分离。JSP 模型 2 具有清晰的页面角色和开发角色划分,可以使团队开发人员的优势得到充分发挥,在大型 Web 应用开发中表现出色。

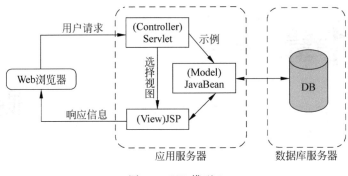

图 1-2 JSP 模型 2

JSP 模型 2 充分体现了 MVC(模型、视图、控制器)的设计模式,目前应用广泛的 Struts 框架就是基于 JSP 模型 2 的。

1.2 MVC 设计模式

MVC 设计模式把一个应用软件系统划分成相对独立又协同工作的 3 部分,即模型 (Model)、视图(View)和控制器(Controller),这 3 部分分别担负着不同的功能。这样划分的好处是:当软件系统需求发生变化时,只需改变其中的一层就可以满足这种变化要求,使系统的维护变得容易。图 1-3 显示了 3 个层次之间的关系及各自的功能。

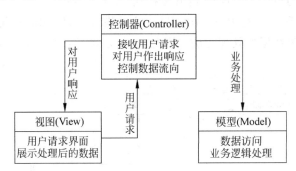

图 1-3 MVC 设计模式

(1) 模型:用于存储数据,提供了设置和获取数据的方法,即属性的 set 和 get 方法。

(2) 视图:用于向控制器提供必要的数据和显示保存在 JavaBean 中的数据,也就是用户的交互界面。

(3) 控制器:用于负责具体的业务逻辑操作,接收用户请求,对数据进行处理,并将处理的结果保存在 JavaBean 中,即负责处理模型和视图之间的数据流向和转换。

1.3 Struts 2 框架的定义

Struts 2 框架是 MVC 设计模式的具体实现,是创建企业级 Java Web 应用的优雅的、可扩展的框架。Struts 2 框架设计的目标贯穿整个开发周期,从构建到部署再到应用维护的整个过程。

Struts 2 框架实质上是在 WebWork 2 框架的基础上发展而来的，和 Struts 1 框架有着很大的不同。对于熟悉 WebWork 框架技术的开发人员来说，使用 Struts 2 框架进行 Web 应用开发会更得心应手，而对于熟悉 Struts 1 框架技术的开发人员来说，使用 Struts 2 框架进行 Web 应用开发则是一个全新的体验。

图 1-4 是 Struts 2 框架的体系结构图。

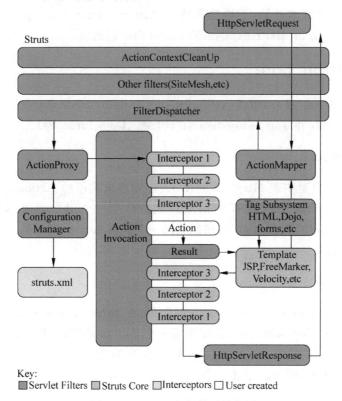

图 1-4　Struts 2 框架体系结构图

由图 1-4 可以看出，Struts 2 框架由 Servlet Filters、Struts Core、Interceptors 和 User created 四大模块组成。各个模块的介绍如下。

1.3.1　Servlet Filters

Servlet Filters(Servlet 过滤器)包括 ActionContextCleanUp、Other filters(如 SiteMesh 等)和 FilterDispatcher 等。

1. ActionContextCleanUp 过滤器

该过滤器主要用来和 FilterDispatcher 过滤器一起集成其他的第三方过滤器，如 SiteMesh 等。该过滤器为可选项，并且在 Struts 2.1.3 之后不推荐使用。

2. Other filters 过滤器

一般是指第三方过滤器框架。该过滤器是可选项。

3. FilterDispatcher 过滤器

该过滤器是 Struts 2 框架的核心控制器，它负责拦截所有的用户请求，当用户请求到达时，该过滤器就会对用户的请求进行过滤。如果用户的请求是以 action 作为后缀，则该请求将被转入 Struts 2 框架进行处理。FilterDispatcher 过滤器主要通过 AcionMapper 来决定调用哪一个 Action。该过滤器为必选项。

从 Struts 2.1.3 版本开始，不再推荐使用 ActionContextCleanUp 和 FilterDispatcher 过滤器，而是推荐使用 StrutsPrepareAndExecuteFilter 过滤器。

1.3.2 Struts Core

Struts Core(Struts 核心)包括 ActionProxy、ConfigurationManager、ActionInvocation、ActionMapper、Result 和 Tag Subsystem(如 HTML、Dojo、forms 等)。

1. ActionMapper

ActionMapper 负责实现 HttpServletRequest 和请求调用的 Action 之间的映射，它实现了 Aciton 类和 Servlet API 之间的彻底解耦。对于一个用户请求，如果没有匹配的 Action，ActionMapper 会返回 null，否则就返回一个描述 Action 调用的 ActionMapping 给 Struts 2 框架使用。

2. ActionProxy

ActionProxy 负责创建 ActionInvocation 实例，在 Struts 2 框架和真正的 Action 之间充当代理的角色。在 ActionMapper 决定激活一个 Action 的时候，FilterDispatcher 会将控制权委托给 ActionProxy。ActionProxy 查阅框架的配置文件管理器（ConfigurationManager）。然后，ActionProxy 创建一个 ActionInvocation。

3. ActionInvocation

当一个请求被调用时，ActionInvocation 确定如何执行 Action，包括在激活 Action 本身之前激活任意的拦截器(Interceptors)。并且在 Action 执行完成后，还负责根据 Action 的返回结果码(Result)在 struts.xml 配置文件中查找对应的视图资源。

4. Result

Action 的返回结果是一个字符串，字符串的类型用于在 struts.xml 配置文件中选择配置的 Result 元素。

5. ConfigurationManager

ConfigurationManager 使用 struts.xml 文件进行初始化，负责配置文件的初始化及管理。

6. Tag Subsystem

Tag Subsystem 提供了不依赖于视图层的标签库，并且标签库中的大部分标签都可以用于 JSP、FreeMarker 和 Velocity 等模板。

1.3.3 Interceptors

Interceptors(拦截器)是 Struts 2 框架的核心内容之一,拦截器能够在 Action 被调用的前后执行代码。Struts 2 框架的大多数核心功能是通过拦截器实现的。例如像重复提交保护,类型转换,数据校验,文件上传等这些特性都是借助于拦截器实现的。每一个拦截器都是可插拔的,因此,开发人员可以根据 Action 的需要决定使用哪些特性。这样可以实现业务逻辑处理代码和技术保障代码之间的解耦。另外,定义在拦截器栈中的拦截器指定了执行的顺序,在某种情况下,拦截器栈中拦截器的顺序是非常重要的。

从图 1-4 还可以看出,对于同一个 Action 配置的多个拦截器,在 Action 被调用之前和 Action 被调用之后都会被执行,并且这些拦截器在 Action 被调用后的执行顺序和 Action 被调用前执行的顺序相反。

1.3.4 User Created

User Created(用户创建的代码)包括 struts.xml、Action 和模板(如 JSP、FreeMarker、Velocity 等)。在进行基于 Struts 2 框架的 Web 应用开发时,需要开发人员创建的代码。

其中:

(1) struts.xml 文件是 Struts 2 框架的核心配置文件,用于配置和管理 Web 应用中 Action 的定义、Action 返回值对应的视图资源以及命名空间信息等内容。

(2) Template 用于视图的输出,在进行输出的时候,模板中可以使用 Struts 2 框架提供的 Struts 标签。

1.4 基本开发环境

工欲善其事必先利其器,说的就是要做好一件事,准备工作非常重要。在学习本书内容的时候,要进行编码实践进行测试、验证。在这里首先对本书使用的最基本开发环境进行简要介绍。由于每种软件都有很多种版本,在使用过程中,会发现"最新"版本与"最好用"版本会出现不一致的现象,我们建议,在使用过程中,选择"最好用"版本。

本书所使用的基本开发环境为 JDK 6、MyEclipse 5.1、Tomcat 6.0 和 MySQL 5.0。

1.4.1 JDK 的下载与安装

1. JDK 的下载

JDK(Java Development Kit)是免费的 Java 开发工具包,可以从 Oracle 公司的官方网站下载。

在撰写本章内容的时候,JDK 的最新版本为 JDK 7。本书使用的是 JDK 6,JDK 6 的下载步骤如下:

(1) 打开网址为 http://www.oracle.com/technetwork/java/的页面。在打开的页面

的右侧单击 Java SE Support 链接，如图 1-5 所示。

（2）在打开的页面底部，单击 Java product archive 链接，如图 1-6 所示。

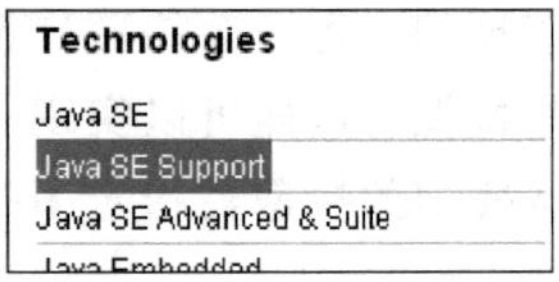

图 1-5　Java SE Support 链接 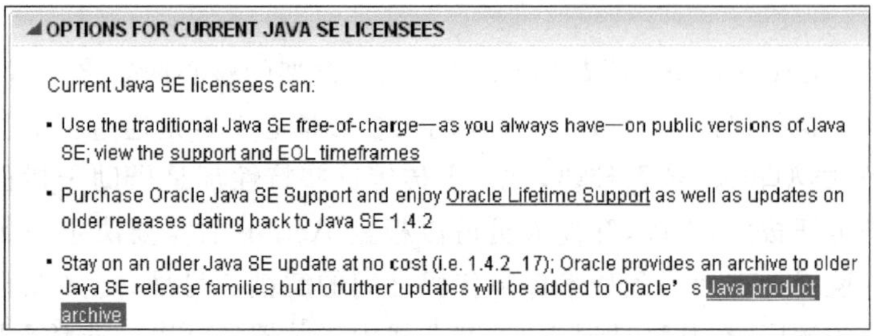　　　　图 1-6　Java product archive 链接

（3）在打开的页面中，单击 Java SE 6 链接，如图 1-7 所示。

（4）在打开的页面中是 JDK 6 的多个不同版本的下载链接，如图 1-8 所示。

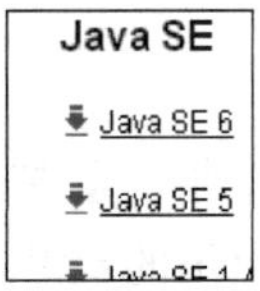

图 1-7　Java SE 6 链接 　　　　图 1-8　JDK 6 不同版本的下载链接

（5）本书使用的是 Java SE Development Kit 6u10 这个版本，用户根据自己的需要下载相应的版本。单击 Java SE Development Kit 6u10 链接后，界面如图 1-9 所示。

（6）选中图 1-9 上方的 Accept License Agreement 单选按钮，然后单击 jdk-6u10-windows-i586-p.exe 链接，即可下载，文件名为 jdk-6u10-windows-i586-p.exe。

2．JDK 的安装与配置

JDK 的安装很简单，只需双击下载的 JDK 安装文件（jdk-6u10-windows-i586-p.exe）即可。在安装过程中，可以更改 JDK 的安装位置，例如将 JDK 安装位置设置为 E:\Java\jdk16。JDK 安装完成后，需要配置如下的环境变量。

（1）JAVA_HOME：该环境变量用于设置 JDK 的安装路径，其值是 JDK 在电脑中的

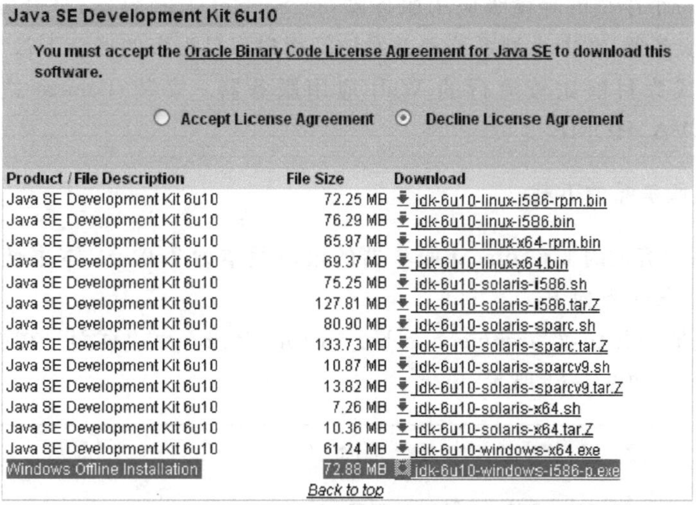

图 1-9　Java SE Development Kit 6u10 下载页面

安装位置,例如：E:\Java\jdk16。

（2）PATH：该环境变量用于设置命令的搜索路径。在原有 PATH 值的最后添加内容如下：

```
;E:\java\jdk16\bin;E:\java\jre6\bin
```

（3）CLASSPATH：该环境变量用于设置 Java 加载类的路径,其内容如下：

```
;E:\java\jdk16\lib\dt.jar;E:\java\jdk16\lib\tools.jar
```

3. JDK 的测试

JDK 安装是否成功,可以通过在命令提示符窗口中输入 java-version 命令进行测试。如果显示了 Java 的版本信息,则表明 JDK 安装及环境变量设置正确,如图 1-10 所示。

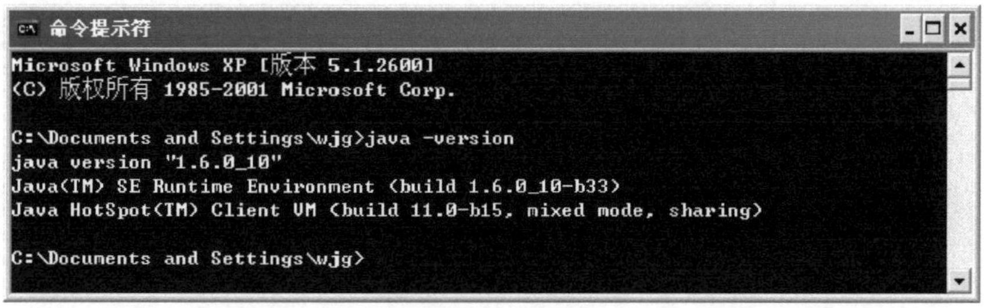

图 1-10　JDK 的测试

1.4.2　Tomcat 服务器的下载与安装

Tomcat 是 Apache 软件基金会的 Jakarta 项目中的一个核心项目,是由 Apache、Sun 和其他一些公司及个人共同开发完成并进行维护的。

Tomcat 是一个小型的轻量级应用服务器,是 Servlet/JSP 容器,是开发和调试 JSP 程序的首选 Web 服务器,在中小型系统和并发访问用户不是很多的场合下被普遍使用。

Tomcat 已成为目前比较流行的 Web 应用服务器。安装 Tomcat 之前必须先安装 JDK,并设置 JAVA_HOME 变量。

1. Tomcat 服务器的下载

在撰写本章内容的时候,Tomcat 服务器的最新版本为 7.0.x。本书使用的是 Tomcat 6,Tomcat 6 的下载步骤如下:

(1) 打开网址为 http://archive.apache.org/dist/tomcat/的页面。在打开的页面中单击"tomcat-6"链接,如图 1-11 所示。

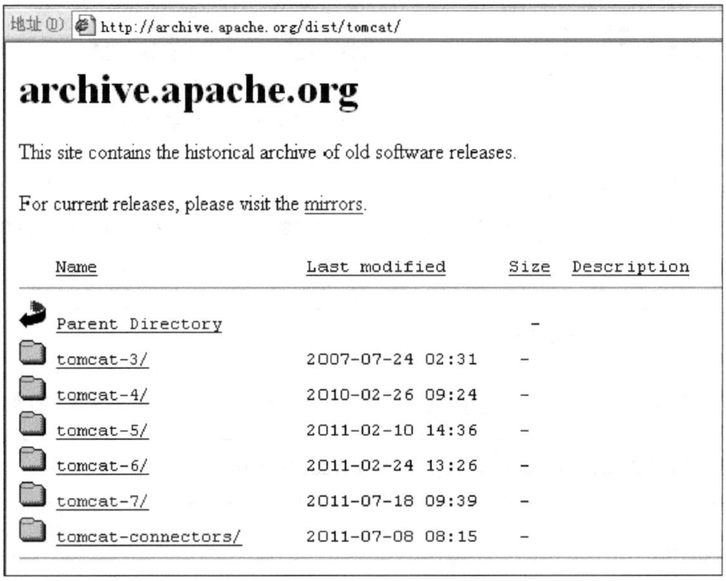

图 1-11 tomcat-6 链接

(2) 本书使用的是 6.0.18 版本。在打开的页面中单击"v6.0.18/"链接,如图 1-12 所示。

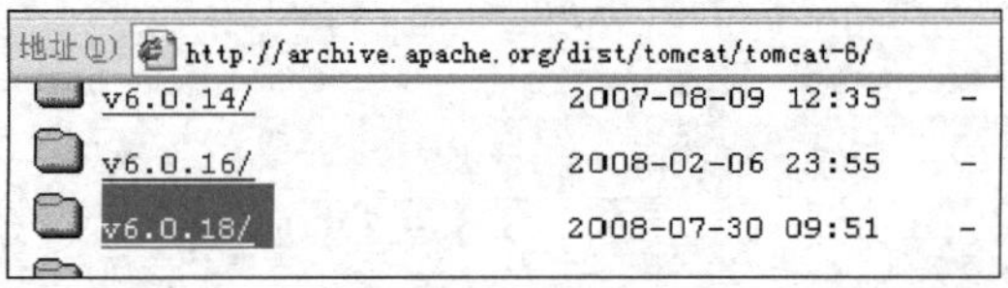

图 1-12 v6.0.18/链接

(3) 在打开的页面中单击"bin/"链接。在新打开的页面中可以下载 Tomcat-6.0.18 软件,如图 1-13 所示。其中 apache-tomcat-6.0.18.exe 是安装版软件,安装后才能使用;apache-tomcat-6.0.18.zip 是压缩版软件,只要解压后就可以使用。

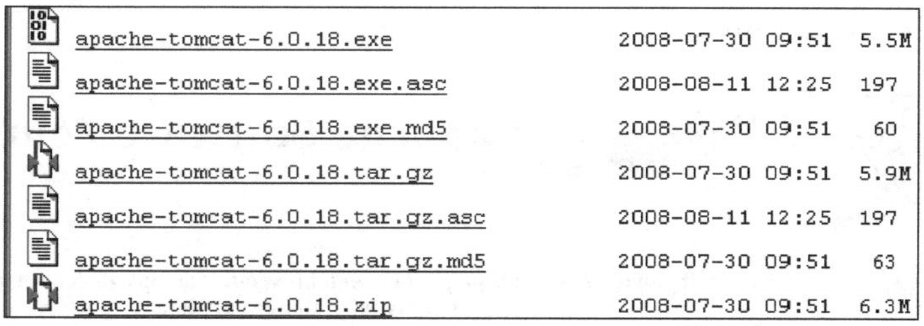

第1章 概述

图 1-13　Tomcat-6.0.18 下载

2. Tomcat 的安装与测试

本书使用的是压缩版软件。

（1）将下载的 apache-tomcat-6.0.18.zip 文件解压缩到指定的位置，如图 1-14 所示。

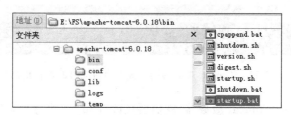

图 1-14　解压后的目录结构

（2）运行解压目录中 bin 文件夹下的 startup.bat 文件，启动 Tomcat 服务器，如图 1-15 所示。

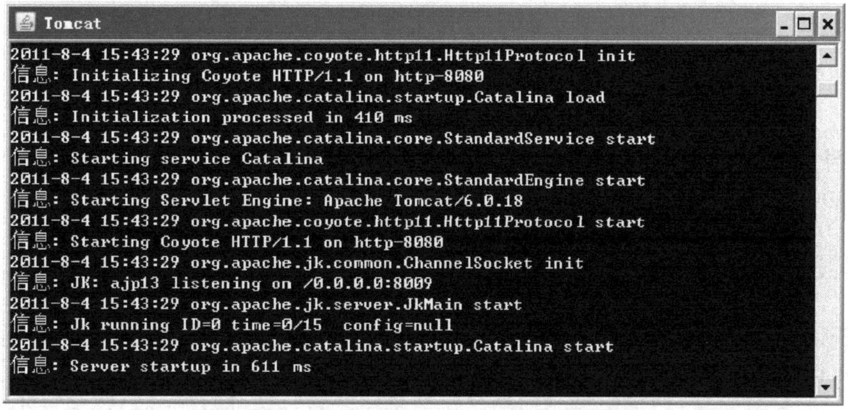

图 1-15　Tomcat 服务器启动后的窗口信息

（3）打开 Web 浏览器，输入 http://localhost:8080/，正常情况下，能够得到如图 1-16 所示的页面。

3. 配置

如果 Tomcat 服务器默认的 8080 端口已经被占用，则在启动 Tomcat 服务器的时候出

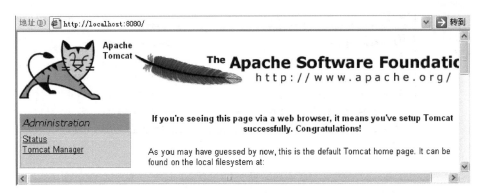

图 1-16　Tomcat 服务器启动后的 Web 页面

现如图 1-17 所示的提示信息。

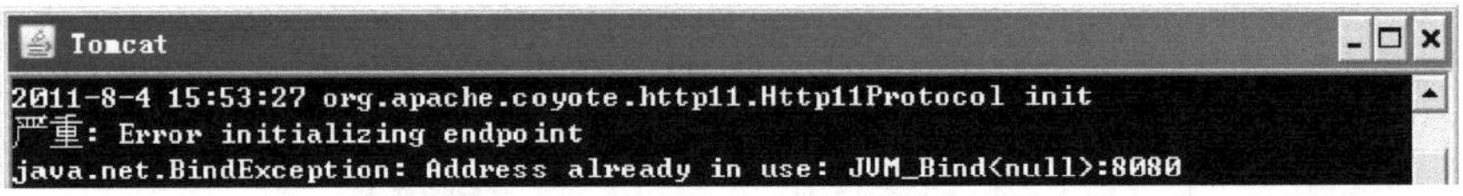

图 1-17　Tomcat 服务器端口被占用启动时的窗口信息

为了能够成功启动 Tomcat 服务器，就需要更改 Tomcat 服务器运行的端口号。具体操作步骤如下：

(1) 打开 Tomcat 安装目录中 conf 文件夹下的 server.xml 文件，找到如图 1-18 所示的内容。

```
<Connector port="8080" protocol="HTTP/1.1"
           connectionTimeout="20000"
           redirectPort="8443" />
```

图 1-18　Tomcat 服务器端口配置

(2) 将 port 的属性值改为"9090"后保存，然后重新启动 Tomcat 服务器，输入 http://localhost:9090/ 进行测试。

1.4.3　MyEclipse 的下载与安装

MyEclipse 是一款商业软件，需要注册才能使用，借助于该软件用户可以在数据库和 JavaEE 的开发、发布以及应用程序服务器的整合方面极大地提高工作效率。它是功能丰富的 JavaEE 集成开发环境，包括了完备的编码、调试、测试和发布功能，完整支持 HTML、Struts、JSF、CSS、JavaScript、SQL 和 Hibernate。

在浏览器地址栏中输入网址 http://www.myeclipseide.com，在打开的页面中能够看到 MyEclipse 的各个版本。可以根据提示下载。另外，读者也可以不通过 MyEclipse 官方网站下载，在 Google 中进行搜索，一般能够很方便地下载到安装文件。

虽然 MyEclipse 已经推出了高的版本，但是综合考虑系统速度和开发需求，本书选择了 MyEclipse 5.1，读者也可以根据需要选择更高的版本，使用起来没有太大区别。下载之后，得到一个可执行文件，文件名为 MyEclipseEnterpriseWorkbenchInstaller_5.1.0GA_E3.2.1.exe。

1. 安装

（1）双击下载的可执行文件，根据提示进行安装，安装成功后，启动 MyEclipse。

（2）在启动过程中，会出现如图 1-19 所示的界面。该界面用于设置工程文件存放的路径，通过单击 Browse 按钮可以更改默认的路径。

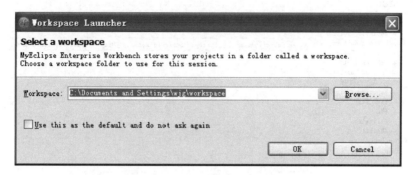

图 1-19　设置工程存放的路径

（3）由于 MyEclipse 是商业软件，因此需要注册才能使用。启动成功后，单击菜单 Window→Preferences，打开如图 1-20 所示的窗口。

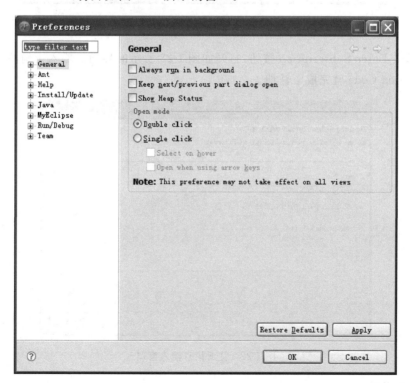

图 1-20　Preferences 窗口

（4）在图 1-20 所示的窗口中选择 MyEclipse→Subscription，打开如图 1-21 所示的窗口。在该窗口中可以看到，如果不进行注册则只能有 30 天的试用期。

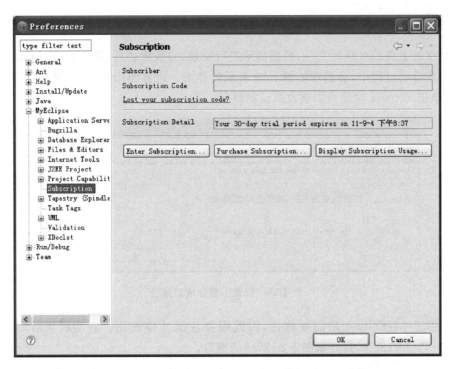

图 1-21　Subscription 窗口

（5）单击 Enter Subscription 按钮，打开如图 1-22 所示的界面。输入正确的 Subscriber 和 Subscription Code 就完成了注册。

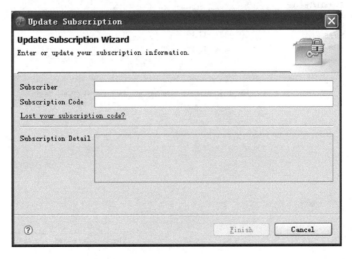

图 1-22　注册信息输入窗口

2．配置 JDK

（1）MyEclipse 启动成功后，单击菜单 Window→Preferences，打开如图 1-23 所示的窗口。

（2）在打开的窗口中单击 Java→Installed JREs，如图 1-23 所示。

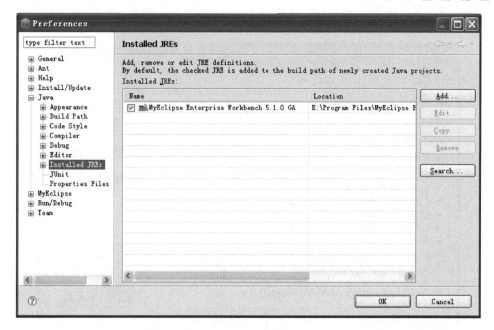

图 1-23　Installed JREs 窗口

（3）由图 1-23 可以看到，MyEclipse 已经和 JDK 进行了绑定，该 JDK 的版本可能不是用户所需要的 JDK 版本。为了使用其他版本的 JDK，可以单击 Add 按钮添加所需版本的 JDK，也可以选中已经绑定的 JDK，然后单击 Edit 按钮进行更改。单击 Add 按钮后，打开如图 1-24 所示的界面。

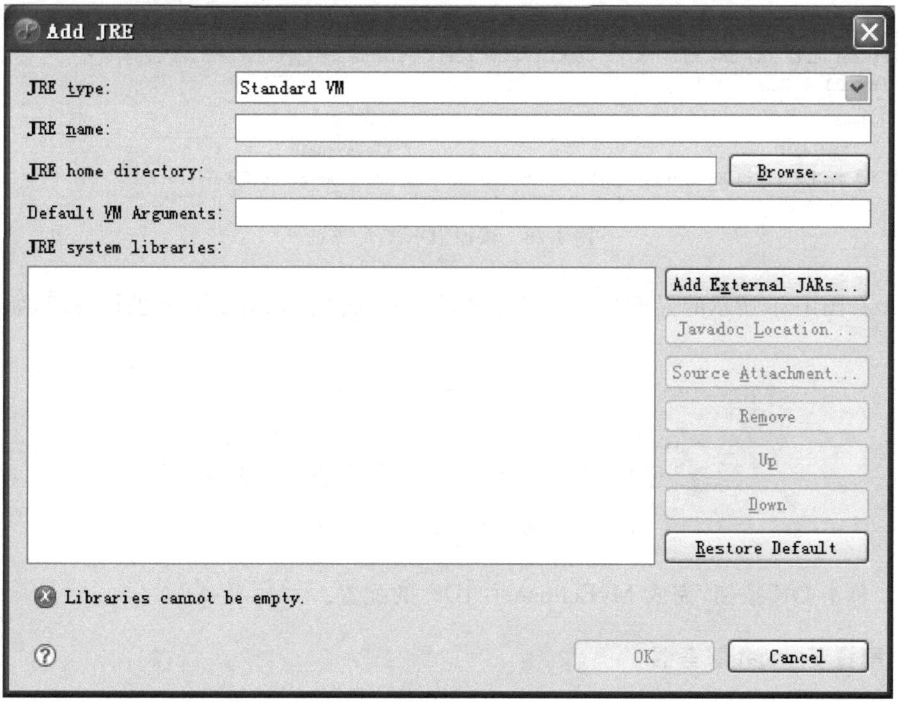

图 1-24　Add JRE 界面

（4）单击 Browse 按钮，选择所需版本 JDK 的安装路径，如图 1-25 所示。

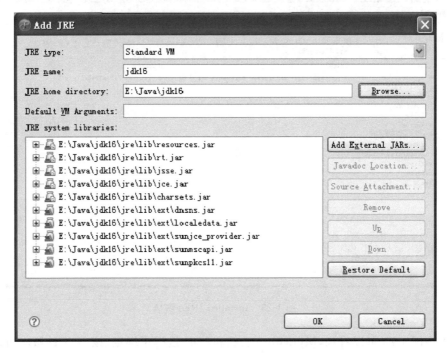

图 1-25　选择 JDK 安装路径的界面

（5）单击 OK 按钮后，结果如图 1-26 所示。

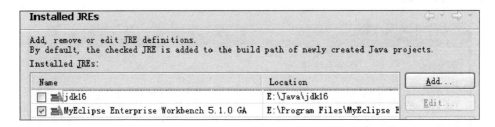

图 1-26　添加 JDK 后的界面

（6）在图 1-26 所示的界面中勾选新添加的 JDK，将之和 MyEclipse 进行绑定，如图 1-27 所示。

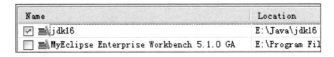

图 1-27　将 JDK 6 和 MyEclipse 绑定

（7）单击 OK 按钮，完成 MyEclipse 中 JDK 的配置。

3．配置 Tomcat 服务器

（1）单击菜单 Window→Preferences，或者单击工具栏中如图 1-28 所示的服务器按钮，在弹出的菜单中选择 Configure Server 选项,打开 Preferences 窗口。

(2) 在打开窗口的左侧依次单击 MyEclipse→Application Servers,出现图 1-29 所示的界面。选中 Enable 选项,然后单击第一个 Browse 按钮,选择 Tomcat 安装的路径,最后单击 OK 按钮关闭图 1-29 所示界面。

图 1-28　配置 Server 的工具栏按钮

(3) 单击图 1-29 左侧的 Tomcat 选项,展开 Tomcat 选项,在展开的列表中单击 JDK,然后在窗口的右侧选择 MyEclipse 中绑定的 JDK,如图 1-30 所示。

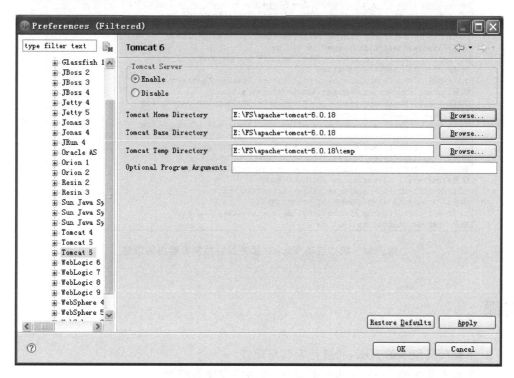

图 1-29　配置 Tomcat 服务器

图 1-30　选择 MyEclipse 绑定的 JDK

(4) 单击 OK 按钮,完成 MyEclipse 中 Tomcat 的配置。

(5) 单击图 1-28 中服务器按钮(左侧第二个按钮)的下拉黑三角,按如图 1-31 所示依次选择 Tomcat 6→Start 选项,启动 Tomcat 服务器。

(6) 启动成功后控制台的信息如图 1-32 所示。

至此,进行 Web 应用开发的基本环境搭建完成。

图 1-31　启动 Tomcat 服务器

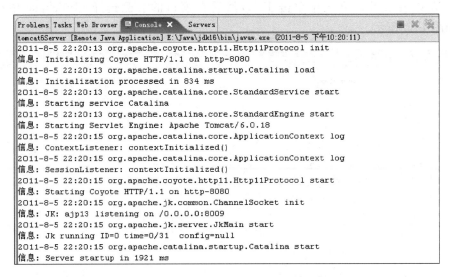

图 1-32　成功启动 Tomcat 服务器后控制台输出的信息

习题

1. 下载较高版本的 JDK，并进行安装与配置。
2. 下载安装版的 Tomcat 服务器软件，并进行安装与配置。
3. 安装 MyEclipse 软件，并进行 JDK 和 Tomcat 配置。

第 2 章

Struts 2框架技术入门

本章首先介绍了 Struts 2 框架的下载与安装过程，然后在讲解了如何搭建基于 Struts 2 框架的开发环境后给出了一个示例，最后对基于 Struts 2 框架的 Web 应用开发流程进行了总结。

2.1 Struts 2 框架的下载与安装

2.1.1 Struts 2 框架的下载

截至作者撰写本章内容时，Struts 2 框架的最新版本是 Struts 2.2.3。Struts 2 框架下载步骤如下：

（1）打开网址为 http://struts.apache.org/ 的页面，找到 Struts 2 框架的最新版本的链接，如图 2-1 所示。

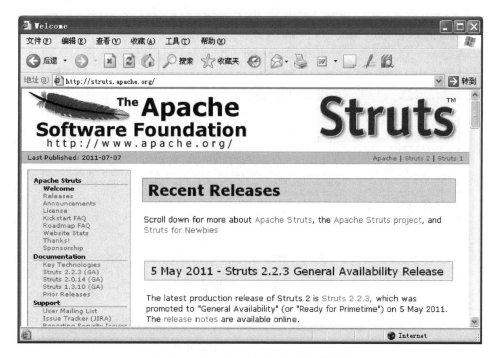

图 2-1　Struts 2 框架最新版本链接的页面

（2）单击图 2-1 中左侧的 Struts 2.2.3(GA)链接或者图 2-1 中右上角的 Struts 2 链接，打开如图 2-2 所示的页面。

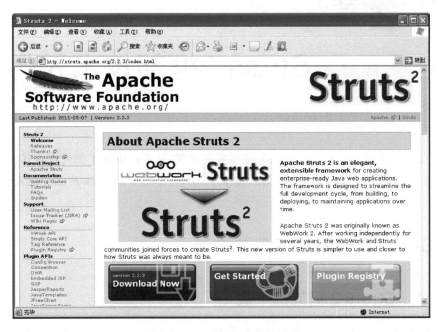

图 2-2　Struts 2.2.3 下载页面

（3）单击图 2-2 中的 Download Now 按钮或者图 2-1 中右侧的 Struts 2.2.3 链接，进入如图 2-3 所示的页面。

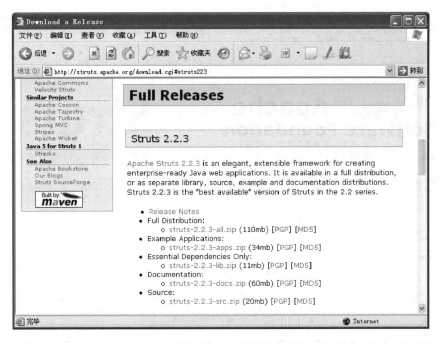

图 2-3　Struts 2.2.3 下载列表

（4）从图 2-3 所示的页面中可以看到，Struts 2.2.3 有多个压缩包可以下载。每个压缩包的作用如下：

- Full Distribution 压缩包

该压缩包是 Struts 2 框架的完整版本，包含 Jar 文件、源代码、示例应用和文档。本书在后面章节内容的讲解时使用的是这个完整版的压缩文件，建议读者下载该版本的压缩包文件。

- Example Applications 压缩包

该压缩包提供了基于 Struts 2 框架的一些 Web 应用示例，这些示例对于初学者学习基于 Struts 2 框架的 Web 开发很有帮助。这个压缩包中的内容已经被包含在完整版压缩包的 apps 文件夹中。

- Essential Dependencies Only 压缩包

该压缩包提供了 Struts 2 框架的核心类库及它所依赖的类库。这个压缩包中的内容已经被包含在完整版压缩包的 lib 文件夹中。

- Documention 压缩包

该压缩包提供了 Struts 2 框架的相关文档，包含 Struts 2 框架的使用文档、参考手册和 API 文档等。这个压缩包中的内容已经被包含在完整版压缩包的 docs 文件夹中。

- Source 压缩包

该压缩包提供了 Struts 2 框架的源代码，通过阅读源代码，可以使读者更好地理解 Struts 2 框架的结构及运行机制。这个压缩包中的内容已经被包含在完整版压缩包的 src 文件夹中。

（5）下载 Struts 2 的完整版压缩包，解压缩后，其目录结构如图 2-4 所示。

图 2-4　完整版压缩包解压缩后的目录结构

其中 lib 文件夹中包含的内容就是基于 Struts 2 框架开发 Web 应用时所需要的 Jar 文件。

2.1.2　Struts 2 框架的安装

Struts 2 框架的安装就是将基于 Struts 2 框架的 Web 应用所需要的 Jar 文件拷贝或者发布到 Web 服务器中 Web 应用的 lib 文件夹中，并对 Struts 2 框架进行配置。下面通过 Struts 2 框架压缩包中自带的 Web 应用示例进行讲解，具体步骤如下：

（1）将 Struts 2 框架压缩包中 apps 文件夹下的 struts2-blank.war 文件复制到 Tomcat 服务器安装路径下的 webapps 文件夹中，如图 2-5 所示。

图 2-5　struts2-blank.war 文件在 Tomcat 服务器中的位置

（2）启动 Tomcat 服务器，struts2-blank.war 文件将被解压并发布，如图 2-6 所示；Tomcat 控制台信息如图 2-7 所示。由图 2-7 中的最后一行信息可以看出，struts2-blank.war 文件正在被发布。

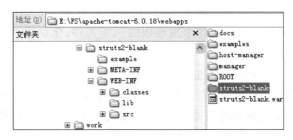

图 2-6　struts2-blank.war 文件发布后的示意图

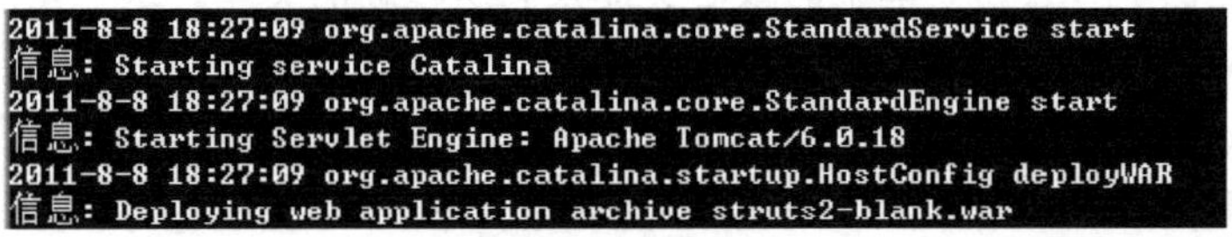

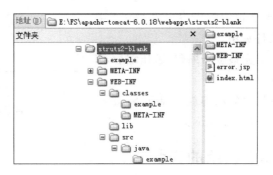

图 2-7　Tomcat 控制台信息

（3）struts2-blank 文件夹中的内容如图 2-8 所示。其中 classes 和 lib 文件夹必须在 WEB-INF 文件夹下。

图 2-8　struts2-blank 文件夹中的内容

（4）lib 文件夹中保存了基于 Struts 2 框架的 Web 应用所需要的 Jar 文件，如图 2-9 所示。

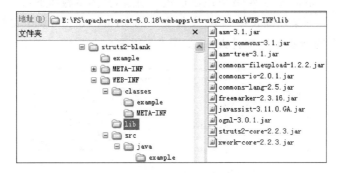

图 2-9　lib 文件夹中的内容

（5）打开浏览器，输入网址 http://localhost:8080/struts2-blank，结果如图 2-10 所示。

图 2-10　struts2-blank 应用访问界面

2.2　搭建基于 Struts 2 框架的 Web 应用开发环境

本节主要介绍在 MyEclipse 开发工具中如何添加对 Struts 2 框架的支持。具体操作步骤如下：

（1）启动 MyEclipse，并在 MyEclipse 中新建一个 Web 项目 Chapter2。

（2）在新建的项目 Chapter2 上单击鼠标右键，在弹出的菜单项中依次选择 Build Path→Configure Build Path，打开如图 2-11 所示的窗口。

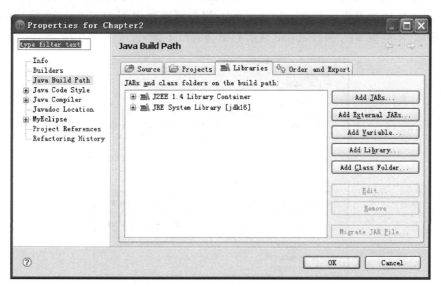

图 2-11　Java Build Path 窗口

（3）选择图 2-11 中左侧的 Java Build Path，然后选择右侧的 Libraries 标签，单击该标签页的 Add External JARs 按钮，打开 JAR Selection 对话框，如图 2-12 所示。

找到 Struts 2 的完整版解压后 lib 文件夹的位置，选择如下的 Jar 文件（参考图 2-9 中 lib 文件夹中的内容）：asm-3.1.jar、asm-commons-3.1.jar、asm-tree-3.1.jar、commons-fileupload-1.2.2.jar、commons-io-2.0.1.jar、commons-lang-2.5.jar、freemarker-2.3.16.jar、javassist-3.11.0.GA.jar、ognl-3.0.1.jar、struts2-core-2.2.3.jar 和 xwork-core-2.2.3.jar。

这 11 个 Jar 文件是基于 Struts 2 框架进行 Web 应用开发时最少依赖的 Jar 文件。

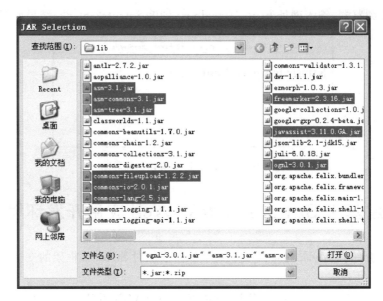

图 2-12　Jar 文件选择窗口

（4）单击"打开"按钮，JAR Selection 对话框关闭，然后单击图 2-11 界面中的 OK 按钮，完成导入 Jar 文件的操作。在 MyEclipse 的左侧包浏览器界面中 Chapter2 项目的内容如图 2-13 所示（不同版本的 MyEclipse 显示界面会有所不同）。

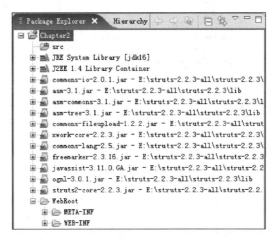

图 2-13　添加 Struts 2 框架支持的 Web 项目

至此，支持 Struts 2 框架的 Web 应用开发环境搭建完成。

2.3　基于 Struts 2 框架的 Web 应用示例

在学习 Struts 2 框架相关技术之前，先以一个简单的 Web 应用为例，介绍使用 Struts 2 框架进行开发的步骤。通过本节内容的学习使初学者对使用 Struts 2 框架进行 Web 应用开发有一个初步的认识。

例 2-1　基于 Struts 2 框架的 Web 应用示例。

任务描述：单击如图 2-14 所示界面中的 Submit 按钮后，输出所输入的信息。

需要完成的内容包括：①创建 3 个视图（JSP 页面），分别是信息输入页面、有信息输入的跳转页面和无信息输入的跳转页面。②创建用于处理 Web 用户请求的业务控制器类（Action）。③创建并配置 Struts 2 框架的核心配置文件 struts.xml。④编辑 web.xml 文件。

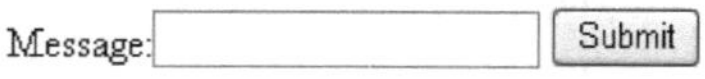

图 2-14　信息输入界面

2.3.1　创建视图

本节完成 3 个视图的创建，分别是：信息输入页面（inputMessage.jsp）、有信息输入的跳转页面（showMessage.jsp）和无信息输入的跳转页面（noMessage.jsp）。

1. 信息输入页面

inputMessage.jsp

```
<%@ page language="java" contentType="text/html; charset=UTF-8"
    pageEncoding="UTF-8"%>
<html>
<head>
<meta http-equiv="Content-Type" content="text/html; charset=UTF-8">
<title>Input message page</title>
</head>
<body>
    <form Action="helloWorld.action" method="post">
        Message:<input name="message" type="text">
        <input type="submit" value="Submit">
    </form>
</body>
</html>
```

<form>表单标签的 action 属性值也可以不使用扩展名，即直接使用"helloWorld"。

2. 有信息输入的跳转页面

showMessage.jsp

```
<%@ page language="java" contentType="text/html; charset=UTF-8"
    pageEncoding="UTF-8"%>
<%@ taglib prefix="s" uri="/struts-tags" %>
<html>
<head>
<meta http-equiv="Content-Type" content="text/html; charset=UTF-8">
<title>Show message page</title>
</head>
```

```
<body>
    HelloWorld, the input message is : <s:property value = "message"/>
</body>
</html>
```

在该页面中使用了 Struts 2 框架的标签,标签的具体使用方法将在第 6 章进行介绍。

3. 无信息输入的跳转页面

noMessage.jsp

```
<%@ page language = "java" contentType = "text/html; charset = UTF - 8"
    pageEncoding = "UTF - 8" %>
<html>
<head>
<meta http - equiv = "Content - Type" content = "text/html; charset = UTF - 8">
<title>No message page</title>
</head>
<body>
    HelloWorld, no message was input!
</body>
</html>
```

2.3.2 创建业务控制器类

业务控制器类(Action 类)是用于处理 Web 用户请求的关键组件,开发人员根据所需要完成的功能编写代码。业务控制器就是一个普通的 Java 类,代码如下:

HelloWorld.java

```java
package example.struts2;
public class HelloWorld {
    private String message;
    public String getMessage() {
        return message;
    }
    public void setMessage(String message) {
        this.message = message;
    }
    public String execute(){
        if(getMessage().isEmpty()){
            return "error";
        }
        else{
            return "success";
        }
    }
}
```

在业务控制器类中需要定义一个execute()方法,在该方法中编写业务逻辑处理代码。

2.3.3　创建struts.xml文件

struts.xml文件是Struts 2框架的核心配置文件,该文件中包含了包空间及其命名空间信息的配置、action的定义、action返回的结果码和对应的视图资源等内容的配置。

该示例中struts.xml文件内容如下:

struts.xml

```xml
<?xml version="1.0" encoding="UTF-8"?>
<!DOCTYPE struts PUBLIC
    "-//Apache Software Foundation//DTD Struts Configuration 2.0//EN"
    "http://struts.apache.org/dtds/struts-2.0.dtd">
<struts>
    <package name="default" namespace="/" extends="struts-default">
        <Action name="helloWorld" class="example.struts2.HelloWorld">
            <result name="success">/showMessage.jsp</result>
            <result name="error">/noMessage.jsp</result>
        </action>
    </package>
</struts>
```

该文件保存在Web项目Chapter2中的src文件夹下,当Web项目发布后,struts.xml文件将被保存在Web应用程序的WEB-INF/classes文件夹中。

在struts.xml配置文件中:

(1) 使用＜package＞元素定义了包空间,其名称为default,并且该包继承了Struts 2框架的默认包struts-default。

(2) 使用＜action＞元素定义了action,其名称为helloWorld,该名称要和form表单标签中action属性的值一致;class属性用于指定action的完整实现类。

(3) 使用＜result＞元素定义了action返回的视图资源。

2.3.4　编辑web.xml文件

在web.xml文件中使用＜filter＞元素注册了StrutsPrepareAndExecuteFilter过滤器作为Struts 2框架的核心控制器,使用＜filter-mapping＞元素配置了核心控制器所要拦截的URL。

web.xml文件的内容如下:

web.xml

```xml
<?xml version="1.0" encoding="UTF-8"?>
<web-app version="2.4"
    xmlns="http://java.sun.com/xml/ns/j2ee"
    xmlns:xsi="http://www.w3.org/2001/XMLSchema-instance"
    xsi:schemaLocation="http://java.sun.com/xml/ns/j2ee
```

```
            http://java.sun.com/xml/ns/j2ee/web-app_2_4.xsd">
  <filter>
    <filter-name>Struts 2</filter-name>
    <filter-class>
        org.apache.struts2.dispatcher.ng.filter.StrutsPrepareAndExecuteFilter
    </filter-class>
  </filter>
  <filter-mapping>
    <filter-name>Struts 2</filter-name>
    <url-pattern>/*</url-pattern>
  </filter-mapping>
</web-app>
```

2.3.5　Web 项目的发布与测试

1. 发布 Web 项目

具体操作步骤如下：

(1) 选中 Web 项目 Chapter2,如图 2-15 所示。

(2) 单击工具栏中的发布按钮(图 2-15 中右侧第二个按钮),打开发布窗口,如图 2-16 所示。

(3) 单击图 2-16 中的 Add 按钮,打开 Web 服务器选择窗口,如图 2-17 所示。

图 2-15　选中 Web 项目

(4) 在图 2-17 所示的窗口中选择要发布到的 Web 服务器(本例中选择了 MyEclipse 中绑定的 Tomcat 6 Web 服务器)。单击 Finish 按钮,返回如图 2-18 所示的界面。

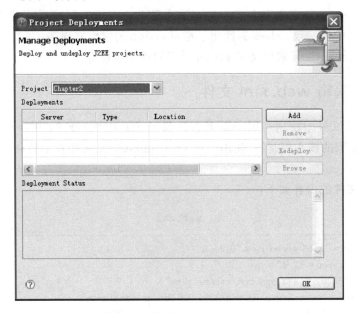

图 2-16　发布 Web 项目窗口

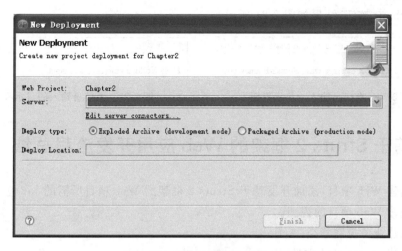

图 2-17　Web 服务器选择窗口

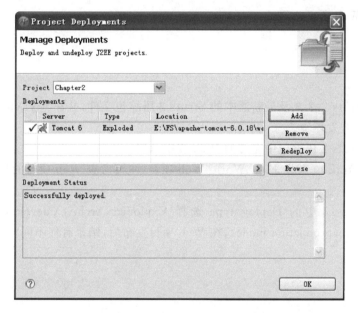

图 2-18　Web 项目发布成功后的界面

(5) 关闭图 2-18 所示界面,完成 Web 项目的发布。

2．测试 Web 项目

(1) 启动 Tomcat 服务器。

(2) 在浏览器地址栏中输入网址 http://localhost:8080/Chapter2/inputMessage.jsp,结果如图 2-19 所示。

(3) 在图 2-19 的输入框中输入信息后,单击 Submit 按钮,结果如图 2-20 所示。

(4) 在图 2-19 所示的输入框中不输入信

图 2-19　信息输入界面

息,单击 Submit 按钮,结果如图 2-21 所示。

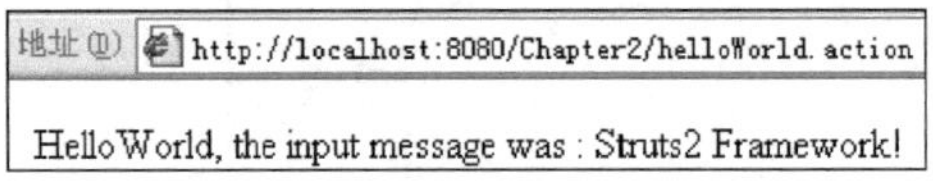

图 2-20 有信息输入的显示界面 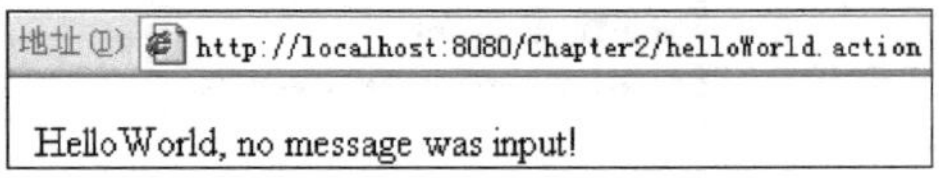　　　图 2-21 无信息输入的显示界面

2.4 基于 Struts 2 框架的 Web 应用开发流程总结

(1) 创建 Web 项目,添加开发基于 Struts 2 框架的 Web 项目所需的 Jar 包。
(2) 创建 Web 项目的视图。
(3) 创建业务控制器类(Action)。
(4) 创建 struts.xml 文件。
(5) 编辑 web.xml 文件。
(6) 发布 Web 项目并测试。

习题

1. 下载 Struts 2.2.3 及最新版本的 Struts 2 压缩包文件,并在 MyEclipse 中搭建基于 Struts 2 框架的开发环境。
2. 测试 Struts 2 压缩包中自带的 Web 应用示例。
3. 使用 UltraEdit 开发 HelloWorld Web 项目(主要考查 Web 项目的目录结构及如何发布)。
4. 比较图 2-17 中的 Deploy type 选择 Exploded Archive(development mode)和 Packaged Archive(production mode),在 Web 项目发布后,结果有何不同?

第 3 章

Struts 2框架的配置

本章首先讲解了 Struts 2 框架的工作原理,然后介绍了 Struts 2 框架工作时所需要的配置文件的结构。

3.1 Struts 2 框架的体系结构与工作原理

在学习 Struts 2 框架的工作原理之前,首先介绍 Struts 2 框架的体系结构。

3.1.1 Struts 2 框架的体系结构

在 Struts 2 的官方文档中(参见 http://struts.apache.org/2.2.3/docs/big-picture.html)提供了 Struts 2 框架的体系结构示意图,如图 3-1 所示。

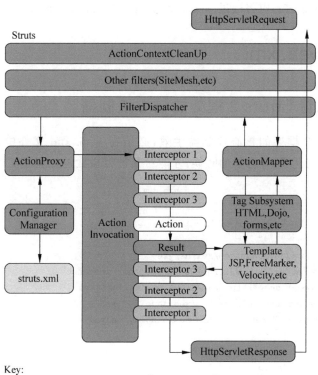

图 3-1 Struts 2 框架的体系结构

由官方文档可知，Struts 2 框架包含 Servlet Filters、Struts Core、Interceptors 和 User created 四个模块。

3.1.2 Struts 2 框架的工作原理

接下来结合图 3-1 对 Struts 2 框架的工作原理进行讲解。

当初始的请求到达 Servlet 容器(例如 Jetty,Resin 或者 Tomcat)的时候,该请求:

(1) 首先要经过 Struts 2 框架的核心控制器 StrutsPrepareAndExecuteFilter。

(2) 然后核心控制器依次查找 ActionMapper 来决定这个请求是否调用 action。如果 ActionMapper 决定调用 action,StrutsPrepareAndExecuteFilter 就把请求的处理控制权委托给 ActionProxy。

(3) ActionProxy 通过 Configuration Manager 查询 Struts 2 框架的核心配置文件 (struts.xml),找到所要调用的 Action 类。

(4) 接下来,ActionProxy 创建一个 ActionInvocation 实例,该实例负责命令模式的实现,包括在激活 action 本身之前调用配置的任意拦截器。

(5) 在 action 执行完成之后,ActionInvocation 实例负责根据 action 的返回结果码在 struts.xml 配置文件中查找要输出的结果(Result)。该结果通常是 JSP 或者 FreeMarker 模板,也可能是 action。在输出的时候,模板可以使用框架提供的 Struts 2 标签。

(6) 然后为 action 配置的拦截器将被再次执行(顺序和 action 被调用之前的执行顺序相反)。

(7) 最后,服务器的响应将通过 web.xml 文件中配置的 Struts 2 框架核心控制器返回。

3.2 配置 web.xml

3.2.1 配置 Struts 2 框架的核心控制器

要想使 Struts 2 框架的核心控制器生效,必须在 web.xml 文件中进行配置,这样当 HttpRequest 请求到达 Web 容器的时候,才能对其进行拦截并处理。基本配置如下:

web.xml

```
<?xml version = "1.0" encoding = "UTF-8"?>
<web-app version = "2.4"
    xmlns = "http://java.sun.com/xml/ns/j2ee"
    xmlns:xsi = "http://www.w3.org/2001/XMLSchema-instance"
    xsi:schemaLocation = "http://java.sun.com/xml/ns/j2ee
    http://java.sun.com/xml/ns/j2ee/web-app_2_4.xsd">
    <display-name>Struts 2 Framework</display-name>
    <filter>
        <filter-name>struts2</filter-name>
        <filter-class>
            org.apache.struts2.dispatcher.ng.filter.StrutsPrepareAndExecuteFilter
```

```xml
            </filter-class>
        </filter>
        <filter-mapping>
            <filter-name>struts2</filter-name>
            <url-pattern>/*</url-pattern>
        </filter-mapping>
        <welcome-file-list>
            <welcome-file>index.html</welcome-file>
        </welcome-file-list>
</web-app>
```

3.2.2 配置第三方过滤器框架

如果在激活action之前想调用第三方过滤器框架,例如SiteMesh,那么该如何配置web.xml配置文件呢?其实,StrutsPrepareAndExecuteFilter过滤器可以被拆分成StrutsPrepareFilter和StrutsExecuteFilter两个过滤器,可以在这两个过滤器之间加上所需要的第三方过滤器。配置如下:

web.xml

```xml
<?xml version="1.0" encoding="UTF-8"?>
<web-app version="2.4"
    xmlns="http://java.sun.com/xml/ns/j2ee"
    xmlns:xsi="http://www.w3.org/2001/XMLSchema-instance"
    xsi:schemaLocation="http://java.sun.com/xml/ns/j2ee
    http://java.sun.com/xml/ns/j2ee/web-app_2_4.xsd">
<display-name>Struts 2 Framework</display-name>
<filter>
        <filter-name>StrutsPrepareFilter</filter-name>
        <filter-class>
            org.apache.struts2.dispatcher.ng.filter.StrutsPrepareFilter
        </filter-class>
</filter>
<filter>
        <filter-name>sitemesh</filter-name>
        <filter-class>
            com.opensymphony.module.sitemesh.filter.PageFilter
        </filter-class>
</filter>
<filter>
        <filter-name>StrutsExecuteFilter</filter-name>
        <filter-class>
            org.apache.struts2.dispatcher.ng.filter.StrutsExecuteFilter
        </filter-class>
</filter>
<filter-mapping>
```

```xml
            <filter-name>struts2</filter-name>
            <url-pattern>/*</url-pattern>
        </filter-mapping>
        <welcome-file-list>
            <welcome-file>index.html</welcome-file>
        </welcome-file-list>
</web-app>
```

3.2.3 配置初始化参数

在配置 Struts 2 框架的核心控制器时，还可以配置初始化参数。这些参数具体如下：

1. config 参数

config 参数用于配置 Struts 2 框架要加载的 XML 配置文件，如果有多个配置文件，它们之间要用逗号分隔。如果省略该参数，则使用默认的配置文件。例如有 struts-config1.xml 和 struts-config2.xml 两个配置文件要加载，其配置如下：

web.xml

```xml
<?xml version="1.0" encoding="UTF-8"?>
<web-app version="2.4"
    xmlns="http://java.sun.com/xml/ns/j2ee"
    xmlns:xsi="http://www.w3.org/2001/XMLSchema-instance"
    xsi:schemaLocation="http://java.sun.com/xml/ns/j2ee
    http://java.sun.com/xml/ns/j2ee/web-app_2_4.xsd">
    ...
    <filter>
        <filter-name>struts2</filter-name>
        <filter-class>
            org.apache.struts2.dispatcher.ng.filter.StrutsPrepareAndExecuteFilter
        </filter-class>
        <init-param>
            <param-name>config</param-name>
            <param-value>
                struts-config1.xml, struts-config2.xml
            </param-value>
        </init-param>
    </filter>
    ...
</web-app>
```

当配置文件的名称或者位置发生变化时，通常配置 config 参数来加载配置文件。

2. ActionPackages 参数

ActionPackages 参数用于指定包空间，如果有多个包空间，则它们之间使用逗号分隔。Struts 2 框架将会扫描这些包空间下的 Action 类。例如，如果设置的包空间是 struts2.actions1

和 struts2.actions2,则将在这两个包空间下搜索 Action 类,其配置如下:

web.xml

```xml
<?xml version = "1.0" encoding = "UTF-8"?>
<web-app version = "2.4"
    xmlns = "http://java.sun.com/xml/ns/j2ee"
    xmlns:xsi = "http://www.w3.org/2001/XMLSchema-instance"
    xsi:schemaLocation = "http://java.sun.com/xml/ns/j2ee
    http://java.sun.com/xml/ns/j2ee/web-app_2_4.xsd">
    ...
    <filter>
        <filter-name>struts2</filter-name>
        <filter-class>
            org.apache.struts2.dispatcher.ng.filter.StrutsPrepareAndExecuteFilter
        </filter-class>
        <init-param>
            <param-name>ActionPackages</param-name>
            <param-value>
                struts2.actions1,struts2.actions2
            </param-value>
        </init-param>
    </filter>
    ...
</web-app>
```

3. configProviders 参数

configProviders 参数用于设置实现了 ConfigurationProvider 接口的 Java 类列表,它们之间用逗号分隔。例如,在 Struts 2 包中的 MyConfigurationProvider1 和 MyConfigurationProvider2 两个 Java 类实现了 ConfigurationProvider 接口,其配置方式如下:

web.xml

```xml
<?xml version = "1.0" encoding = "UTF-8"?>
<web-app version = "2.4"
    xmlns = "http://java.sun.com/xml/ns/j2ee"
    xmlns:xsi = "http://www.w3.org/2001/XMLSchema-instance"
    xsi:schemaLocation = "http://java.sun.com/xml/ns/j2ee
    http://java.sun.com/xml/ns/j2ee/web-app_2_4.xsd">
    ...
    <filter>
        <filter-name>struts2</filter-name>
        <filter-class>
            org.apache.struts2.dispatcher.ng.filter.StrutsPrepareAndExecuteFilter
        </filter-class>
        <init-param>
            <param-name>configProviders</param-name>
            <param-value>
```

```xml
            struts2.MyConfigurationProvider1,struts2.MyConfigurationProvider2
        </param-value>
    </init-param>
</filter>
    ...
</web-app>
```

4. loggerFactory 参数

loggerFactory 参数用于指明 LoggerFactory 实现类的类名。例如,如果 LoggerFactory 实现类的类名是 MyLoggerFactory,则其配置如下:

web.xml

```xml
<?xml version="1.0" encoding="UTF-8"?>
<web-app version="2.4"
    xmlns="http://java.sun.com/xml/ns/j2ee"
    xmlns:xsi="http://www.w3.org/2001/XMLSchema-instance"
    xsi:schemaLocation="http://java.sun.com/xml/ns/j2ee
    http://java.sun.com/xml/ns/j2ee/web-app_2_4.xsd">
    ...
    <filter>
        <filter-name>struts2</filter-name>
        <filter-class>
            org.apache.struts2.dispatcher.ng.filter.StrutsPrepareAndExecuteFilter
        </filter-class>
        <init-param>
            <param-name>loggerFactory</param-name>
            <param-value>MyLoggerFactory</param-value>
        </init-param>
    </filter>
    ...
</web-app>
```

5. *

"*"是指任意的被当作 Struts 2 框架常量的任何其他的参数。例如,要指定 Struts 2 框架处理的 HTTP 请求的扩展名为"do",则其配置如下:

web.xml

```xml
<?xml version="1.0" encoding="UTF-8"?>
<web-app version="2.4"
    xmlns="http://java.sun.com/xml/ns/j2ee"
    xmlns:xsi="http://www.w3.org/2001/XMLSchema-instance"
    xsi:schemaLocation="http://java.sun.com/xml/ns/j2ee
    http://java.sun.com/xml/ns/j2ee/web-app_2_4.xsd">
    ...
```

```xml
<filter>
    <filter-name>struts2</filter-name>
    <filter-class>
        org.apache.struts2.dispatcher.ng.filter.StrutsPrepareAndExecuteFilter
    </filter-class>
    <init-param>
        <param-name>struts.action.extension</param-name>
        <param-value>do</param-value>
    </init-param>
</filter>
    ...
</web-app>
```

3.3 配置 struts.xml

struts.xml 文件是 Struts 2 框架的核心配置文件,该文件中包含了包空间及其命名空间信息的配置、action 的定义、action 返回的结果码和对应的视图资源等内容。

struts.xml 文件的基本结构如下:

```xml
<?xml version="1.0" encoding="UTF-8" ?>
<!DOCTYPE struts PUBLIC
    "-//Apache Software Foundation//DTD Struts Configuration 2.0//EN"
    "http://struts.apache.org/dtds/struts-2.0.dtd">
<struts>
    ...
</struts>
```

struts.xml 文件是 XML 格式的文件,其结构由两大部分组成:文件序言和文件主体两个大的部分。其中<struts>元素之前的部分是文件序言部分,<struts>…</struts>部分是文件的主体部分。

文件序言部分的第一行是 XML 文件必须要声明的,并且必须位于 XML 文件的第一行,其主要作用是告诉 XML 解析器如何工作。其中,version 属性必填,用于指定该文件所使用的版本号。encoding 属性可选,用于指定该 XML 文件内容所使用的编码类型;如果省略该属性,则该 XML 文件中的内容使用 Unicode 编码,建议不要省略。

文件序言部分的"<!DOCTYPE"语句表示开始声明应用外部 DTD 文件,其中"struts"表示该 DTD 文件的根元素的名称,"struts-2.0.dtd"是外部 DTD 文件的名称。

文件主体部分的元素、元素的属性、元素的顺序及元素能够包含的内容等必须符合"struts-2.0.dtd"文件中的定义。

3.3.1 struts.xml 文件的基本框架

要想了解 struts.xml 文件的基本框架,就必须熟悉 struts-2.0.dtd 文件,该 DTD 文件位于 struts2-core-2.2.3.jar 文件中,如图 3-2 所示。

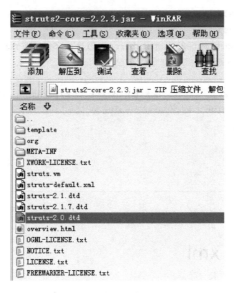

图 3-2 struts-2.0.dtd 文件位置

选中 struts-2.0.dtd 文件,然后单击图 3-2 中的"查看"按钮将显示该文件内容。

为了方便讲解,将 struts.dtd 文件进行分解介绍。struts-2.0.dtd 文件的开始部分内容如下:

```
<?xml version = "1.0" encoding = "UTF-8"?>
...
<!--
    Struts configuration DTD.
    Use the following DOCTYPE

    <!DOCTYPE struts PUBLIC
     "-//Apache Software Foundation//DTD Struts Configuration 2.0//EN"
     "http://struts.apache.org/dtds/struts-2.0.dtd">
-->

<!ELEMENT struts (package|include|bean|constant) *>
```

其中:

(1) 第一行告诉 XML 解析器将如何工作。

(2) 在"<!--"和"-->"标记符之间的内容是注释内容,其中"Use the following DOCTYPE"这一行内容指出在 struts.xml 文件的文件序言部分第一行之后,要写上如下的内容来应用外部 DTD 文件。

```
<!DOCTYPE struts PUBLIC
  "-//Apache Software Foundation//DTD Struts Configuration 2.0//EN"
  "http://struts.apache.org/dtds/struts-2.0.dtd">
```

(3) <！ELEMENT struts（package|include|bean|constant）*＞语句指出配置文件的根元素是 struts,该元素可以包含的子元素有 package、include、bean 和 constant,并且这些子元素可以不出现,也可以出现多次。另外,这些子元素的关系是并列的,并且这 4 个子元素使用的时候,必须按 struts-2.0.dtd 文件中描述的顺序进行配置。

由上面的描述可知,struts.xml 文件的最基本框架如下:

```
<?xml version = "1.0" encoding = "UTF - 8" ?>
<!DOCTYPE struts PUBLIC
    " - //Apache Software Foundation//DTD Struts Configuration 2.0//EN"
    "http://struts.apache.org/dtds/struts - 2.0.dtd">
<struts>
</struts>
```

下面将分别对各个子元素进行介绍。

3.3.2 package 及其包含的子元素

1. package 子元素

在 struts-2.0.dtd 文件中 package 子元素的描述如下:

```
<!ELEMENT package (result - types?, interceptors?, default - interceptor - ref?, default -
action - ref?, default - class - ref?, global - results?, global - exception - mappings?,
action * )>
<!ATTLIST package
    name CDATA #REQUIRED
    extends CDATA #IMPLIED
    namespace CDATA #IMPLIED
    abstract CDATA #IMPLIED
    externalReferenceResolver NMTOKEN #IMPLIED
>
```

package 子元素用于配置包的相关信息。该子元素具有 name、extends、namespace、abstract 和 externalReferenceResolver 五个属性。并且,该子元素还可以包含 0 个或者多个 action 子元素及一些可选的子元素: result-types、interceptors、default-interceptor-ref、default-action-ref、default-class-ref、global-results、global-exception-mappings。这些子元素的关系是并列的,并且这些子元素在使用的时候,必须按 struts-2.0.dtd 文件中描述的顺序进行配置,并且要作为 package 元素的子元素使用。

(1) package 子元素的属性如表 3-1 所示。

表 3-1 package 子元素的属性

属 性 名 称	功　　能	必　选
name	用于配置包的名称	是
extends	当该包继承其他包时,通过将该包的 extends 属性值配置成所要继承包的包名(name 属性值)即可	否

续表

属性名称	功　能	必选
namespace	用于配置包的命名空间。使用命名空间可以避免 action 名称之间的冲突,在不同的命名空间中的 action 可以使用相同的名称。如果没有配置此属性,则表明该包使用的是默认的名称空间	否
abstract	用于配置包是否为抽象包,属性的默认值是 false。如果属性值设置为 true,则表明该包为抽象包,在该抽象包中不能定义 action,只能作为父包	否
externalReferenceResolver	用于配置外部引用解析器,可以用来整合第三方类库,如 Spring	否

(2) package 子元素的属性配置示例。

例 3-1　抽象包及子包的定义。

struts.xml

```
<?xml version = "1.0" encoding = "UTF-8" ?>
<!DOCTYPE struts PUBLIC
    "-//Apache Software Foundation//DTD Struts Configuration 2.0//EN"
    "http://struts.apache.org/dtds/struts-2.0.dtd">

<struts>
    <package name = "childPackage" namespace = "/subSpace" extends = "parentPackage">
        ...
    </package>
    <package name = "parentPackage" abstract = "true" extends = "struts-default">
        ...
    </package>
</struts>
```

在该示例中配置了两个 package,其中名称为"parentPackage"的 package 是抽象包,该抽象包继承了"struts-default"包。名称为"childPackage"的 package 继承了"parentPackage"包,其命名空间是"/subSpace"。例如,如果用户的请求是/subSpace/sub.action,那么系统将会在/subSpace 命名空间中搜索名称为 sub 的 action。如果找到了,就会调用该 action,否则将会去默认的命名空间中搜索名称为 sub 的 action。

2. result-types 子元素

result-types 子元素用于设置视图的类型。在 struts-2.0.dtd 文件中 result-types 子元素的描述如下：

```
<!ELEMENT result-types (result-type+)>

<!ELEMENT result-type (param*)>
<!ATTLIST result-type
    name CDATA #REQUIRED
```

```
        class CDATA #REQUIRED
        default (true|false) "false"
>
<!ELEMENT param (#PCDATA)>
<!ATTLIST param
        name CDATA #REQUIRED
>
```

result-types 子元素包含一个或多个 result-type 子元素,但不具有属性。

result-type 子元素具有 name、class 和 default 三个属性,并且该子元素包含 0 个或多个 param 子元素。param 子元素用来给 result-type 子元素传递参数,参数的名称由 param 子元素的 name 属性指定。

(1) result-type 子元素的属性如表 3-2 所示。

表 3-2 result-type 子元素的属性

属性名称	功　能	必　选
name	用于配置结果类型的名称	是
class	用于配置处理该结果类型的完整实现类	是
default	用于配置结果类型是不是 Struts 2 框架的默认结果类型,其值是逻辑值 true 或 false。默认值是 false	否

(2) result-types 子元素的配置示例。

例 3-2　result-types 及其子元素 result-type 的配置。

```
<result-types>
    <result-type name="chain" class="com.opensymphony.xwork2.ActionChainResult"/>
    <result-type name="dispatcher"
            class="org.apache.struts2.dispatcher.ServletDispatcherResult" default="true"/>
</result-types>
```

在该示例中,result-types 子元素下又配置了两个 result-type 子元素。其中名称为 chain 的结果类型的实现类是 com.opensymphony.xwork2.ActionChainResult;名称为 dispatcher 的结果类型的实现类是 org.apache.struts2.dispatcher.ServletDispatcherResult,并且该结果类型是 Struts 2 框架默认的结果类型。

3. interceptors 子元素

interceptors 子元素用于设置与拦截器和拦截器栈有关的内容。在 struts-2.0.dtd 文件中 interceptors 子元素的描述如下:

```
<!ELEMENT interceptors (interceptor|interceptor-stack)+>
```

interceptors 子元素包含一个或多个 interceptor 及 interceptor-stack 子元素,但不具有属性。

(1) interceptor 子元素。

interceptor 子元素用来配置拦截器的相关内容,其定义如下:

```
<!ELEMENT interceptor (param * )>
<!ATTLIST interceptor
    name CDATA #REQUIRED
    class CDATA #REQUIRED
>
```

interceptor 子元素包含 0 个或多个 param 子元素,param 子元素用来给 interceptor 子元素传递参数。interceptor 子元素具有如表 3-3 所示的属性。

表 3-3　interceptor 子元素的属性

属 性 名 称	功　　能	必　选
name	用于配置拦截器(interceptor)的名称	是
class	用于配置拦截器的实现类	是

(2) interceptor-stack 子元素。

interceptor-stack 子元素用来配置拦截器栈的相关内容,其定义如下:

```
<!ELEMENT interceptor-stack (interceptor-ref * )>
<!ATTLIST interceptor-stack
    name CDATA #REQUIRED
>

<!ELEMENT interceptor-ref (param * )>
<!ATTLIST interceptor-ref
    name CDATA #REQUIRED
>
```

interceptor-stack 子元素具有一个 name 属性,必选,用于配置拦截器栈的名称,并且该子元素包含 0 个或多个 interceptor-ref 子元素。interceptor-ref 子元素用于配置所要引用的拦截器或拦截器栈,其中所要引用的拦截器或拦截器栈的名称由 name 属性指定。

(3) interceptors 子元素配置示例。

例 3-3　interceptors 及其子元素的配置。

```
<interceptors>
    <interceptor name = "logger"
            class = "com.opensymphony.xwork2.interceptor.LoggingInterceptor"/>
    <interceptor name = "timer"
            class = "com.opensymphony.xwork2.interceptor.TimerInterceptor"/>
    <interceptor-stack name = "loggerTimer">
        <interceptor-ref name = "logger"/>
        <interceptor-ref name = "timer"/>
    </interceptor-stack>
</interceptors>
```

在该示例中,定义了"logger"和"timer"两个拦截器,每个拦截器的实现类由对应的 class 属性指定。另外,还定义了一个名称为"loggerTimer"的拦截器栈,在该拦截器栈的定义中使用了 interceptor-ref 子元素引用了已经定义的"logger"和"timer"拦截器。如果在调用 action 的时候,需要使用"logger"和"timer"两个拦截器,则可以通过使用"loggerTimer"拦截器栈来完成,这样可以减少代码的重复。

4．default-interceptor-ref 子元素

default-interceptor-ref 子元素用于配置所要引用的默认的拦截器或拦截器栈。在 struts-2.0.dtd 文件中 default-interceptor-ref 子元素的描述如下:

```
<!ELEMENT default-interceptor-ref (#PCDATA)>
<!ATTLIST default-interceptor-ref
    name CDATA #REQUIRED
>
```

default-interceptor-ref 子元素不包含其他子元素,只具有一个 name 属性。

(1) default-interceptor-ref 子元素的属性如表 3-4 所示。

表 3-4 default-interceptor-ref 子元素的属性

属性名称	功能	必选
name	用于配置所要引用的默认的拦截器或拦截器栈的名称,即 interceptor 子元素或 interceptor-stack 子元素的 name 属性设置的值	是

(2) default-interceptor-ref 子元素的示例。

例 3-4 配置默认的拦截器。

```
<default-interceptor-ref name="defaultStack"/>
```

在该示例中,将名称为"defaultStack"的拦截器栈设置成了默认的拦截器栈。这样在整个包范围内的所有 action 都可以应用这个默认的拦截器栈信息。

5．default-action-ref 子元素

default-action-ref 元素用于配置所要引用的默认的 action。在 struts-2.0.dtd 文件中 default-action-ref 子元素的描述如下:

```
<!ELEMENT default-action-ref (#PCDATA)>
<!ATTLIST default-action-ref
    name CDATA #REQUIRED
>
```

default-action-ref 子元素不包含其他子元素,只具有一个 name 属性。

(1) default-action-ref 子元素的属性如表 3-5 所示。

表 3-5 default-action-ref 子元素的属性

属性名称	功能	必选
name	用于配置所要引用的默认的 action 的名称,即 action 子元素的 name 属性设置的值	是

（2）default-action-ref 子元素的示例。

例 3-5　配置的默认 action。

```
<default-action-ref name="defaultAction"></default-action-ref>
```

在该示例中，将名称为"defaultAction"的 action 设置成了默认的 action。这样当 Struts 2 框架发现请求中不包含 action 名称或者请求中的 action 名称不存在的时候，则 Struts 2 框架将使用 default-action-ref 子元素配置的默认 action 进行处理，本例中的默认 action 是 "defaultAction"。

6. default-class-ref 子元素

default-class-ref 子元素用于配置所要引用的默认的 class。在 struts-2.0.dtd 文件中 default-class-ref 子元素的描述如下：

```
<!ELEMENT default-class-ref (#PCDATA)>
<!ATTLIST default-class-ref
    class CDATA #REQUIRED
>
```

default-class-ref 子元素不包含其他子元素，只具有一个 class 属性。

（1）default-class-ref 子元素的属性如表 3-6 所示。

表 3-6　default-class-ref 子元素的属性

属性名称	功　　能	必　选
class	用于配置所引用的默认 class 的完整实现类的名称	是

（2）default-class-ref 子元素的示例。

例 3-6　配置默认的 class。

```
<default-class-ref class="com.opensymphony.xwork2.ActionSupport" />
```

在该示例中，配置了默认的完整实现类为"com.opensymphony.xwork2.ActionSupport"。当 struts.xml 文件中的 action 省略 class 属性的时候，就会调用 default-class-ref 子元素配置的类。

7. global-results 子元素

global-results 子元素为当前包配置全局 result，其可以应用于当前包中所有的 action。在 struts-2.0.dtd 文件中 global-results 子元素的描述如下：

```
<!ELEMENT global-results (result+)>

<!ELEMENT result (#PCDATA|param)*>
<!ATTLIST result
```

```
        name CDATA #IMPLIED
        type CDATA #IMPLIED
>
```

global-results 子元素包含一个或者多个 result 子元素,不具有任何属性。result 子元素包含 0 个或多个 param 子元素或者字符串,用于配置视图资源路径,并且还具有 name 和 type 两个属性。

(1) result 子元素的属性如表 3-7 所示。

表 3-7　result 子元素的属性

属性名称	功　　能	必　选
name	用于配置 result 子元素的名称	否
type	用于配置 result 子元素的类型。如果不配置 type 属性,Struts 2 框架则使用默认的 result 类型 dispatcher。Struts 2 框架支持的内置 result 类型是在 struts-default.xml 配置文件中定义的	否

(2) global-results 和 result 子元素的示例。

例 3-7　配置全局的 result。

```
<global-results>
    <result name="success">/success.jsp</result>
    <result name="error">
        <param name="location">/fail.jsp</param>
    </result>
</global-results>
```

该示例配置了全局的 result,其中定义了两个 result 子元素,名称分别为 success 和 error。当 action 执行完毕后,根据 action 返回的结果码(success 或者 error)调用相应的视图资源。其中名称为"success"的 result 子元素使用字符串指定了视图资源的路径,而名称为"error"的 result 子元素使用 param 子元素指定了视图资源的路径。

8. global-exception-mappings 子元素

global-exception-mappings 子元素为当前包配置全局异常映射,其可以应用于当前包中的所有 action。在 struts-2.0.dtd 文件中 global-exception-mappings 子元素的描述如下:

```
<!ELEMENT global-exception-mappings (exception-mapping+)>

<!ELEMENT exception-mapping (#PCDATA|param)*>
<!ATTLIST exception-mapping
    name CDATA #IMPLIED
    exception CDATA #REQUIRED
    result CDATA #REQUIRED
>
```

global-exception-mappings 子元素包含一个或者多个 exception-mapping 子元素,不具

有任何属性。exception-mapping 子元素包含 0 个或多个 param 子元素或者字符串，并且还具有 name、exception 和 result 三个属性。

（1）result 子元素的属性如表 3-8 所示。

表 3-8　result 子元素的属性

属性名称	功　　能	必　选
name	用于配置 exception-mapping 子元素的名称	否
exception	用于配置 exception-mapping 子元素所对应的异常类	是
result	当 action 抛出异常时，配置 action 返回的 result 的名称	是

（2）global-exception-mappings 和 exception-mapping 子元素的示例。

例 3-8　配置全局异常映射。

```
<global-exception-mappings>
    <exception-mapping exception="java.lang.Exception" result="Exception"/>
</global-exception-mappings>
```

该示例配置了全局异常映射，其中定义了省略名称的 exception-mapping 子元素。当 action 执行过程中抛出"java.lang.Exception"异常时，将查找名称为"Exception"的 result 配置，找到后，显示对应的视图资源的内容。

9. action 子元素

action 子元素用于配置 action 映射。在 struts-2.0.dtd 文件中 action 子元素的描述如下：

```
<!ELEMENT action (param|result|interceptor-ref|exception-mapping)*>
<!ATTLIST action
    name CDATA #REQUIRED
    class CDATA #IMPLIED
    method CDATA #IMPLIED
    converter CDATA #IMPLIED
>
```

action 子元素具有 name、class、method 和 converter 四个属性，并且该子元素还可以包含 0 个或者多个 param、result、interceptor-ref 和 exception-mapping 子元素。action 子元素中的 result 和 exception-mapping 两个子元素，其作用范围只局限在该 action 子元素中，因此，action 中的 result 子元素配置称为局部 result 配置，exception-mapping 子元素配置称为局部 exception 配置。

（1）action 子元素的属性如表 3-9 所示。

表 3-9　action 子元素的属性

属性名称	功　　能	必　选
name	action 的逻辑名称，用于匹配用户的请求	是
class	配置 action 的完整实现类的名称	否

续表

属 性 名 称	功　　能	必　选
method	配置所要调用的 action 中的方法名称	否
converter	配置 action 类型转换器的完整实现类名称	否

（2）action 子元素的示例。

例 3-9　定义 action。

```
<action name = "helloWorld" class = "example.struts2.HelloWorld">
    …
</action>
```

在该示例的 action 定义中，action 名称为"helloWorld"，其完整实现类的名称为"example.struts2.HelloWorld"。当请求的 action 名称是"helloWorld"时，Struts 2 框架将执行"example.struts2.HelloWorld"进行处理。

```
<action name = "helloWorld" class = "example.struts2.HelloWorld" method = "hello">
    …
</action>
```

在该示例的 action 定义中，配置了 method 属性。当请求的 action 名称是"helloWorld"，并且请求的方法是"hello"时，Struts 2 框架将执行"example.struts2.HelloWorld"类中的"hello"方法进行处理。

3.3.3　include 子元素

include 子元素用于包含其他的配置文件。在 struts-2.0.dtd 文件中 include 子元素的描述如下：

```
<!ELEMENT include (#PCDATA)>
<!ATTLIST include
    file CDATA #REQUIRED
>
```

include 子元素只具有一个必选的 file 属性。

1．include 子元素的属性（如表 3-10 所示）

表 3-10　include 子元素的属性

属 性 名 称	功　　能	必　选
file	用于配置被包含的配置文件的名称	是

2．include 子元素的示例

例 3-10　在 struts.xml 文件中包含 strutsConfig.xml 配置文件。

struts.xml

```xml
<?xml version="1.0" encoding="UTF-8" ?>
<!DOCTYPE struts PUBLIC
    "-//Apache Software Foundation//DTD Struts Configuration 2.0//EN"
    "http://struts.apache.org/dtds/struts-2.0.dtd">
<struts>
    <include file="strutsConfig.xml" />
</struts>
```

在该示例中,符合 struts-2.0.dtd 定义的 strutsConfig.xml 文件被包含到了 struts.xml 配置文件中。如果 struts.xml 配置文件中还要使用 package 子元素,则必须将 package 子元素放置在 include 子元素之前:

struts.xml

```xml
<?xml version="1.0" encoding="UTF-8" ?>
<!DOCTYPE struts PUBLIC
    "-//Apache Software Foundation//DTD Struts Configuration 2.0//EN"
    "http://struts.apache.org/dtds/struts-2.0.dtd">
<struts>
    <package name="strutsPackage">
        ...
    </package>
    <include file="strutsConfig.xml" />
</struts>
```

3.3.4　bean 子元素

bean 子元素用于对象注入和值注入。在 struts-2.0.dtd 文件中 bean 子元素的描述如下:

```
<!ELEMENT bean (#PCDATA)>
<!ATTLIST bean
    type CDATA #IMPLIED
    name CDATA #IMPLIED
    class CDATA #REQUIRED
    scope CDATA #IMPLIED
    static CDATA #IMPLIED
    optional CDATA #IMPLIED
>
```

bean 子元素不包含其他子元素,但具有 type、name、class、scope、static 和 optional 六个属性。

(1) bean 子元素的属性如表 3-11 所示。

(2) bean 子元素的示例。

由于 Struts 2 框架提供的默认 bean 配置已经能够满足 Web 应用开发,因此,通常不在

struts.xml 文件中配置 bean。Struts 2 框架中配置 bean 的默认文件是 struts-default.xml。如果在 Web 应用中需要程序员自己开发组件，则要在 Struts 2 框架的配置文件中进行配置来扩展或者替换 Struts 2 框架中组件的定义。

表 3-11　bean 子元素的属性

属 性 名 称	功　　能	必　选
type	用于配置 bean 类实现的接口	否
name	用于配置 bean 实例的名称。在多个 bean 的配置中，如果 type 属性具有相同的值，则 name 属性值不能相同	否
class	用于配置 bean 的完整实现类名称	是
scope	用于配置 bean 的作用范围，其可选的范围值包括：default、singleton、request、session 和 thread	否
static	用于配置 bean 是否使用静态方法注入，其值为 true 或 false。如果配置了 type 属性，则该属性值必须为 false	否
optional	用于配置 bean 是否为可选的 bean	否

例 3-11　配置 bean。

示例代码如下：

struts.xml

```
<?xml version = "1.0" encoding = "UTF-8"?>
<!DOCTYPE struts PUBLIC
    "-//Apache Software Foundation//DTD Struts Configuration 2.0//EN"
    "http://struts.apache.org/dtds/struts-2.0.dtd">
<struts>
    <bean type = "com.opensymphony.xwork2.ObjectFactory" name = "mybean"
          class = "myapp.ObjectFactory"/>
</struts>
```

在该示例中，使用一个自定义的 ObjectFactory 对象替换了 Struts 2 框架中内置的 ObjectFactory。

3.3.5　constant 子元素

constant 子元素用于配置常量。在 struts-2.0.dtd 文件中 bean 子元素的描述如下：

```
<!ELEMENT constant (#PCDATA)>
<!ATTLIST constant
    name CDATA #REQUIRED
    value CDATA #REQUIRED
>
```

constant 子元素不包含其他子元素，但具有 name 和 value 两个属性。

（1）constant 子元素的属性如表 3-12 所示。

表 3-12 constant 子元素的属性

属 性 名 称	功 能	必 选
name	用于配置常量的名称	是
value	用于配置常量的值	是

(2) constant 子元素的示例。

例 3-12 在 struts.xml 配置文件中使用 constant 子元素配置 Struts 2 框架的属性。

任务描述：在第 2 章 Chapter2 项目的基础上进行操作，在 struts.xml 文件中配置 struts.action.extension 常量，使 Struts 2 框架能够处理的请求的后缀是 struts2。

具体操作步骤如下：

① 编辑 struts.xml 文件，在 package 子元素的结束标记之后添加 constant 子元素的配置，内容如下：

struts.xml

```xml
<?xml version="1.0" encoding="UTF-8"?>
<!DOCTYPE struts PUBLIC
    "-//Apache Software Foundation//DTD Struts Configuration 2.0//EN"
    "http://struts.apache.org/dtds/struts-2.0.dtd">

<struts>
    <package name="default" namespace="/" extends="struts-default">
        <action name="helloWorld" class="example.struts2.HelloWorld">
            <result name="success">/showMessage.jsp</result>
            <result name="error">/noMessage.jsp</result>
        </action>
    </package>
    <constant name="struts.action.extension" value="struts2"/>
</struts>
```

注意：在 struts.xml 配置文件中，package 和 constant 子元素是并列的，并且是有先后顺序的。

② 编辑 inputMessage.jsp 页面，修改<form>表单元素的 action 属性值，如下：

```
<form action="/Chapter2/helloWorld.struts2" method="post">
```

③ 发布 Web 项目 Chapter2，启动 Tomcat 服务器，并输入如下的网址访问 inputMessage.jsp 页面。

```
http://localhost:8080/Chapter2/inputMessage.jsp
```

④ 在打开的页面中单击 Submit 按钮，结果如图 3-3 所示。

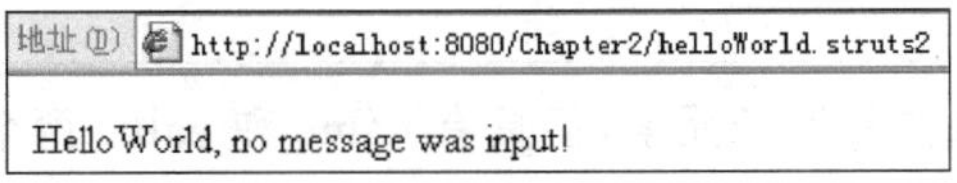

图 3-3 单击 Submit 按钮后的页面

由图 3-3 可以看出，用户的请求的扩展名已经变成了"struts2"。

3.4 配置 struts.properties

例 3-12 通过在 struts.xml 文件中使用 constant 子元素配置了 Struts 2 框架的 struts.action.extension 常量，更改了用户请求的扩展名。其实，Struts 2 框架在 struts2-core-2.2.3.jar 文件的 org.apache.struts2 目录下使用 default.properties 文件配置了 struts.action.extension 常量的值，如下：

```
struts.action.extension=action,,
```

这个配置表明了 Struts 2 框架能够处理的用户请求的后缀是 action 或者为空，也就是说，在第 2 章中的 Chapter2 项目中，将 inputMessage.jsp 页面的＜form＞表单元素的 action 属性值写成如下的形式也是正确的。

```
<form action="/Chapter2/helloWorld" method="post">
```

由于在 struts.xml 配置文件中，使用 constant 子元素为 struts.action.extension 常量配置了新的值，因此 Struts 2 框架处理的用户请求的扩展名发生了变化。由此可知，struts.xml 文件中配置的常量优先 default.properties 文件中配置的常量加载。也就是说，如果在 struts.xml 和 default.properties 文件中有同名的常量，则 struts.xml 文件中配置的常量生效。

由于 default.properties 文件是 Struts 2 框架中默认的常量（或属性）配置文件，因此，不能在该文件中修改其内容。如果想要修改 default.properties 文件中的属性配置，必须在 struts.xml 文件中使用 constant 子元素进行配置吗？其实，default.properties 文件中的这些属性还可以在 struts.properties 文件中进行配置，其配置格式同 default.properties 文件中属性配置格式。struts.properties 和 default.properties 文件都是标准的属性配置文件，由每行一条形如"key=value"格式的内容组成，其中，"key"表示 Struts 2 框架的属性，"value"表示 Struts 2 框架属性的值。

由于 struts.properties 是由开发人员创建的，因此，为了使 Struts 2 框架能够加载该文件，必须将其保存在 Web 应用的"WEB-INF/classes"文件夹下。

下面针对 default.properties 文件中的一些常用属性进行介绍。

3.4.1 开发模式属性

struts.devMode、struts.configuration.xml.reload 和 struts.i18n.reload 三个属性通常应用于 Web 应用开发阶段。

1. struts.devMode

该属性用于配置基于 Struts 2 框架的 Web 应用是否使用开发模式。该属性有 true 和 false 两个可选值，默认值是 false，用于 Web 应用发布阶段。在开发阶段，通常将该属性值设置为 true,这样当 Web 应用出错时,则可以显示更多、更友好的出错提示信息。

2. struts.configuration.xml.reload

该属性用于配置当 struts.xml 文件改变后,是否自动重新被系统加载。该属性有 true 和 false 两个可选值,默认值是 false,用于 Web 应用发布阶段。在开发阶段,通常将该属性值设置为 true,这样当 struts.xml 配置文件被修改后,系统能够自动加载该文件,使得程序开发人员能够实时看到 struts.xml 文件的配置效果。

3. struts.i18n.reload

该属性用于配置是否每次 HTTP 请求到达的时候,系统都会重新加载国际化资源文件。该属性有 true 和 false 两个可选值,默认值是 false,用于 Web 应用发布阶段。在开发阶段,通常将该属性值设置为 true,这样在每次请求时都可以重新加载国际化资源文件,从而可以让程序开发人员能够实时看到开发效果。

在 Web 应用发布阶段,将上述三个属性值设置为 false,是为了提高系统的响应性能。

4. 配置示例

例 3-13 开发模式属性的配置。

struts.properties 中的属性可以在 web.xml、struts.properties、struts.xml、struts-plugin.xml 和 struts-default.xml 等文件中进行配置。如果在这些配置文件中同时配置了同名属性,则 Struts 2 框架将使用后加载的配置文件中的同名属性覆盖先加载的配置文件中的属性。

(1) 在 web.xml 文件中配置属性。

```xml
<filter>
    ...
    <init-param>
        <param-name>struts.devMode</param-name>
        <param-value>true</param-value>
    </init-param>
    <init-param>
        <param-name>struts.configuration.xml.reload</param-name>
        <param-value>true</param-value>
    </init-param>
    <init-param>
        <param-name>struts.i18n.reload</param-name>
        <param-value>true</param-value>
    </init-param>
</filter>
```

(2) 在 struts.properties 文件中配置属性。

```
struts.devMode = true
struts.configuration.xml.reload = true
struts.i18n.reload = true
```

（3）在 struts.xml 文件中配置属性。

```
<struts>
    ...
    <constant name = "struts.devMode" value = "true" />
    <constant name = "struts.configuration.xml.reload" value = "true" />
    <constant name = "struts.i18n.reload" value = "true" />
</struts>
```

在 struts-default.xml 和 struts-plugin.xml 两个配置文件中进行属性配置，和在 struts.xml 文件中进行属性配置方法一样，都是使用<constant>子元素。由于开发人员不能修改 struts-default.xml 和 struts-plugin.xml 这两个配置文件，因此，对属性的配置采用前三种方法来完成。通常情况下，为了保持 struts.xml 文件和 web.xml 文件的独立性，最好将属性配置放置在 struts.properties 文件中。

3.4.2 国际化属性

struts.locale、struts.i18n.encoding 和 struts.custom.i18n.resources 三个属性用于资源的国际化。

1. struts.locale

该属性用于配置默认的 locale，其值格式为："语言_国家"。例如，配置中国中文为默认的 locale，则 struts.locale 属性的配置如下：

```
struts.locale = zh_CN
```

2. struts.i18n.encoding

该属性用于配置默认的字符编码，其默认值为 UTF-8。

3. struts.custom.i18n.resources

该属性用于配置要加载的国际化资源文件，如果有多个资源文件要加载，则用它们之间逗号分隔。

```
struts.custom.i18n.resources = res1,res2,res3
```

配置的资源文件名称只写主文件名，省略扩展名".properties"。

3.4.3 文件上传属性

struts.multipart.parser、struts.multipart.saveDir 和 struts.multipart.maxSize 三个属性用于文件上传。

1. struts.multipart.parser

该属性用于配置处理 MIME 的类型为 enctype = "multipart/form-data"，且 HTTP 请

求为 POST 的解析器。属性值可以选择 cos、pell 和 jakarta 等值,默认值为 jakarta,即使用 JakartaMultiPartRequest 解析器,该解析器使用 Common-FileUpload 文件上传框架对请求进行解析,然后将解析结果保存。

2. struts.multipart.saveDir

该属性用于配置上传后的文件临时存储目录,该属性的默认值是 javax.servlet.context.tempdir,其值和 Tomcat 服务器的安装位置有关。假设 Tomcat 安装的目录为"d:\tomcat",则 javax.servlet.context.tempdir 的值为"d:\tomcat\work\Catalina\localhost\web 应用名称",其中 Catalina 是 Tomcat 服务器默认的引擎名称,localhost 是 Catalina 引擎下虚拟主机的名称。

3. struts.multipart.maxSize

该属性用于配置允许客户端上传文件的最大字节数,默认值是 2097152,即 2M 字节。

3.4.4 模板和主题属性

模板和主题是 Struts 2 框架中所有 UI 标签的核心。通过使用模板和主题,开发人员可以快速开发出便于维护的、美观的界面。struts.ui.theme、struts.ui.templateDir 和 struts.ui.templateSuffix 三个属性用于配置模板和主题的类型。

1. struts.ui.theme

该属性用于标准 UI 主题的配置,可选值有 simple、xhtml、css_xhtml 和 ajax,其默认值是 xhtml。

2. struts.ui.templateSuffix

该属性用于配置默认模板的类型,可选值有 ftl、vm 和 jsp 三个可选值,其默认值是 ftl。其中 ftl 表示使用 FreeMarker 编写的模板,vm 表示使用 Velocity 编写的模板,jsp 表示使用 Java Server Page 编写的模板。

3. struts.ui.templateDir

该属性用于配置 UI 主题的模板文件存放目录,其默认值是 template。在设置了 struts.ui.templateDir 属性后,Struts 2 框架将如何搜索设置的目录呢?下面举例说明,假设属性值使用的是默认值 template。

(1) 首先,在 Web 应用程序中查找名称为 template 的目录(和 WEB-INF 目录具有相同的父目录),如果找到,则使用该目录下的模板文件。注意:具体的路径要在模板名称后加上主题的名称。

例如,这三个属性都使用的是默认值,则其路径为"Web 应用/template/xhtml/"。

(2) 如果在 Web 应用程序中没有找到名称为 template 目录,则 Struts 2 框架会在 classpath 中进行查找。

例如,这三个属性都使用的是默认值,则其路径为"classpath/template/xhtml/"。

3.4.5 url 属性

struts.url.http.port、struts.url.https.port 和 struts.url.includeParams 三个属性用于 Struts 2 框架的 url 标签。

1. struts.url.http.port

当 Struts 2 框架要生成 url 时,该属性用于提供 Web 应用的服务端口。

2. struts.url.https.port

当 Struts 2 框架要生成 url 时,该属性用于提供 Web 应用的加密服务端口。

3. struts.url.includeParams

该属性用于配置在生成 url 时,是否包含请求参数,可选值有 none、get 和 all,默认值为 none。三个属性值的意义如表 3-13 所示。

表 3-13 none、get 和 all 三个属性值的意义

属 性 值	意 义
none	在生成 url 时,不包含请求提交的参数
get	在生成 url 时,仅包含 get 请求提交的参数
all	在生成 url 时,包含所有请求(get 和 post)提交的参数

例 3-14 struts.url.includeParams 属性示例。

(1) 创建 urlTest.jsp 页面:

urlTest.jsp

```
<%@ page language="java" contentType="text/html; charset=UTF-8"
    pageEncoding="UTF-8" %>
<%@ taglib uri="/struts-tags" prefix="s" %>
<s:url>
    <s:param name="number" value="1" />
</s:url>
```

(2) 启动 Tomcat 服务器,访问如下的网址,结果如图 3-4 所示。

```
http://localhost:8080/Chapter3/urlTest.jsp?id=123
```

(3) 修改 struts.url.includeParams 属性的值为 get,重新启动 Tomcat 服务器,并访问 urlTest.jsp 页面,结果如图 3-5 所示。

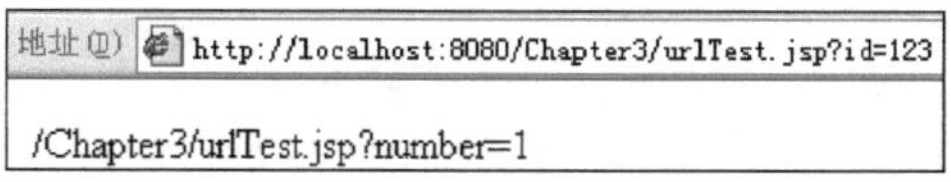

图 3-4 struts.url.includeParams 属性值为 none 的结果

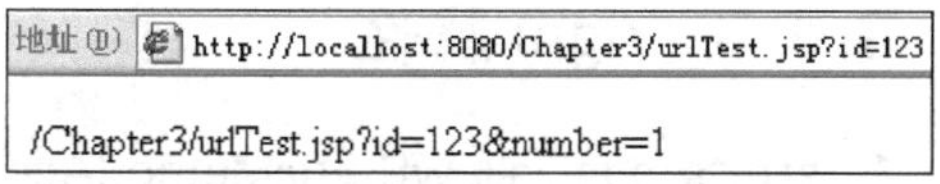

图 3-5 struts.url.includeParams 属性值为 get 的结果

在图 3-4 和图 3-5 的结果中可以看出 struts.url.includeParams 属性设置为不同值的区别。

3.4.6 freemarker 属性

1. struts.freemarker.manager.classname

该属性配置 Struts 2 框架所要使用的 FreeMarker 管理器的类名称,其默认值是 org.apache.struts2.views.freemarker.FreemarkerManager,它是 Struts 2 框架内建的 FreeMarker 管理器。

2. struts.freemarker.templatesCache

该属性配置是否启用缓存 FreeMarker 模板,有 true 和 false 两个可选值,默认值为 false。

3. struts.freemarker.beanwrapperCache

该属性配置是否启用 BeanWrapper 的缓存功能,有 true 和 false 两个可选值,默认值为 false。

4. struts.freemarker.wrapper.altMap

该属性有 true 和 false 两个可选值,默认值是 true。

5. struts.freemarker.mru.max.strong.size

该属性用于配置 FreeMarker 缓存的大小。对于高负荷的系统,该值通常为非 0,否则会导致频繁重新加载、解析 FreeMarker 模板。

3.4.7 velocity 属性

1. struts.velocity.configfile

该属性用于配置 Velocity 框架的 velocity.properties 文件的位置,其默认值为 velocity.properties。

2. struts.velocity.contexts

该属性用于配置 Velocity 框架的 Context 位置。如果 Velocity 框架有多个 Context,则它们之间用逗号分隔。

3. struts.velocity.toolboxlocation

该属性用于配置 Velocity 框架的 toolbox 的位置。

3.4.8 ognl 属性

1. struts.ognl.allowStaticMethodAccess

该属性用于配置是否允许在 OGNL 表达式中访问静态方法,有 true 和 false 两个可选值,默认值为 false。

如果将该属性值设置为 true,则可以在基于 Struts 2 框架的 JSP 页面中使用 OGNL 表达式访问类的静态方法,其格式为:@[类全名]@[方法名 | 值名]。

例 3-15 struts.ognl.allowStaticMethodAccess 属性示例。

(1) 创建 ognlTest.jsp。

ognlTest.jsp

```
<%@ page language="java" contentType="text/html; charset=UTF-8"
    pageEncoding="UTF-8" %>
<%@ taglib uri="/struts-tags" prefix="s" %>

<s:property value="@java.lang.String@format('Struts 2 %s', 'ognl test!')"/>
```

(2) 启动 Tomcat 服务器,访问 ognlTest.jsp 页面,不输出任何内容。

(3) 将 struts.ognl.allowStaticMethodAccess 属性值设置为 true。重新启动 Tomcat 服务器,并访问 ognlTest.jsp 页面,结果如图 3-6 所示。

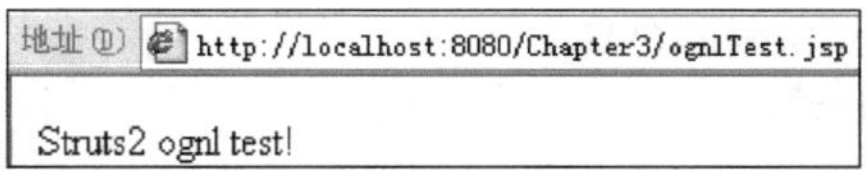

地址(D) http://localhost:8080/Chapter3/ognlTest.jsp
Struts2 ognl test!

图 3-6 struts.ognl.allowStaticMethodAccess 属性值为 true 的结果

2. struts.ognl.logMissingProperties

该属性用于配置是否记录警告属性日志,有 true 和 false 两个可选值,默认值为 false。

3. struts.ognl.enableExpressionCache

该属性用于配置是否缓存解析过的 OGNL 表达式,有 true 和 false 两个可选值,默认值为 false。如果设置为 true,则进行缓存,这种情况下,当 Web 应用生成大量不同的表达式时会导致内存泄漏。

3.4.9 其他属性

1. struts.enable.DynamicMethodInvocation

该属性用于配置 Struts 2 框架是否支持动态方法调用,有 true 和 false 两个可选值,默认值是 true。

例 3-16 struts.enable.DynamicMethodInvocation 属性示例。

(1) 创建 inputMessage.jsp 页面。

inputMessage.jsp

```
<%@ page language="java" contentType="text/html; charset=UTF-8"
    pageEncoding="UTF-8" %>
```

```
<form action = "/Chapter3/helloWorld!en.action" method = "post">
    Message:<input name = "message" type = "text">
    <input type = "submit" value = "Submit">
</form>
```

（2）创建 HelloWorld.java 文件。

HelloWorld.java

```java
package example.struts2;
public class HelloWorld {
    private String message;
    public String getMessage() {
        return message;
    }
    public void setMessage(String message) {
        this.message = message;
    }
    public String en(){
        if(getMessage().isEmpty()){
            return "error";
        }
        else{
            return "success";
        }
    }
}
```

（3）创建 struts.xml 文件。

struts.xml

```xml
<?xml version = "1.0" encoding = "UTF-8" ?>
<!DOCTYPE struts PUBLIC
    "-//Apache Software Foundation//DTD Struts Configuration 2.0//EN"
    "http://struts.apache.org/dtds/struts-2.0.dtd">
<struts>
    <package name = "default" namespace = "/" extends = "struts-default">
        <action name = "helloWorld" class = "example.struts2.HelloWorld">
            <result name = "success">/showMessage.jsp</result>
            <result name = "error">/noMessage.jsp</result>
        </action>
    </package>
</struts>
```

（4）发布 Web 项目 Chapter3，启动 Tomcat 服务器，访问 inputMessage.jsp 页面，单击 Submit 按钮，能够成功执行。

（5）将 struts.enable.DynamicMethodInvocation 属性设置为 false。重新启动 Tomcat 服务器，访问 inputMessage.jsp 页面，单击 Submit 按钮后，结果如图 3-7 所示。

图 3-7　struts.ognl.allowStaticMethodAccess 属性值为 false 的结果

2．struts.enable.SlashesInActionNames

该属性用于配置是否允许在 Action 的名称中使用斜线，有 true 和 false 两个可选值，默认值是 false。

例 3-17　struts.enable.SlashesInActionNames 属性示例。

在例 3-16 的基础上进行操作。

（1）修改 inputMessage.jsp 页面。

inputMessage.jsp

```
<%@ page language = "java" contentType = "text/html; charset = UTF-8"
    pageEncoding = "UTF-8" %>
<form action = "/Chapter3/HelloWorld/en.action" method = "post">
    Message:<input name = "message" type = "text">
    <input type = "submit" value = "Submit">
</form>
```

（2）修改 HelloWorld.java 文件。

HelloWorld.java

```
package example.struts2;
public class HelloWorld {
    private String message;
    public String getMessage() {
        return message;
    }
    public void setMessage(String message) {
        this.message = message;
    }
    public String en(){
        if(getMessage().isEmpty()){
            return "error";
        }
        else{
            return "success";
        }
    }
    public String zh(){
        if(getMessage().isEmpty()){
            return "error";
```

```
            }
            else{
                return "success";
            }
        }
    }
```

(3) 修改 struts.xml 文件。

struts.xml

```
<?xml version = "1.0" encoding = "UTF-8" ?>
<!DOCTYPE struts PUBLIC
    "-//Apache Software Foundation//DTD Struts Configuration 2.0//EN"
    "http://struts.apache.org/dtds/struts-2.0.dtd">
<struts>
    <package name = "default" namespace = "/" extends = "struts-default">
        <action name = "*/*" method = "{2}" class = "example.struts2.{1}">
            <result name = "success">/show{2}Message.jsp</result>
            <result name = "error">/no{2}Message.jsp</result>
        </action>
    </package>
</struts>
```

(4) 创建信息显示页面。

showenMessage.jsp

```
<%@ page language = "java" contentType = "text/html; charset = UTF-8"
    pageEncoding = "UTF-8" %>
<%@ taglib prefix = "s" uri = "/struts-tags" %>
HelloWorld, the input message was : <s:property value = "message"/>
```

showzhMessage.jsp

```
<%@ page language = "java" contentType = "text/html; charset = UTF-8"
    pageEncoding = "UTF-8" %>
<%@ taglib prefix = "s" uri = "/struts-tags" %>
输入信息是: <s:property value = "message"/>
```

noenMessage.jsp

```
<%@ page language = "java" contentType = "text/html; charset = UTF-8"
    pageEncoding = "UTF-8" %>
<%@ taglib prefix = "s" uri = "/struts-tags" %>
HelloWorld, no message was input!
```

nozhMessage.jsp

```
<%@ page language = "java" contentType = "text/html; charset = UTF-8"
    pageEncoding = "UTF-8" %>
```

```
<%@ taglib prefix = "s" uri = "/struts-tags" %>
HelloWorld,没有输入信息!
```

(5)发布 Web 项目 Chapter3,启动 Tomcat 服务器,访问 inputMessage.jsp,单击 Submit 按钮后出错,提示信息如图 3-8 所示。

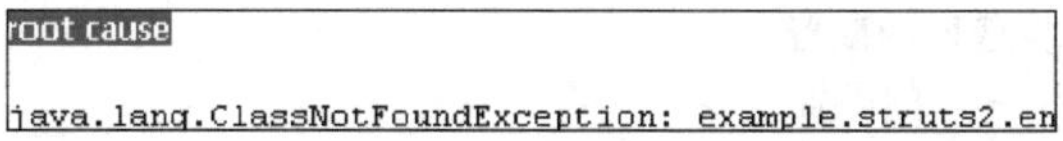

```
root cause
java.lang.ClassNotFoundException: example.struts2.en
```

图 3-8　访问 inputMessage.jsp 页面的出错信息

(6)将 struts.enable.SlashesInActionNames 属性设置为 true,重新启动 Tomcat 服务器,访问 inputMessage.jsp,单击 Submit 按钮后结果如图 3-9 所示。

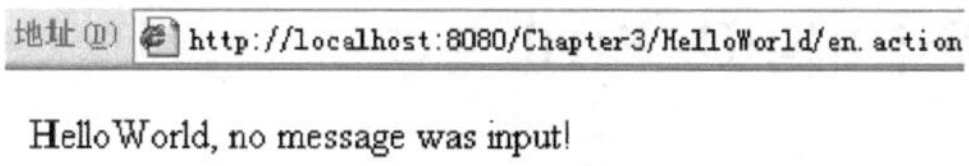

```
地址(D)  http://localhost:8080/Chapter3/HelloWorld/en.action
HelloWorld, no message was input!
```

图 3-9　struts.enable.SlashesInActionNames 属性值为 true 的结果

3. struts.action.extension

该属性用于配置 Struts 2 框架处理的用户请求的后缀,其默认值是 action 和空字符串,即所有没有后缀和后缀是 action 的用户请求都由 Struts 2 框架处理。可以同时配置多个后缀,它们之间用逗号分隔。

4. struts.custom.properties

该属性用于配置基于 Struts 2 框架的 Web 应用所要加载的用户自定义的属性文件,该属性文件中配置的属性不会覆盖 struts.properties 文件中配置的属性。如果需要加载多个属性文件,它们之间用逗号分隔。

3.5　配置 struts-default.xml

struts-default.xml 配置文件包含在 struts2-core-2.2.3.jar 文件中,是 Struts 2 框架默认加载的基本配置文件。struts-default.xml 配置文件提供了标准配置项的设置,包括 Struts 2 框架的一些核心的 bean、拦截器、拦截器栈和所有的默认的结果类型,这些配置会被自动包含到 struts.xml 配置文件中,而不用复制这些配置项。struts-default.xml 配置文件是 Struts 2 框架中自带的,只能使用,不能修改。

struts-default.xml 文件的基本结构如下:

```
<?xml version = "1.0" encoding = "UTF-8" ?>
<!DOCTYPE struts PUBLIC
    "-//Apache Software Foundation//DTD Struts Configuration 2.1.7//EN"
    "http://struts.apache.org/dtds/struts-2.1.7.dtd">
```

```xml
<struts>
    <!-- bean 的定义 -->
    <package name="struts-default" abstract="true">
                <!-- 结果类型的定义 -->
        <!-- 拦截器的定义 -->
        <!-- 拦截器栈的定义 -->
        <!-- 默认拦截器的配置 -->
        <!-- 默认 class 的配置 -->
    </package>
</struts>
```

struts-2.1.7.dtd 文档描述文件定义了 struts-default.xml 文件的结构。

3.6 配置 struts-plugin.xml

Struts 2 框架可以通过插件(plugin)机制来扩展、替换其自身的功能,并且通过插件这种方式扩展的功能,都与 Struts 2 框架的核心程序(struts2-core-2.2.3.jar)保持独立性。

Struts 2 框架的插件是以 JAR 包的形式存在,其包含了扩展、替换或者添加到 Struts 2 框架中的类及配置。为了配置插件,插件 JAR 包中应该包含一个 struts-plugin.xml 文件,该文件的格式和 struts.xml 文件的格式相同。在 struts-plugin.xml 文件中可以使用 results,interceptors 和 action 子元素定义新的包(packages),可以覆盖 Struts 2 框架的常量,添加新的扩展点实现类。

Struts 2 框架中很多流行的但可选的特性都是以插件的形式发布的。一个应用可以保留发布时所有的插件,或者仅包含它使用的插件。

插件不是以特定的顺序加载的,插件之间不应该有依赖关系。插件可以依赖 Struts 2 框架核心 JAR 包中的类,但是不应该依赖其他插件中的类。Struts 2 框架首先加载默认的配置文件(struts-default.xml),然后加载 classpath 中所有插件 JAR 包中的 struts-plugin.xml 配置文件,最后加载 struts.xml 配置文件。由于 struts.xml 配置文件最后被加载,因此,它可以使用插件中提供的任何资源。

struts2-spring-plugin-2.2.3.jar 文件是和 Struts 2.2.3 框架一起发布的集成 Spring 框架的插件,其 struts-plugin.xml 文件的配置如下:

struts-plugin.xml

```xml
<?xml version="1.0" encoding="UTF-8" ?>
<!DOCTYPE struts PUBLIC
    "-//Apache Software Foundation//DTD Struts Configuration 2.0//EN"
    "http://struts.apache.org/dtds/struts-2.0.dtd">

<struts>
    <bean type="com.opensymphony.xwork2.ObjectFactory" name="spring"
          class="org.apache.struts2.spring.StrutsSpringObjectFactory" />
    <!-- Make the Spring object factory the automatic default -->
```

```xml
<constant name = "struts.objectFactory" value = "spring" />
<constant name = "struts.class.reloading.watchList" value = "" />
<constant name = "struts.class.reloading.acceptClasses" value = "" />
<constant name = "struts.class.reloading.reloadConfig" value = "false" />
    <package name = "spring-default">
        <interceptors>
            <interceptor name = "autowiring" class = "com.opensymphony.xwork2.spring
                                    .interceptor.ActionAutowiringInterceptor"/>
        </interceptors>
    </package>
</struts>
```

习题

1. 阅读 struts-2.0.dtd 文件。
2. 修改 struts.xml 配置文件中子元素的顺序，查看运行结果。
3. 在 3.3.5 节的示例中，省略步骤(2)，查看运行结果。
4. 测试 struts.xml 和 struts.properties 文件中常量的加载顺序。
5. 阅读 struts-2.1.7.dtd 文件和 struts-default.xml 文件。

第 4 章 Struts 2框架进阶

本章详细介绍了 struts.xml 配置文件中 result 及 action 的配置,并且介绍了 ActionSupport 类的使用。

为了方便讲解本章内容,首先创建 Web 项目 Chapter4,并添加 Struts 2 框架的支持。Chapter4 项目的目录结构如图 4-1 所示。

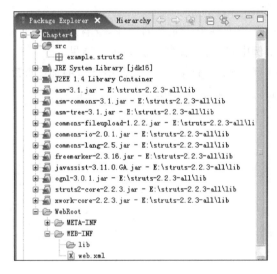

图 4-1 Chapter4 项目目录结构

编辑 web.xml 文件,内容如下:

web.xml

```
<?xml version = "1.0" encoding = "UTF - 8"?>
< web - app version = "2.4"
    xmlns = "http://java.sun.com/xml/ns/j2ee"
    xmlns:xsi = "http://www.w3.org/2001/XMLSchema - instance"
    xsi:schemaLocation = "http://java.sun.com/xml/ns/j2ee
    http://java.sun.com/xml/ns/j2ee/web - app_2_4.xsd">

    < filter >
      < filter - name > Struts 2 </filter - name >
      < filter - class >
```

```
            org.apache.struts2.dispatcher.ng.filter.StrutsPrepareAndExecuteFilter
        </filter-class>
    </filter>
    <filter-mapping>
        <filter-name>Struts 2</filter-name>
        <url-pattern>/*</url-pattern>
    </filter-mapping>
</web-app>
```

4.1 result 配置

一个 action 类的方法执行完成后，将会返回一个字符串，这个字符串的值用于选择 result 元素。在 struts.xml 配置文件中的 action 定义中，通常会有多个用于表示不同可能输出的 result 配置。

result 元素有以下两个作用：

（1）提供一个逻辑名称。action 在不知道任何其他实现细节的情况下，能够返回一个类似于"success"或"error"的标记。

（2）提供一个 result 类型。大多数 result 元素简单地转发服务器页面或模板，但是其他的 result 类型能够实现更复杂的功能。

result 元素的定义如下：

```
<!ELEMENT result (#PCDATA|param) * >
<!ATTLIST result
    name CDATA  #IMPLIED
    type CDATA  #IMPLIED
>
<!ELEMENT param (#PCDATA)>
<!ATTLIST param
    name CDATA  #REQUIRED
>
```

由定义可以看出，result 元素具有 name 和 type 两个属性。name 属性用于配置 result 元素的名称，和 action 返回的结果码进行匹配；type 属性用于指定 result 元素的类型。

另外，result 元素还可以包含 0 个或多个 param 子元素，该子元素只有一个 name 属性，用于指定参数的名称。param 子元素用于设置为 result 对象传递的参数，不同的 result 类型可以设置的参数也有所不同。

4.1.1 Struts 2 框架中内置的 result 类型

Struts 2 框架提供了一些实现 com.opensymphony.xwork2.Result 接口的 result 类型，并在 struts-default.xml 配置文件中进行了定义，内容如下：

```
<result-types>
    <result-type name="chain" class="com.opensymphony.xwork2.ActionChainResult"/>
```

```xml
<result-type name="dispatcher"
        class="org.apache.struts2.dispatcher.ServletDispatcherResult" default="true"/>
<result-type name="freemarker"
        class="org.apache.struts2.views.freemarker.FreemarkerResult"/>
<result-type name="httpheader"
        class="org.apache.struts2.dispatcher.HttpHeaderResult"/>
<result-type name="redirect"
        class="org.apache.struts2.dispatcher.ServletRedirectResult"/>
<result-type name="redirectAction"
        class="org.apache.struts2.dispatcher.ServletActionRedirectResult"/>
<result-type name="stream" class="org.apache.struts2.dispatcher.StreamResult"/>
<result-type name="velocity" class="org.apache.struts2.dispatcher.VelocityResult"/>
<result-type name="xslt" class="org.apache.struts2.views.xslt.XSLTResult"/>
<result-type name="plainText" class="org.apache.struts2.dispatcher.PlainTextResult"/>
</result-types>
```

在名称为 dispatcher 的 result 类型定义中,使用了 default="true" 设置,表明 dispatcher 是默认的 result 类型。因此,如果在 result 配置中没有指定 type 属性时,就默认使用 dispatcher。

这些内置的 result 类型可以直接在 Web 应用中使用。下面对内置的 result 类型进行介绍。

1. chain result

chain result 用于 action 的链式操作。这种 result 调用另外一个完整的 action,包括它的拦截器栈和 result。它包含如表 4-1 所示的 4 个参数。

表 4-1 chain result 包含的 4 个参数

名称	作用
actionName	默认参数,用于配置链接到的 action 的名称,即被调用的 action 的名称
namespace	用于配置被调用的 action 所在的命名空间,如果为空,则默认是当前的命名空间
method	用于指定目标 action 中被调用的其他方法,如果为空,默认调用 execute 方法
skipActions	用于配置被链式调用的 action 的名称列表,之间用逗号分隔,该参数可选

例 4-1 chain 配置示例。

```xml
<package name="public" extends="struts-default">
    <action name="createUser" class="…">
        <result type="chain">login</result>
    </action>
    <action name="login" class="…">
        <result type="chain">
            <param name="actionName">dashboard</param>
            <param name="namespace">/secure</param>
        </result>
    </action>
</package>
```

在名称为"createUser"的 action 定义中，使用默认的参数，将 createUser 链接到了 login。在名称为"login"的 action 定义中，使用 namespace 和 actionName 两个参数，将 login 链接到了"/secure"命名空间中的 dashboard。

2．dispatcher result

dispatcher result 用于包含或转发到某一视图（通常是 JSP 页面），即 Web 资源集成。它包含如表 4-2 所示的两个参数。

表 4-2　dispatcher result 包含的两个参数

名称	作用
location	默认参数，用于配置当 action 执行完成后所转发的位置
parse	默认值 true。如果设置为 false，则不解析 location 参数中的 OGNL 表达式

例 4-2　dispatcher 配置示例。

```
<result name="success" type="dispatcher">
    <param name="location">foo.jsp</param>
</result>
```

该配置表示当 action 执行完成后，如果返回的结果码是"success"，则使用 dispatcher 类型转向页面 foo.jsp。

由于 type 默认值是 dispatcher，另外 location 参数也是默认的参数，因此，上述配置可以写成如下的形式：

```
<result name="success">foo.jsp</result>
```

3．freemarker result

freemarker result 用于集成 FreeMarker，即使用 Freemarker 模板引擎输出一个视图。它包含如表 4-3 所示的 4 个参数。

表 4-3　freemarker result 包含的 4 个参数

名称	作用
location	默认参数，要处理的模板的位置
parse	默认值 true。如果设置为 false，则不解析 location 参数中的 OGNL 表达式
contentType	默认值"text/html"，可以设置为其他值
writeIfCompleted	默认值 false，表示在处理模板时如果没有任何错误则写入流

例 4-3　freemarker 配置示例。

```
<result name="success" type="freemarker">foo.ftl</result>
```

4．httpheader result

httpheader result 用于控制具体的 HTTP 行为。它包含如表 4-4 所示的 5 个参数。

表 4-4　httpheader result 包含的 5 个参数

名称	作用
status	在一个响应中设置 http servlet 响应状态码
parse	默认值 true。如果设为 false,则不解析 headers 参数中的 OGNL 表达式
headers	用于设置 header 值
error	在一个响应中设置 http servlet 响应错误码
errorMessage	指响应的错误消息,如果设置了"error"参数,则要设置该参数

例 4-4　httpheader 配置示例。

```
<result name="proxyRequired" type="httpheader">
    <param name="error">305</param>
    <param name="errorMessage">This action must be accessed through a proxy!</param>
</result>
```

该配置表示当 action 执行完成后,如果返回的结果码是 proxyRequired,则使用 httpheader 类型对 HTTP 行为进行控制,设置的响应错误码是 error,响应的错误消息是 "This action must be accessed through a proxy!"。

5. redirect result

redirect result 用于重定向到其他的 Web 资源(URL),重定向时将调用 HttpServletResponse 的 sendRedirect(String)方法。它包括如表 4-5 所示的 3 个参数。

表 4-5　redirect result 包含的 3 个参数

名称	作用
location	默认参数,用于配置当 action 执行完成后重定向的位置
parse	默认值 true。如果设为 false,则不解析 headers 参数中的 OGNL 表达式
anchor	为结果指定一个 anchor,该参数可选

例 4-5　redirect 配置示例。

```
<result name="success" type="redirect">
    <param name="location">foo.jsp</param>
    <param name="parse">false</param>
    <param name="anchor">FRAGMENT</param>
</result>
```

redirect result 在重定向时调用 HttpServletResponse 的 sendRedirect(String)方法,该操作意味着在重定向的 Web 资源中,刚刚被调用的 action 实例将不能再被获得,如果要为重定向的 Web 资源传递数据,则要通过 session 属性、web 参数或者使用 OGNL 表达式。

dispatcher result 则可以在包含或转发的视图中继续使用刚刚被调用的 action 实例。

例 4-6　redirect result 和 dispatcher result 的比较。

在 Chapter4 项目中进行操作,具体操作步骤如下:

（1）创建 HelloWorld.java、inputMessage.jsp、showMessage.jsp、noMessage.jsp 及 struts.xml 文件，内容同第 2 章例 2-1 中同名文件的内容。

（2）编辑 struts.xml 文件，为名称为 success 的 result 配置 type 属性，内容如下：

```
<result name="success" type="redirect">/showMessage.jsp</result>
```

（3）发布 Web 项目，启动 Tomcat 服务器，并访问 inputMessage.jsp 页面，输入信息"test"（或其他信息），如图 4-2 所示。

（4）单击 Submit 按钮，结果如图 4-3 所示。

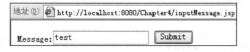
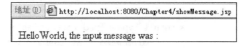

图 4-2　inputMessage.jsp 输入页面　　　　图 4-3　redirect 配置输出页面

那么如何才能输出输入的信息呢？上面提到可以通过 session 属性、web 参数或者使用 OGNL 表达式。下面给出通过 Web 参数的方式输出信息。

例 4-7　通过 Web 参数的方式输出信息。

具体操作步骤如下：

（1）编辑 struts.xml 文件，修改名称为"success"的 result 配置，内容如下：

```
<result name="success" type="redirect">/showMessage.jsp?message=${message}</result>
```

（2）编辑 showMessage.jsp 文件，内容如下：

```
HelloWorld, the input message was : <%=request.getParameter("message")%>
```

（3）重新启动 Tomcat 服务器，访问 inputMessage.jsp 页面，输入信息"test"（或其他信息）后单击 Submit 按钮，结果如图 4-4 所示。

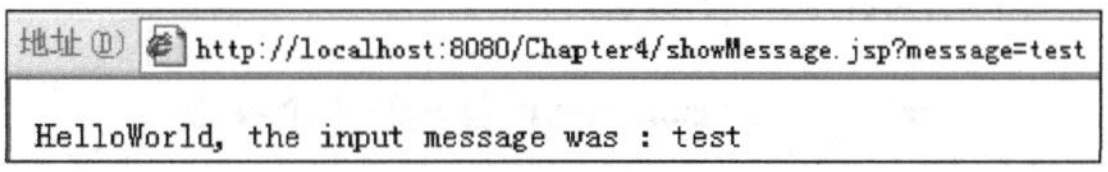

图 4-4　使用 Web 参数输出信息

6. redirectAction result

redirectAction result 使用 ActionMapperFactory 提供的 ActionMapper 将浏览器重定向到指定的命名空间（可选）中的 action。redirectAction result 比 ServletRedirectResult（redirect result）好的原因是：前者不要求开发人员对 ActionMapper 要处理的 URL 模式在 struts.xml 文件中进行编码，这就意味着开发人员可以在应用的任何位置改变 URL 模式。因此，如果想要重定向到其他的 action，则强烈建议使用 redirectAction result 而不是标准的 redirect result。

redirectAction 包含如表 4-6 所示的 3 个参数。

表 4-6　redirectAction 包含的 3 个参数

名称	作用
actionName	默认参数,用于配置将被重定向到的 action 的名称
namespace	用于配置被重定向到的 action 所在的命名空间,如果为空,则默认是当前的命名空间
supressEmptyParameters	可以防止无值参数被包含到重定向的 URL 中,默认值为 false

例 4-8　redirectAction 配置示例。

```xml
<package name="public" extends="struts-default">
    <action name="login" class="...">
        <result type="redirectAction">
            <param name="actionName">dashboard</param>
            <param name="namespace">/secure</param>
        </result>
    </action>
</package>

<package name="secure" extends="struts-default" namespace="/secure">
    <action name="dashboard" class="...">
        <result>dashboard.jsp</result>
        <result name="error" type="redirectAction">error</result>
    </action>
    <action name="error" class="...">
        <result>error.jsp</result>
    </action>
</package>
```

7. stream result

stream result 通过 InputStream 直接发送原始数据给 HttpServletResponse,通常用于文件下载。它包含如表 4-7 所示的 7 个参数。

表 4-7　stream result 包含的 7 个参数

名称	作用
contentType	发送给 Web 浏览器的流的 MIME 类型,默认值是 text/plain
contentLength	流的长度,以字节为单位(浏览器在状态栏显示一个进度条)
contentDisposition	设置响应的 Content-Disposition 头,指定下载文件的名称
inputName	action 链中 InputStream 属性的名称,默认值是 inputStream
bufferSize	用于文件下载的缓存大小,默认值 1024 字节
allowCaching	是否允许客户端缓存下载的内容,默认值 true
contentCharSet	设置字符集编码

例 4-9　stream 配置示例。

```xml
<result name="success" type="stream">
    <param name="contentType">application/pdf</param>
```

```
    <param name = "inputName"> inputStream </param>
    <param name = "contentDisposition"> attachment;filename = "document.pdf"</param>
</result>
```

该示例中配置了要下载的文件类型是 pdf 文件,文件名为 document.pdf。

8. velocity result

velocity result 用于集成 Velocity。它使用 Servlet 容器的 JspFactory 模仿 JSP 执行环境,然后显示 Velocity 模板。它包含如表 4-8 所示的两个参数。

表 4-8 velocity result 包含的两个参数

名 称	作 用
location	默认参数,要处理的模板的位置
parse	默认值 true。如果设置为 false,则不解析 location 参数中的 OGNL 表达式

例 4-11   velocity 配置示例。

```
<result name = "success" type = "velocity"> foo.vm </result>
```

9. xslt result

xslt result 用于集成 XML/XSLT。它包含如表 4-9 所示的 3 个参数。

表 4-9 xslt result 包含的 3 个参数

名 称	作 用
location	默认参数,用于配置当 action 执行完成后转到的位置
parse	默认值 true。如果设置为 false,则不解析 location 参数中的 OGNL 表达式
struts.xslt.nocache	和 struts.properties 有关的配置,默认值为 false。如果设置为 true,会禁用 stylesheet 缓冲,有利于开发,不利于发布

例 4-11   xslt 配置示例。

```
<result name = "success" type = "xslt"> foo.xslt </result>
```

10. plainText result

plainText result 用于显示一个特定页面(如 JSP,HTML)的原始内容。它包含如表 4-10 所示的两个参数。

表 4-10 plainText result 包含的两个参数

名 称	作 用
location	默认参数,被显示文件(JSP/HTML)的位置
charSet	设置所使用的字符集,可选参数

例 4-12 plainText 配置示例。

```xml
<action name = "displayJspRawContent">
    <result type = "plaintext">
        <param name = "location">/jspFile.jsp</param>
        <param name = "charSet">UTF-8</param>
    </result>
</action>
```

4.1.2 缺省配置

Struts 2 框架在 struts-default.xml 配置文件的 struts-default 包中定义了可用的 result 类型。在每一个包中都可以设置一个默认的 result 类型,这样,如果在 result 元素中没有指定 type 属性,就会使用默认的 result 类型。如果一个包继承了其他的包,则子包既可以使用继承自父包的默认的 result 类型,也可以设置自己的默认的 result 类型。默认的 result 类型需要使用 default="true"进行设置:

```xml
<result-type name = "dispatcher"
        class = "org.apache.struts2.dispatcher.ServletDispatcherResult"
        default = "true"/>
```

这样,如果在 result 元素中没有指定 type 属性,Struts 2 框架将会使用默认的 dispatcher 类型,转发到其他的 Web 资源。

1. 无缺省的 result 元素配置

```xml
<action name = "hello">
    <result name = "success" type = "dispatcher">
        <param name = "location">/result.jsp</param>
    </result>
</action>
```

该配置中,result 的名称为 success,类型为 dispatcher,转发到的 Web 资源是根目录下的 JSP 页面 result.jsp。

2. 部分缺省的 result 元素配置

```xml
<action name = "hello">
    <result>
        <param name = "location">/result.jsp</param>
    </result>
</action>
```

该配置中省略了 result 元素的 name 和 type 属性,这时将使用默认值,即 name 的默认值为 success,type 的默认值为 dispatcher。

3. 简单的 result 元素配置

```
<action name="hello">
    <result>/result.jsp</result>
</action>
```

该配置省略了 result 元素的属性，并且省略了 param 子元素的配置，这时将/result.jsp 作为 dispatcher 的默认参数(location)的值，进行转发。

4. result 元素的混合配置

```
<action name="hello">
    <result>/result.jsp</result>
    <result name="error">/error.jsp</result>
    <result name="input">/input.jsp</result>
</action>
```

4.1.3 "其他"result 配置

通过将 result 元素 name 属性的值设置为"*"，可以配置特殊的"其他"result。当没有匹配名称的 result 时，才会选择该 result。配置如下：

```
<action name="hello">
    <result>/result.jsp</result>
    <result name="error">/error.jsp</result>
    <result name="input">/input.jsp</result>
    <result name="*">/other.jsp</result>
</action>
```

上述配置中的 name="*" 不是通配符模式，它是一个特殊的名称，只有在没有精确的匹配的情况下才会被选择。

4.1.4 动态 result 配置

有时，直到执行时才能知道返回的 result，这样就无法在配置阶段设置具体的下一个 Web 资源。结果码可以从对应的 Action 实现中使用 OGNL 表达式查找到，EL 表达式访问 Action 的属性，就像 Struts 2 的标签库。

例 4-13 通过如图 4-5 所示页面输入要转到的 Web 资源的名称，单击 Go 按钮后，显示相应的 Web 资源。

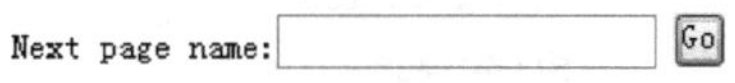

图 4-5　Web 资源名称输入界面

具体操作步骤如下：
(1) 创建 JSP 页面。

dynamicResult.jsp

```html
<form action="nextPage.action" method="post">
    Next page name:<input name="nextPage" type="text">
    <input type="submit" value="Go">
</form>
```

(2) 创建 DynamicResult.java 文件。

DynamicResult.java

```java
package example.struts2;
public class DynamicResult {
    private String nextPage;
    public String getNextPage() {
        return nextPage;
    }
    public void setNextPage(String nextPage) {
        this.nextPage = nextPage;
    }
    public String execute(){
        if((nextPage == null) || (getNextPage().isEmpty())){
            return "input";
        }
        else{
            return "next";
        }
    }
}
```

(3) 编辑 struts.xml 文件。

struts.xml

```xml
<?xml version="1.0" encoding="UTF-8" ?>
<!DOCTYPE struts PUBLIC
    "-//Apache Software Foundation//DTD Struts Configuration 2.0//EN"
    "http://struts.apache.org/dtds/struts-2.0.dtd">
<struts>
    <package name="default" namespace="/" extends="struts-default">
        <action name="nextPage" class="example.struts2.DynamicResult">
            <result name="next">/${nextPage}</result>
            <result name="input">/dynamicResult.jsp</result>
        </action>
    </package>
</struts>
```

(4) 发布 Web 项目，启动 Tomcat 服务器，并访问 dynamicResult.jsp 页面，结果如图 4-5 所示。如果不输入任何内容，单击 Go 按钮，则仍显示如图 4-5 所示界面。如果在 Web 服务器中找到了和输入内容对应的 Web 资源，则显示该 Web 资源。如果在 Web 服务器中没有找到和输入的内容对应的 Web 资源,则显示如图 4-6 所示的界面。

> 地址(D) http://localhost:8080/Chapter4/nextPage.action
> HTTP Status 404 - /Chapter4/test.jsp

图 4-6 错误提示界面

4.1.5 局部和全局 result 配置

局部 result 是指 result 元素嵌套在 action 元素中,即＜result＞作为＜action＞的子元素。局部 result 的作用范围局限在该 action 的定义中。

全局 result 是指 result 元素嵌套在 global-results 元素中,即＜result＞作为＜global-results＞的子元素。全局 result 对当前包中所有的 action 都起作用,如果 action 需要共享 result,则可以为每一个包定义全局的 result。

Struts 2 框架将首先查找嵌套在 action 中的局部 result,如果没有匹配的局部 result,则查找全局 result。如果局部 result 和全局 result 有同名的,则局部 result 优先匹配。

4.2 action 配置

action 定义是 Struts 2 框架的基本工作单元,其实质是在标识符和处理类之间建立一个映射。当到达的请求匹配 action 的名称(标识符)时,Struts 2 框架使用映射来决定如何处理该请求。

在第 3 章的 3.3.2 节中介绍了 action 子元素的结构、属性及其包含的子元素的作用,接下来对该子元素的详细使用进行介绍。

4.2.1 默认类

action 子元素有 4 个属性,其中 name 属性是必选项。如果 action 子元素只配置 name 属性,则 Struts 2 框架使用默认的类对请求进行处理。该默认的类在 struts-default.xml 配置文件中使用 default-class-ref 子元素配置,其值为 com.opensymphony.xwork2.ActionSupport。ActionSupport 中的 execute 方法内容如下:

```
public String execute() throws Exception {
    return SUCCESS;
}
```

该方法中只有一条 return 语句,返回的结果码是 SUCCESS,即不做任何处理。

例 4-14 默认类使用示例。

具体操作步骤如下:

(1) 在 Chapter4 项目中创建 struts.xml 文件,内容如下:

struts.xml

```xml
<?xml version="1.0" encoding="UTF-8"?>
<!DOCTYPE struts PUBLIC
    "-//Apache Software Foundation//DTD Struts Configuration 2.0//EN"
    "http://struts.apache.org/dtds/struts-2.0.dtd">
<struts>
    <package name="default" namespace="/" extends="struts-default">
        <action name="helloWorld">
            <result name="success">/showMessage.jsp</result>
            <result name="error">/noMessage.jsp</result>
        </action>
    </package>
</struts>
```

(2) 创建 JSP 页面。

inputMessage.jsp

```html
<form action="helloWorld.action" method="post">
    Message:<input name="message" type="text">
    <input type="submit" value="Submit">
</form>
```

showMessage.jsp

```
HelloWorld, the message was input!
```

noMessage.jsp

```
HelloWorld, no message was input!
```

(3) 发布 Web 项目,启动 Tomcat 服务器,访问 inputMessage.jsp 页面,可以发现,无论是否输入信息,当单击 Submit 按钮后,结果都如图 4-7 所示。

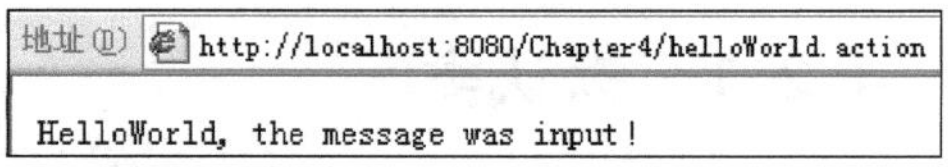

图 4-7 单击 Submit 按钮后的结果

这是由于名称为 helloWorld 的 action 配置中没有配置 class 属性,因此,将实例化默认的 com.opensymphony.xwork2.ActionSupport 类,并执行其中的 execute 方法。由于该方法中只有一条返回 SUCCESS 的 return 语句,因此,无论是否输入信息,提交后返回的结果码都一样。

4.2.2 method 属性

1. 设置不同的访问点

Action 实现类中的默认入口方法是 execute 方法,也就是说,在调用某个 action 的时候,将自动查找并执行 execute 方法。有时,开发人员可能会为一个 action 创建多个入口点(方法),例如,在数据访问的 action 中,开发人员希望把增(create)、删(delete)、改(update)、查(retrieve)的访问点分开,这样,不同的访问点可以使用 method 属性指定。

例 4-15 增、删、改、查示例。

具体操作步骤如下:

(1) 创建 DataCrud.java 文件。

DataCrud.java

```java
package example.struts2;
public class DataCrud {
    public String create(){
        System.out.println("Create method was called!");
        return "success";
    }
    public String delete(){
        System.out.println("Delete method was called!");
        return "success";
    }
    public String update(){
        System.out.println("Update method was called!");
        return "success";
    }
    public String retrieve(){
        System.out.println("Retrieve method was called!");
        return "success";
    }
}
```

(2) 编辑 struts.xml 文件,内容如下:

struts.xml

```xml
<?xml version="1.0" encoding="UTF-8" ?>
<!DOCTYPE struts PUBLIC
    "-//Apache Software Foundation//DTD Struts Configuration 2.0//EN"
    "http://struts.apache.org/dtds/struts-2.0.dtd">
<struts>
    <package name="default" namespace="/" extends="struts-default">
        <global-results>
            <result name="success">/okMessage.jsp</result>
        </global-results>
        <action name="createData" class="example.struts2.DataCrud" method="create"/>
        <action name="retrieveData" class="example.struts2.DataCrud"
                                                        method="retrieve"/>
        <action name="updateData" class="example.struts2.DataCrud" method="update"/>
        <action name="deleteData" class="example.struts2.DataCrud" method="delete"/>
    </package>
</struts>
```

（3）创建 JSP 页面。

dataCrud.jsp

```
<form action = "createData.action" method = "post">
    <input type = "submit" value = "Data Create">
</form>
<form action = "retrieveData.action" method = "post">
    <input type = "submit" value = "Data Retrieve">
</form>
<form action = "updateData.action" method = "post">
    <input type = "submit" value = "Dada Update">
</form>
<form action = "deleteData.action" method = "post">
    <input type = "submit" value = "Data Delete">
</form>
```

okMessage.jsp

```
Operate successfully!
```

（4）发布 Web 项目，启动 Tomcat 服务器，并访问 dataCrud.jsp 页面，结果如图 4-8 所示。

（5）单击图 4-8 中的各个按钮后的结果如图 4-9 所示，控制台输出如图 4-10 所示。

图 4-8　访问 dataCrud.jsp 页面的结果

图 4-9　访问 dataCrud.jsp 页面的结果

```
Create method was called!
Retrieve method was called!
Update method was called!
Delete method was called!
```

图 4-10　控制台输出结果

2. 通配符方法

在很多时候，多个 action 定义共享同一种模式，例如，例 4-15 中的增、删、改、查示例，在这个示例的 struts.xml 文件中，为每一个使用同一种模式的 action 类配置了一个 action 定义，这种配置方式有较多的代码冗余，这个问题可以通过使用通配符映射进行解决。

例 4-16　通配符方法示例。

具体操作步骤如下：

（1）编辑 struts.xml 文件。

struts.xml

```
<?xml version = "1.0" encoding = "UTF-8" ?>
<!DOCTYPE struts PUBLIC
```

```
    " - //Apache Software Foundation//DTD Struts Configuration 2.0//EN"
    "http://struts.apache.org/dtds/struts - 2.0.dtd">
<struts>
    <package name = "default" namespace = "/" extends = "struts - default">
        <global - results>
            <result name = "success">/okMessage.jsp</result>
        </global - results>
        <action name = " * Data" class = "example.struts2.DataCrud" method = "{1}"/>
    </package>
</struts>
```

(2) 重新启动 Tomcat 服务器,并访问 dataCrud.jsp 页面,然后单击页面中的各个按钮,结果如图 4-9 和图 4-10 所示。

4.2.3 动态方法调用

Struts 2 框架中有一个内嵌的特性,就是除了调用 execute 方法外,还可以使用"!"符号来调用一个方法,称之为动态方法调用。在动态方法调用中,使用 action 名称中"!"符号之后的字符串表示要调用的方法(非 execute 方法),例如,"Category! create.action"表示匹配名称为 Category 的 action 定义,并调用 create 方法。

例 4-17 动态方法调用示例。

具体操作步骤如下:

(1) 编辑 struts.xml 文件。

struts.xml

```
<?xml version = "1.0" encoding = "UTF - 8" ?>
<!DOCTYPE struts PUBLIC
    " - //Apache Software Foundation//DTD Struts Configuration 2.0//EN"
    "http://struts.apache.org/dtds/struts - 2.0.dtd">
<struts>
    <package name = "default" namespace = "/" extends = "struts - default">
        <global - results>
            <result name = "success">/okMessage.jsp</result>
        </global - results>
        <action name = "dataCrud" class = "example.struts2.DataCrud"/>
    </package>
</struts>
```

(2) 编辑 dataCrud.jsp 页面。

dataCrud.jsp

```
<form action = "dataCrud!create.action" method = "post">
    <input type = "submit" value = "Data Create">
</form>
<form action = "dataCrud!retrieve.action" method = "post">
```

```
            <input type = "submit" value = "Data Retrieve">
    </form>
    <form action = "dataCrud!update.action" method = "post">
            <input type = "submit" value = "Dada Update">
    </form>
    <form action = "dataCrud!delete.action" method = "post">
            <input type = "submit" value = "Data Delete">
    </form>
```

(3) 重新启动 Tomcat 服务器,并访问 dataCrud.jsp 页面,然后单击页面中的各个按钮,控制台输出的结果如图 4-10 所示,浏览器端结果如图 4-11 所示。

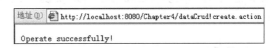

图 4-11 浏览器端输出结果

单击不同的按钮,浏览器端的 url 会有所不同。

动态方法调用的本质是,代码扫描 action 的名称,查找"!"符号,如果找到,则以该符号之后的字符串作为方法的名称,并欺骗 Struts 2 框架调用该方法,而不是调用 execute 方法。虽然调用了非 execute 方法,但是使用了和 execute 方法相同的配置,包括输入校验配置。

使用动态方法调用,Struts 2 框架"相信"它自己正在调用名称为 Category 的 action,且该 action 具有 execute 方法。这和通配符方法调用是不同的,在通配符方法调用中,Struts 2 框架"相信"它自己正在执行 Category!create 这个 action,并且"知道"它正在执行对应的 action 类中的 create 方法,这样,程序员就可以为通配符方法的 action 定义添加它自己的输入校验,信息资源及类型转换器。

4.2.4 默认的 action

通常,如果请求了一个 action,并且 Struts 2 框架不能够为该请求找到匹配的 action,结果通常是"404 - Page not found"错误。但是,如果开发员愿意使用一个总括的 action 来处理没有匹配的请求,就可以通过指定一个默认的 action 来实现。如果没有其他匹配的 action,就使用默认的 action。

对于默认的 action 来说,没有特殊的要求,每一个包(package)都能够拥有自己默认的 action,但是,在每一个命名空间中只能有一个默认的 action。

例 4-18 默认的 action 示例。

具体操作步骤如下:

(1) 编辑 struts.xml 配置文件。

struts.xml

```
<?xml version = "1.0" encoding = "UTF - 8" ?>
<!DOCTYPE struts PUBLIC
    " - //Apache Software Foundation//DTD Struts Configuration 2.0//EN"
```

```
        "http://struts.apache.org/dtds/struts-2.0.dtd">
<struts>
    <package name="default" namespace="/" extends="struts-default">
        <default-action-ref name="underConstruction"/>
        <action name="underConstruction">
            <result name="success">/underConstruction.jsp</result>
        </action>
    </package>
</struts>
```

（2）创建 underConstruction.jsp 页面。

struts.xmlunderConstruction.jsp

```
Sorry, the requested page is under construction!
```

（3）重新启动 Tomcat 服务器，并向服务器提出请求，如果该请求在 struts.xml 文件中没有匹配的 action，则结果如图 4-12 所示。

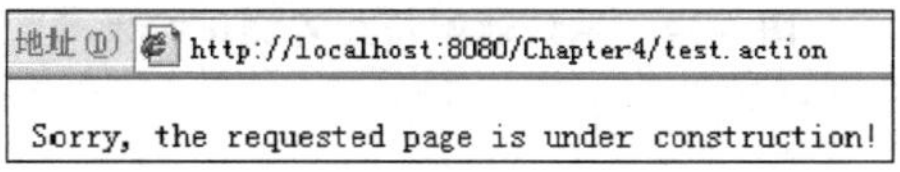

图 4-12　默认 action 配置测试结果

4.2.5　默认的通配符

使用通配符是定义默认 action 的另一种方法，在配置的最后可以使用通配符 action 捕捉没有匹配的查询。

例 4-19　默认的通配符配置示例。

具体操作步骤如下：

（1）编辑 struts.xml 配置文件。

struts.xml

```
<?xml version="1.0" encoding="UTF-8"?>
<!DOCTYPE struts PUBLIC
    "-//Apache Software Foundation//DTD Struts Configuration 2.0//EN"
    "http://struts.apache.org/dtds/struts-2.0.dtd">
<struts>
    <package name="default" namespace="/" extends="struts-default">
        <action name="*">
            <result name="success">/{1}Message.jsp</result>
        </action>
    </package>
</struts>
```

这里使用通配符匹配所有的请求，由于通配符能够匹配所有的请求，因此，当有其他的

action 映射时,一定要把默认的通配符配置放到配置文件的最后。

另外,在 result 子元素的定义中使用了占位符{1},其作用是用被通配符(*)所匹配的结果进行替换,例如,如果通配符匹配了用户的请求 test,则占位符{1}用"test"字符串进行替换。

(2) 重新启动 Tomcat 服务器,并向服务器提出请求,结果如图 4-13 所示。

图 4-13　默认通配符配置测试结果

如图 4-13 所示,当用户提出的请求是 show.action 时,则显示 showMessage.jsp 页面的内容。如果替换后的 JSP 页面不存在,将显示错误提示信息,如图 4-14 所示。

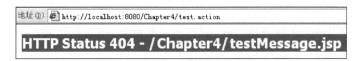

图 4-14　默认通配符配置测试结果

4.2.6　使用 param 子元素为 action 传递参数

action 可以使用 param 子元素为其传递参数,每一个 param 子元素都对应着 action 实现类的一个属性。staticParams 拦截器负责将 param 子元素配置的参数映射到 action 实现类中的属性。

例 4-20　使用 param 子元素为 action 传递参数示例。

在例 4-13 的基础上进行操作,具体操作步骤如下:

(1) 启动 Tomcat 服务器,并访问 http://localhost:8080/Chapter4/nextPage.action,结果如图 4-5 所示。

(2) 编辑 struts.xml 文件:

struts.xml

```xml
<?xml version="1.0" encoding="UTF-8" ?>
<!DOCTYPE struts PUBLIC
    "-//Apache Software Foundation//DTD Struts Configuration 2.0//EN"
    "http://struts.apache.org/dtds/struts-2.0.dtd">
<struts>
    <package name="default" namespace="/" extends="struts-default">
        <action name="nextPage" class="example.struts2.DynamicResult">
            <result name="next">/${nextPage}</result>
            <result name="input">/dynamicResult.jsp</result>
            <param name="nextPage">underConstruction.jsp</param>
        </action>
    </package>
</struts>
```

（3）重新启动 Tomcat 服务器，并访问 http://localhost:8080/Chapter4/nextPage.action，结果如图 4-15 所示。

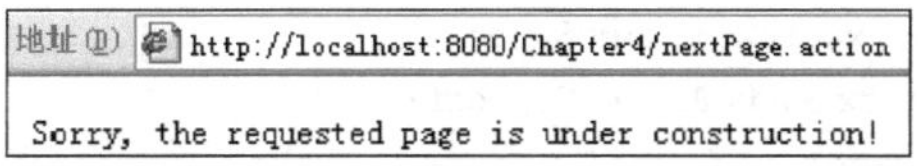

图 4-15　使用 param 子元素为 action 传递参数的结果

4.3　Struts 2 框架中的 Action

4.3.1　ActionSupport 类

由 4.2.1 节可知，Struts 2 框架使用默认类 ActionSupport 对请求进行处理，该类中有一个只有一条 return 语句的 execute 方法，如下：

```
public String execute() throws Exception {
    return SUCCESS;
}
```

该 return 语句返回值（用于查找 struts.xml 文件中的 result）是一个常量，而不是字符串。该常量是在 Action 接口中定义的，ActionSupport 类实现了 Action 接口，因此，在 ActionSupport 类中可以使用该常量。

Action 接口的定义如下：

```
package com.opensymphony.xwork2;
public interface Action {
    public static final String SUCCESS = "success";
    public static final String NONE = "none";
    public static final String ERROR = "error";
    public static final String INPUT = "input";
    public static final String LOGIN = "login";
    public String execute() throws Exception;
}
```

在 Action 接口中，定义了 5 个静态常量和一个 execute() 方法。

由上面的介绍可知，如果想要使用这些常量，开发人员在创建自己的 Action 类时，要实现 Action 接口或者继承 ActionSupport 类。ActionSupport 类除了实现了 Action 接口中的 execute() 方法外，还提供了很多常用的方法，如：数据校验、异常处理、资源的国际化等。因此，开发人员通常直接继承 ActionSupport 类，既可以使用预先定义的常量，又可以很方便地实现业务控制器类。

例 4-21　ActionSupport 类的应用示例。

下面在第 2 章例 2-1 的基础上进行操作，具体操作步骤如下：

（1）修改 HelloWorld.java 文件，如下：

HelloWorld.java

```java
package example.struts2;
import com.opensymphony.xwork2.ActionSupport;
public class HelloWorld extends ActionSupport{
    private String message;
    public String getMessage() {
        return message;
    }
    public void setMessage(String message) {
        this.message = message;
    }
    public String execute(){
        if(getMessage().isEmpty()){
            return ERROR;
        }
        else{
            return SUCCESS;
        }
    }
}
```

（2）启动 Tomcat 服务器，访问 inputMessage.jsp 页面进行测试，效果同例 2-1。

开发人员在实现自己的 Action 时，使用常量而不是使用自己随意书写的字符串的好处是可以规范开发人员的代码。

4.3.2 Action 访问 Servlet API

Struts 2 框架通过拦截器实现了将 Action 和 Servlet API 的分离，因此，Struts 2 框架中的业务控制器类就是普通的 Java 类，降低了框架和 Servlet API 的耦合性，便于 Action 的测试。

那么如何在 Action 中访问 ServletContext、HttpSession、HttpServletRequest、HttpServletResponse 等资源呢？Struts 2 框架为开发人员提供了 ActionContext 和 ServletActionContext 类，借助于这两个类可以访问这些资源。

1. 使用 ActionContext 访问 Servlet API

ActionContext 是 Action 执行时的上下文，其中保存的是 Action 执行时用到的对象，如请求的参数，servlet 的上下文、会话和本地化的一些信息，开发人员可以通过使用 ActionContext 的静态方法 getContext() 来获取当前的 ActionContext 对象。ActionContext 是线程安全的，即在同一个线程里，ActionContext 里的属性是唯一的，这样 Action 就可以在多线程中使用。

例 4-22 使用 ActionContext 访问 Servlet API。

任务描述：在页面中输入信息，提交后，显示输入的信息。

在例 2-1 的基础上操作，具体步骤如下：

（1）修改 HelloWorld.java 文件。

HelloWorld.java

```java
package example.struts2;
import com.opensymphony.xwork2.ActionContext;
import com.opensymphony.xwork2.ActionSupport;
public class HelloWorld extends ActionSupport{
    private String message;
    public String getMessage() {
        return message;
    }
    public void setMessage(String message) {
        this.message = message;
    }
    public String execute(){
        if(getMessage().isEmpty()){
            return ERROR;
        }
        else{
            ActionContext context = ActionContext.getContext();
            context.getSession().put("sessionMessage", message);
            return SUCCESS;
        }
    }
}
```

在这段代码中,通过 ActionContext 取得 HttpSession,取得的 session 是 Map 类型的对象。上面 else 代码块中的代码相当于如下代码:

```
ActionContext context = ActionContext.getContext();
Map session = context.getSession();
session.put("sessionMessage", message);
```

为什么取得的 session 是 Map 类型的对象？这是因为 Struts 2 框架将和 Web 相关的很多对象重新进行了包装,例如本例中将 HttpSession 对象重新包装成了一个 Map 对象,供 Action 使用。这样包装之后,Action 就不用直接和底层的 HttpSession 打交道。正是借助于 Struts 2 框架的包装,实现了 Actoion 和 Servlet API 的解耦。

（2）修改 showMessage.jsp 文件。

showMessage.jsp

```
HelloWorld, the input message was: ${sessionScope.sessionMessage}
```

（3）重新启动 Tomcat 服务器,访问 inputMessage.jsp 页面进行测试。

2. 使用 ServletActionContext 访问 Servlet API

ServletActionContext 类是 ActionContext 类的子类,比 ActionContext 使用更方便。ServletActionContext 和 ActionContext 中的一些功能是重复的,在 Action 中访问 Servlet API 时,如果 ActionContext 能够实现的功能则尽量不要使用 ServletActionContext,即尽

量不要让自己编写的 Action 直接访问 Java Servlet 对象。

例 4-23　使用 ServletActionContext 访问 Servlet API。

在例 4-22 的基础上进行操作,具体操作步骤如下:

(1) 修改 HelloWorld.java 文件。

HelloWorld.java

```java
package example.struts2;
import javax.servlet.http.HttpServletRequest;
import org.apache.struts2.ServletActionContext;
import com.opensymphony.xwork2.ActionSupport;
public class HelloWorld extends ActionSupport{
    private String message;
    private HttpServletRequest request;
    public String getMessage() {
        return message;
    }
    public void setMessage(String message) {
        this.message = message;
    }
    public String execute(){
        if(getMessage().isEmpty()){
            return ERROR;
        }
        else{
            request = ServletActionContext.getRequest();
            request.getSession().setAttribute("sessionMessage", message);
            return SUCCESS;
        }
    }
}
```

(2) 重新启动 Tomcat 服务器,访问 inputMessage.jsp 页面进行测试。

3. 注入方式(IoC)访问 Servlet API

无论使用 ActionContext 还是使用 ServletActionContext 访问 Servlet API,都是非注入方式。Struts 2 框架提供了 Aware 接口,Struts 2 框架在实例化实现了相应 Aware 接口的 Action 时,会将相应的资源通过 Aware 接口方法注入,这种方式叫做注入方式。

Servlet API 中常用的 application、request、response 和 session 对象对应的 Aware 接口分别是 ServletContextAware、ServletRequestAware、ServletResponseAware、SessionAware。实现了这些接口的 Action 必须实现这些接口中的方法。例如:实现了 ServletContextAware 接口的 Action,要实现 setServletContext(ServletContext arg0)方法获取 application 对象,实现了 ServletRequestAware 接口的 Action,要实现 setServletRequest(HttpServletRequest arg0)方法获取 request 对象,实现了 ServletResponseAware 接口的 Action,要实现 setServletResponse(HttpServletResponse arg0)方法获取 response 对象,实现了 SessionAware 接口的 Action,要实现 setSession(Map<String, Object> session)方法获取 session 对象。

IoC 方式使用了 Struts 2 框架的 Aware 拦截器。

例 4-24 使用 IoC 方式访问 Servlet API。

在例 4-23 的基础上进行操作,具体操作步骤如下:

(1) 修改 HelloWorld.java 文件。

HelloWorld.java

```java
package example.struts2;
import java.util.Map;
import org.apache.struts2.interceptor.SessionAware;
import com.opensymphony.xwork2.ActionSupport;
public class HelloWorld extends ActionSupport implements SessionAware{
    private String message;
    private Map session;
    public String getMessage() {
        return message;
    }
    public void setMessage(String message) {
        this.message = message;
    }
    public void setSession(Map<String, Object> session) {
        this.session = session;
    }
    public String execute(){
        if(getMessage().isEmpty()){
            return ERROR;
        }
        else{
            session.put("sessionMessage", message);
            return SUCCESS;
        }
    }
}
```

(2) 重新启动 Tomcat 服务器,访问 inputMessage.jsp 页面进行测试。

习题

1. 熟悉 result 配置和 action 配置。
2. 修改例 4-21,通过实现 Action 接口的方式创建 HelloWorld.java 类。
3. 创建 Web 项目:统计网站被访问的次数,当用户登录成功后,显示该用户访问该网站的次数,及该网站被访问的总次数。

第5章 Struts 2框架的OGNL

本章首先介绍了 OGNL 语法,然后对 OGNL 表达式和 OGNL 中的集合操作进行了介绍,最后介绍了如何在 Struts 2 框架中使用 OGNL。

5.1 OGNL 简介

OGNL(Object Graph Navigation Language,对象图导航语言)是一种功能强大的表达式语言,它提供了访问值栈中对象的统一方式,是 Struts 2 框架默认的表达式语言。

OGNL 可以用来读取和设置 Java 对象的属性,另外,在 Java 中能够实现的大部分功能,OGNL 都能实现,并且还支持集合投影与选择、lambda 表达式等功能。

OGNL 处理的顶级对象是一个 Map(通常被称为 context map 或 context)。在 context 中,OGNL 有一个根对象的概念。OGNL 有且仅有一个根对象,它通过根对象可以访问任何与根对象有关联的对象。

在 OGNL 表达式中,访问根对象的属性时可以不使用特殊的标志符号,但是访问其他对象时要使用井号(♯)标记。

例如,假设在 OGNL 上下文 Map 中有两个对象:"foo" -> foo 和 "bar" -> bar,并且每个对象都有一个 blah 属性,其中 foo 对象被设置为单独的根对象。下面的代码描述了 OGNL 如何处理对象的三种情况:

```
♯foo.blah
♯bar.blah
blah
```

其中,第一个表达式返回 foo 对象 blah 属性的值;第二个表达式返回 bar 对象 blah 属性的值;第三个表达式返回 foo 对象 blah 属性的值,因为 foo 是根对象,所以可以省略井号(♯)标记和根对象。

5.2 OGNL 语法

OGNL 表达式非常简单。例如,为了获取一个对象的 name 属性的值,OGNL 表达式可以简单地表达为 name。为了获取 headline 属性的 text 属性的值,OGNL 表达式可以表达为 headline.text。

OGNL 表达式的基本单位是"导航链",通常简称为"链"。最简单的链由以下几部分组成。

(1) 属性名称:例如上面的 name 和 headline.text。

(2) 方法调用:例如 hashCode() 返回当前对象的 hash 码。

(3) 数组索引:例如 listeners[0] 返回当前对象的 listeners 列表中的第一个元素。

所有的 OGNL 表达式都是在当前对象的上下文中进行计算的。链使用导航链中的前一个链的结果作为当前对象传递给下一个链。另外,链的长度可以任意扩展,如下:

```
name.toCharArray()[0].numericValue.toString()
```

该 OGNL 表达式的计算步骤如下:

(1) 首先提取初始的、根的或者对象(用户通过 OGNL 上下文提供给 OGNL)的 name 属性,并将结果字符串传给下一个链。

(2) 然后调用结果字符串的 toCharArray() 方法,得到字符数组,并将结果数组传给下一个链。

(3) 通过数组索引从结果数组中提取第一个(索引为 0)字符,并将结果字符传给下一个链。

(4) 接下来计算结果字符的 numericValue 属性的值,并将结果传给下一个链。

(5) 最后调用整型结果对象的 toString() 方法。表达式的最终结果是最后调用 toString() 方法返回的字符串。

5.3 OGNL 表达式

这一节介绍 OGNL 表达式的元素。

5.3.1 常量

OGNL 包含的常量类型如下:

(1) 字符串常量:由单引号或双引号进行定界的字符序列,包括所有的转义字符。

例 5-1 "OGNL", 'OGNL', "\'OGNL\' test"。

(2) 字符常量:由单引号定界的一个字符,包括所有的转义字符。

例 5-2 'O', '\''。

(3) 数值常量:除了 Java 中的 int、long、float 和 double 类型,OGNL 还可以让开发人员用"b"或者"B"后缀指定 BigDecimals 类型常量,用"h"或者"H"后缀指定 BigIntegers 类型常量。

例 5-3 53.1f,531h。

(4) 布尔型常量:true 和 false。

(5) null 常量:null。

5.3.2 属性访问

对于不同类型的对象,OGNL 在处理属性访问时是不同的。Map 对象把所有的属性访问都看作元素进行查找或存储,属性的名称作为键(key)。列表(list)和数组(array)对待数值型属性的处理相似,属性的名称作为索引,但是把字符串型属性作为普通对象处理。普通对象(除数值型外)只能够处理字符串型属性,并且使用"get"和"set"方法(或者是"is"和"set"方法)进行属性的操作。

由于属性可以是任意类型,而不仅仅是字符串类型,因此,在访问非字符串型属性时,必须使用索引的概念。

例 5-4 属性访问示例。

为了取得一个数组的长度(数值型),可以使用如下的表达式:

```
array.length
```

但是,为了取得数组的第一个元素,必须使用如下所示的表达式:

```
array[0]
```

5.3.3 操作符

通常,OGNL 的操作符都是从 Java 中借鉴的,并且和 Java 的操作符具有相同的工作方式。这里介绍一下 Java 中没有的 OGNL 操作符。

1. 逗号或序列操作符

这个操作符借鉴于 C 语言,逗号用来分隔两个独立的表达式,第二个表达式的值是逗号表达式的值。

例 5-5 逗号操作符示例。

```
ensureLoaded(), name
```

当计算该逗号表达式时,ensureLoaded() 方法首先被调用,然后使用 name 表达式检索(如果是取值)或替换(如果是设置值)name 值,即两个表达式都被计算,但是整个表达式的值是第二个表达式(name 属性)的值。

2. 大括号操作符

使用大括号将这些值括起来(值之间用逗号分隔),可以创建一个列表。

例 5-6 大括号操作符示例。

```
{null, true, false}
```

该表达式创建了一个有三个元素的列表。

3. in 和 not in 操作符

这是一个包含测试操作符,用于判断一个值是否在一个集合中,返回的结果是一个逻辑值。其中"in"操作符用于判断一个值是否在指定的集合中;"not in"操作符用于判断一个值是否不在指定的集合中。

例 5-7 in 操作符示例。

```
name in {null,"Untitled"} || name
```

该表达式用于输出 name 属性的值。首先判断 name 属性值是 null 还是"Untitled",如果是二者之一,则输出该值。

5.3.4 设置值和检索值

一些可以检索值的表达式不一定可以设置值,这是由表达式的性质决定的。见例 5-8。

例 5-8 设置值和检索值示例。

```
names[0].location
names[0].length + 1
```

第一个表达式是一个可以设置值的表达式,因为表达式的最后组成部分被解析成一个对象中的可设置的属性(location 是一个属性)。

但是第二个表达式不是一个可设置的表达式,因为它们不被解析成一个对象中的可设置的属性。由于它只是一个计算的值,因此,如果尝试着使用 Ognl.setValue()方法计算这个表达式将会失败,并且抛出 InappropriateExpressionException 异常。

5.3.5 访问静态方法和字段

1. 访问静态方法

可以使用如下的语法调用一个静态方法。

```
@class@method(args)
```

(1) 如果指定了 class,则必须给出完整限定类名,例如:

```
@some.package.ClassName@someMethod(args)
```

(2) 如果省略了 class,则默认为 java.lang.Math 类,这将使得调用 min 和 max 方法比较容易,例如:

```
@@min(1,2)
```

另外,如果有一个类的实例,并且想调用类中的静态方法,可以通过对象调用这个静态方法,就像这个静态方法是一个实例的方法一样。如果一个方法被重载了,OGNL 选择调用正确的静态方法,其过程和使用重载实例方法的过程相同。

2. 访问静态字段

可以使用如下语法访问静态字段。

```
@class@field
```

其中 class 必须是完整限定类名。

5.3.6 索引

1. 数组和列表索引

对于数组和列表的索引比较简单,给定一个整数索引值,其元素就是所指的对象。例如:

```
array[0]
{null, true, false}[0]
```

和 Java 中一样,如果索引超出了数组和列表的边界,则将抛出 IndexOutOfBoundsException 异常。

2. JavaBean 的属性索引

JavaBean 支持索引属性的概念,也就是说一个对象有一套遵循如下模式的方法。

```
public PropertyType[] getPropertyName();
public void setPropertyName(PropertyType[] anArray);
public PropertyType getPropertyName(int index);
public void setPropertyName(int index, PropertyType value);
```

其中前两个方法用于数组。通过索引概念 OGNL 能够解释并访问属性,如例 5-9。

例 5-9 属性索引示例。

```
someProperty[2]
```

该表达式能够自动通过正确的属性索引访问方法(getSomeProperty(2) 或者 setSomeProperty(2, value))进行计算。

3. OGNL 对象索引

OGNL 扩展了索引属性的概念,可以使用任意对象进行索引,而不像 JavaBean 索引属性只能使用整数进行索引。当为对象索引找到候选的属性时,OGNL 使用如下的签名查找方法的模式。

```
public PropertyType getPropertyName(IndexType index);
public void setPropertyName(IndexType index, PropertyType value);
```

在对应的 set 和 get 方法中，PropertyType 和 IndexType 必须相互匹配。一个实际的使用对象索引属性的例子是 Servlet API，Session 对象有两个方法用于获取和设置任意的属性，如下：

```
public Object getAttribute(String name)
public void setAttribute(String name, Object value)
```

例 5-10　OGNL 对象索引示例。

一个既可以获取又可以设置属性的 OGNL 表达式形式如下：

```
session.attribute["foo"]
```

该表达式可以在 session 中获取和设置 foo 属性的值。

5.3.7　括号表达式

一个用括号括起来的表达式被作为一个单位进行计算，和任何围绕的操作符分离。这种表达式可以用来强制改变 OGNL 操作符的计算顺序。括号表达式是在一个方法的参数中使用逗号运算符的唯一方式。

5.3.8　链接子表达式

如果在一个点（"."）符号之后使用括号表达式，则在这个点符号处的对象作为括号表达式的当前对象。例如：

```
headline.parent.(ensureLoaded(), name)
```

该表达式将遍历 headline 和 parent 属性，括号表达式中的 ensureLoaded() 方法确保 parent 被加载，然后使用括号表达式中的 name 表达式返回（或者设置）parent 的名称。

顶级表达式也可以使用这种方式进行链接，整个表达式的结果是最右边表达式的值。例如：

```
ensureLoaded(), name
```

该表达式将在根对象上调用 ensureLoaded() 方法，然后获得根对象的 name 属性，并作为表达式的结果。

5.3.9　变量访问

OGNL 有一个简单的变量方案，可以让开发人员存储中间结果并再次使用它们，或者只是取一个名字使得表达式容易理解。OGNL 中的所有变量对整个表达式都是全局的。访问一个变量要在其名称前使用井号（#），如下：

```
#var
```

在计算一个表达式的时候，OGNL 也把在每一个点的当前对象保存在 this 变量中，可以像其他变量一样访问。例如：

```
listeners.size().(#this > 100? 2 * #this : 20 + #this)
```

该表达式中首先计算 listeners 的个数（listeners.size()），并将计算后的 listeners 的个数作为括号表达式的当前对象，this 变量的值是括号表达式的当前对象。#this 是取得 this 变量的值。

为了给一个变量显式地赋值，只要写一条赋值语句即可，如下：

```
#var = 1
```

5.3.10 表达式计算

如果在 OGNL 表达式之后跟随一个在左括号前没有点(".")符号的括号表达式，OGNL 将把第一个表达式的结果作为另一个表达式来计算，并且在计算时使用括号表达式的结果作为根对象。

例如，对于如下表达式：

```
#fact(30H)
```

查找 fact 变量，并将使用 30H 作为根对象的 OGNL 表达式作为变量值。

如果当前对象有一个 fact 属性，而该属性有一个 OGNL 阶乘表达式，就不能使用 fact(30H)进行调用，因为 OGNL 会将之解释为对 fact 方法的调用。这时可以通过使用括号将属性访问括起来的方式进行操作，例如：

```
(fact)(30H)
```

5.3.11 lambda 表达式

OGNL 有一个简单的 lambda 表达式语法，这样开发人员可以写简单的函数。由于 OGNL 中的所有变量都是全局范围，没有闭包，所以它不是一个完善的 lambda 算法。

下面的表达式声明了一个递归的阶乘函数，然后调用它。

```
#fact = :[#this<=1? 1 : #this * #fact(#this-1)], #fact(30H)
```

lambda 表达式是使用方括号括起来的内容。#this 变量代表表达式的参数，初始值是 30H，之后每调用一次表达式，其值就减一。

OGNL 把 lambda 表达式看作常量。

5.4 OGNL 的集合操作

主要包括列表、数组和 Map。

5.4.1 创建集合

1. 创建列表

为了创建列表对象,将一系列表达式用大括号括起来,如例5-11。

例 5-11 创建列表示例。

```
{ "Computer", "Phone" }
```

该表达式创建了一个包含两个元素的列表。

2. 创建数组

OGNL支持创建数组,其创建过程和Java中创建数组的过程一样,需要调用构造函数,但是OGNL创建数组允许使用一个已经存在的列表或者给定数组大小的方式对数组进行初始化,两种方式只能选择其一,如例5-12。

例 5-12 创建数组示例。

```
new int[] { 1, 2, 3 }
new int[5]
```

其中第一个表达式表示使用一个列表对创建的数组进行初始化,数组由1、2、3三个元素组成。为了创建所有数组元素都是null或者0的数组,使用给定数组大小的方式创建数组,第二个表达式表示创建包含5个元素的数组,数组元素初始化为0。

3. 创建Map

使用特殊语法创建Map,语法如下:

```
#{ "Key1" : "Value1", "Key2" : "Value2", ... , "KeyN" : "ValueN" }
```

创建了包含N个元素的Map,每个元素是一个"Key:Value"对,Key和Value之间用冒号分隔。创建Map要使用井号(#)标记,并将"Key:Value"对用大括号括起来,"Key:Value"对之间用逗号分隔。

例 5-13 创建包含两个元素的Map示例。

```
#{ "foo" : "foo_value", "bar" : "bar_value" }
```

如果想选择具体的Map类创建Map,则要在左大括号之前指定该Map类,如例5-14。

例 5-14 创建具体Map类型的Map。

```
#@java.util.LinkedHashMap@{ "foo" : "foo_value", "bar" : "bar_value" }
```

该表达式将会创建JDK 1.4中LinkedHashMap类的一个实例,确保元素的插入顺序被保存。

5.4.2 投影

OGNL 提供了一种简单的"投影"方式,即在一个集合中,对每个元素调用相同的方法或者提取相同的属性,并将结果保存在一个新的集合中。借助于数据库术语,就是从一个表中选择列的一个子集。如例 5-15。

例 5-15 投影示例。

```
leaders.{name}
```

该表达式返回所有领导的名字。

在投影的时候,♯this 变量指向迭代时的当前元素,如例 5-16。

例 5-16 ♯this 变量示例。

```
objects.{ ♯this instanceof String ? ♯this : ♯this.toString()}
```

该表达式从对象列表中生成一个新的元素列表,新列表中的元素是字符串类型。

5.4.3 选择

OGNL 提供了一种简单的"选择"方式,即使用一个表达式从集合中选择一些元素,并将结果保存在新的集合中。借助于数据库术语,就是从一个表中选择行的一个子集。

1. 选择所有匹配的元素

从集合中选择所有匹配的元素的操作符是问号(?)。

例 5-17 问号操作符示例。

```
leaders.{? ♯this.age < 30}
```

该表达式返回所有领导年龄小于 30 的领导的列表。

2. 选择第一个匹配的元素

从集合中选择第一个匹配的元素的操作符是"^"。

例 5-18 "^"操作符示例。

```
leaders.{^ ♯this.age < 30}
```

该表达式返回领导年龄小于 30 的第一个元素。

3. 选择最后一个匹配的元素

从集合中选择最后一个匹配的元素的操作符是"$"。

例 5-19 "$"操作符示例。

```
leaders.{$ ♯this.age < 30}
```

该表达式返回领导年龄小于30的最后一个元素。

5.5 Struts 2 中的 OGNL

Struts 2 框架在 OGNL 的基础上增加的最大功能就是支持值栈(ValueStack)。OGNL 假定只有一个"根"("root"),而 Struts 2 的值栈概念需要有多个"根"。值栈是 Struts 2 框架的核心。

5.5.1 值栈

Struts 2 框架将 OGNL 上下文(context)设置为 ActionContext,并将值栈设置为 OGNL 的根对象。值栈可以包含多个对象,但是对于 OGNL 而言,它表现为一个单一的对象。Struts 2 框架将值栈和其他对象(包括表示 application、session 和 request 等的 Map)一起保存在 ActionContext 中。这些对象和值栈(OGNL 的根)共同存在于 ActionContext 中,如图 5-1 所示。

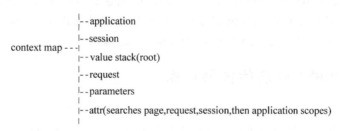

图 5-1 Struts 2 框架中 context map 包含的对象

在 context map 中还有其他的对象,图 5-1 只是一个说明 context map 结构的一个示例。

在 Struts 2 框架中,整个值栈就是上下文的一个根对象。如果从值栈中获取对象,然后再获取属性,并不需要开发人员自己编写表达式(如:peek().blah),Struts 2 框架有一个特殊的 OGNL 属性访问器(PropertyAccessor),它可以自动自上而下查看整个值栈,直到找到具有正在查找的属性的对象,如例 5-20 所示。

例 5-20 值栈示例。

任务描述:假设值栈中包含 Animal 和 Person 两个对象。两个对象都有"name"属性,并且 Animal 对象还有一个"species"属性,Person 对象还有一个"salary"属性。Animal 对象在栈顶,Person 对象在 Animal 对象之下。

```
species
salary
name
```

在这段代码中,第一条代码将会调用 animal.getSpecies()方法返回 Animal 对象的 species 属性的值;第二条代码将会调用 person.getSalary()方法返回 Person 对象的 salary 属性的值;虽然两个对象都有 name 属性,但是由于 Animal 对象在栈顶,因此第三条代码将会调用 animal.getName()方法返回 Animal 对象的 name 属性的值。

5.5.2 索引

在例 5-20 中，由于 Animal 对象在栈顶，因此将调用 animal.getName()方法返回 Animal 对象的 name 属性的值，这也是期望的效果。但是有时想要获取非栈顶对象的属性，如何实现呢？为了能够做到这一点，Struts 2 框架在值栈中增加了索引操作，即使用诸如[0]、[1]、[2]、…、[N]形式的表达式截取值栈。截取值栈操作将返回一个 CompoundRoot 对象，该对象是从值栈的位置 N 开始的部分栈，如例 5-21 所示。

例 5-21 截取值栈的示例。

```
[0].name
[1].name
```

在该示例中，第一条代码中的"[0]"将从值栈的第一个位置对象开始截取值栈，得到的 CompoundRoot 对象包含的内容和原来值栈的内容一样，因此[0].name 将会调用 animal.getName()方法返回 Animal 对象的 name 属性的值。第二条代码中的"[1]"将从值栈的第二个位置对象开始截取值栈，得到的 CompoundRoot 对象包含的内容去除了 Animal 对象，这时，在 Animal 对象之下的 Person 对象成为截取后值栈的栈顶对象，因此[1].name 将会调用 person.getName()方法返回 Person 对象的 name 属性的值。

5.5.3 使用 top 访问栈顶对象

Struts 2 框架支持使用 top 关键字访问栈顶对象，另外和诸如[0]、[1]、[2]、…、[N]形式的表达式相结合可以获取值栈中任意位置的对象(即特定的栈顶对象)，如例 5-22 所示。

例 5-22 获得特定的栈顶对象的示例。

```
[0].top
[1].top
```

在该示例中，第一条代码用于获取值栈中第一个位置的对象，即截取后的栈的栈顶对象(在这种情况下，和 top 表达式一样)。第二条代码用于获取值栈中第二个位置的对象，即截取后的栈的栈顶对象。

5.5.4 访问静态属性

标准 OGNL 支持访问静态属性和静态方法，而 Struts 2 框架的默认配置不允许访问静态属性和静态方法。为了使 Struts 2 框架支持访问静态属性和静态方法，必须设置 struts.ognl.allowStaticMethodAccess 常量的值为 true。

在 Struts 2 框架中，访问静态属性和静态方法不需给出完整限定类名，只需使用"vs"前缀即可，如例 5-23 所示。

例 5-23 "vs"前缀使用示例。

```
@vs@FOO_PROPERTY
@vs@someMethod()
```

在该示例中，"vs"表示"value stack"。需要特别指出的是，如果使用"vs"指定了类名，就会调用栈顶对象的类。

如果想要调用其他位置对象的类，如何实现呢？方法很简单，只要在"vs"字符串后指定一个数字，就会调用栈中指定位置对象的类，如例 5-24 所示。

例 5-24 调用栈中指定位置对象的类的示例。

```
@vs1@FOO_PROPERTY
@vs1@someMethod()
@vs2@BAR_PROPERTY
@vs2@someOtherMethod()
```

5.5.5 Struts 2 框架的命名对象

Struts 2 框架没有将请求参数（parameters）以及 request、session 和 application 等属性存放在 OGNL 的值栈中，因此在访问这些对象时要使用 # 标记。表 5-1 给出对这些对象的访问形式。

表 5-1 Struts 2 框架的命名对象的访问形式

名称	值
#parameters['foo'] #parameters.foo	返回请求参数 foo 的值，相当于调用 request.getParameter("foo")
#request['foo'] #request.foo	返回 request 作用域中 foo 属性的值，相当于 调用 request.getAttribute("foo")
#session['foo'] #session.foo	返回 session 作用域中 foo 属性的值
#application['foo'] #application.foo	返回 application 作用域中 foo 属性的值
#attr['foo'] #attr.foo	如果 PageContext 有效，则访问 PageContext，否则，按照顺序依次搜索 request、session 和 application，返回 foo 属性的值

例 5-25 非根对象访问示例。

```
<s:property value="#session.mySessionPropKey"/>
```

该示例代码使用 Struts 2 标签输出了 session 域的 mySessionPropKey 属性值。

5.5.6 访问 Action 属性

Action 的实例通常压入值栈中。因为 Action 在值栈中，并且值栈是 OGNL 的根，所以访问 Action 的属性时可以省略井号（#）标记，如例 5-26 所示。

例 5-26 访问 Action 属性示例。

```
<s:property value="userPassword"/>
```

该示例代码使用 Struts 2 标签输出了 Action 实例的 userPassword 属性的值。

5.5.7 集合

在 Struts 2 框架中经常要处理集合（Map、List、Set）。下面通过几个示例进行讲解。

1. 为 select 标签创建集合。

例 5-27 使用列表的示例。

```
<s:select label="label" name="name" list="{'name1','name2'}" value="%{'name2'}" />
```

该示例代码创建了一个包含"name1"和"name2"两个字符串的列表，并将"name2"设置为默认值。

例 5-28 使用 Map 的示例。

```
<s:select label="label" name="name" list="#{'foo':'foovalue', 'bar':'barvalue'}" />
```

该示例代码创建了一个 Map，将字符串"foo"映射到字符串"foovalue"，将字符串"bar"映射到字符串"barvalue"。

2. 使用集合判断操作符。

例 5-29 集合判断操作符示例。

```
<s:if test="'foo' in {'foo','bar'}">
    foo
</s:if>
<s:else>
    boo
</s:else>
```

该示例中，如果'foo'是{'foo','bar'}列表中的元素，则输出 foo，否则输出 boo。

5.6 OGNL 应用示例

通过一个简单完整的示例，讲解如何使用 OGNL。

例 5-30 OGNL 应用示例。

具体操作步骤如下：

（1）创建 SchClass 类。在该类中声明 cName（班级名称）和 cStudent（班级学生）两个属性，其中 cStudent 为 Map 类型。

SchClass.java

```
package example.struts2;
import java.util.Map;
public class SchClass {
```

```java
    private String cName;
    private Map<String, String> cStudent;
    public String getCName() {
        return cName;
    }
    public void setCName(String name) {
        cName = name;
    }
    public Map<String, String> getCStudent() {
        return cStudent;
    }
    public void setCStudent(Map<String, String> student) {
        cStudent = student;
    }
    public SchClass(String name, Map<String, String> student) {
        cName = name;
        cStudent = student;
    }
    public SchClass() {
    }
}
```

(2) 创建 Action。

OgnlTest.java

```java
package example.struts2;
import java.util.ArrayList;
import java.util.HashMap;
import java.util.List;
import java.util.Map;
import org.apache.struts2.interceptor.SessionAware;
import com.opensymphony.xwork2.ActionSupport;
public class OgnlTest extends ActionSupport implements SessionAware {
    private Map<String, Object> mSession;
    private List<SchClass> classes = new ArrayList<SchClass>();
    public List<SchClass> getClasses() {
        return classes;
    }
    public void init(){
        Map<String, String> cStudent1 = new HashMap<String, String>();
        cStudent1.put("stuName", "郭一");
        cStudent1.put("stuSex", "男");
        Map<String, String> cStudent2 = new HashMap<String, String>();
        cStudent2.put("stuName", "林一");
        cStudent2.put("stuSex", "女");
        Map<String, String> cStudent3 = new HashMap<String, String>();
        cStudent3.put("stuName", "武一");
        cStudent3.put("stuSex", "男");
```

```
            SchClass classes1 = new SchClass("09 计本 1", cStudent1);
            SchClass classes2 = new SchClass("09 计本 2", cStudent2);
            SchClass classes3 = new SchClass("09 计本 2", cStudent3);
            classes.add(classes1);
            classes.add(classes2);
            classes.add(classes3);
        }
        public String execute() throws Exception {
            init();
            mSession.put("ClassName", classes.get(0).getCName());
            return SUCCESS;
        }
        public void setSession(Map<String, Object> session) {
            this.mSession = session;
        }
    }
```

OgnlTest 类实现了 SessionAware 接口,Struts 2 框架使用该接口向 Action 实例注入 session Map 对象。该接口只有一个 setSession 方法,因此 OgnlTest 类必须实现该方法。由于本例是想演示 session 命名对象的使用,因此,没有给出 mSession 属性的 getter 和 setter 方法。

(3) 创建 struts.xml 配置文件。

struts.xml

```xml
<?xml version="1.0" encoding="UTF-8" ?>
<!DOCTYPE struts PUBLIC
    "-//Apache Software Foundation//DTD Struts Configuration 2.0//EN"
    "http://struts.apache.org/dtds/struts-2.0.dtd">
<struts>
    <package name="default" namespace="/" extends="struts-default">
        <action name="ognlTest" class="example.struts2.OgnlTest">
            <result>/showMessage.jsp?welcome=计算机科学与技术系</result>
        </action>
    </package>
</struts>
```

在 result 子元素中的配置中配置了参数 welcome,主要是为了测试 parameters 命名对象的使用。

(4) 创建 JSP 页面。

showMessage.jsp

```
<%@ page language="java" contentType="text/html; charset=UTF-8"
    pageEncoding="UTF-8" %>
<%@ taglib prefix="s" uri="/struts-tags" %>
-----------Struts 2 框架的命名对象-------------<br>
parameters: <s:property value="#parameters.welcome"/><br>
```

```
session：<s:property value="#session.ClassName"/><br>
-----------------------栈顶对象--------------<br>
学生班级：<s:property value="classes[0].cName"/><br>
学生姓名：<s:property value="classes[0].cStudent.stuName"/><br>
学生性别：<s:property value="classes[0].cStudent.stuSex"/><br>
------------09计本2班学生信息--------------<br>
<s:iterator value="classes.{? #this.cName == '09计本2'}">
    学生姓名：<s:property value="cStudent.stuName"/><br>
    学生性别：<s:property value="#sex =:[#this == '女'.toString() ? 'female' : 'male'],
                                        #sex(cStudent.stuSex)"/><br>
</s:iterator>
```

（5）发布 Web 项目，启动 Tomcat 服务器，并访问 ognlTest，结果如图 5-2 所示。

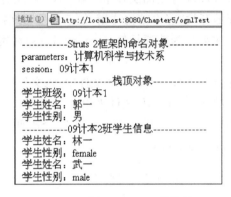

图 5-2 访问 ognlTest 的结果

习题

1. 写出例 5-11、例 5-12、例 5-13 对应的 Java 代码。
2. 使用 lambda 表达式计算斐波那契（Fibonacci）数列的值，描述如下：
fib(0)＝0,fib(1)＝1,fib(n)＝fib(n−1)＋fib(n−2)，其中(n≥2,n∈N*)，输出 fib(10)。
3. 描述 person.relatives.{? #this.gender == 'male'}代码的作用。
4. 在例 5-30 中的 showMessage.jsp 文件中，为什么要使用 toString()方法？

第 6 章 Struts 2框架的标签

Struts 2 框架提供了大量的标签,这些标签支持 OGNL、JSTL、Velocity 等表达式,同时还支持 JSP、FreeMarker、Velocity 等视图层技术。在 Web 应用开发中,使用 Struts 2 框架的标签,可以简化 JSP 页面的开发,程序结构清晰,便于开发与维护。

Struts 2 框架中的标签主要分为普通标签(Generic Tags)和用户界面标签(UI Tags)两大类。

Struts 2 框架的标签包含在 struts-tags.tld 文件中,该文件位于 Struts 2 框架的核心文件(struts2-core-2.2.3.jar)的 META-INF 文件夹中。

在 JSP 页面中使用 Struts 2 框架的标签,必须在 JSP 页面中使用 taglib 指令引入 Struts 2 的标签库。taglib 指令的格式如下:

```
<%@ taglib uri="/struts-tags" prefix="s" %>
```

其中 uri 属性用于指定所要引入的 Struts 2 标签库的位置;prefix 属性用于指定使用标签时的前缀名称。

为了方便讲解本章的内容,首先创建 Web 项目 Chapter6,并添加 Struts 2 框架的支持。编辑 web.xml 配置文件,内容同第 4 章 Chapter4 项目中 web.xml 配置文件的内容。Chapter6 项目目录结构如图 6-1 所示。

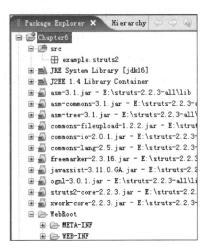

图 6-1　Chapter6 项目目录结构

6.1 普通标签

当页面输出时,普通标签用于控制执行流。如果有内容要输出,则普通标签直接从 tag 简单地输出一些内容。并且普通标签也允许从 action、值栈、国际化、JavaBean 等位置抽取数据。

普通标签分为控制标签(Control Tags)和数据标签(Data Tags)两大类。其中,控制标签提供流的控制,例如 if、else 和 iterator 等;数据标签允许处理或创建数据,例如 bean、push 和 i18n 等。

6.1.1 控制标签

控制标签包含 if、elseif、else、append、generator、iterator、merge、sort、subset 等标签。

1. 条件标签

条件标签用于基本条件控制,包含 if、elseif 和 else 三个标签。if 标签可以单独使用,也可以后面跟随 0 或多个 elseif 标签及 else 标签。if 和 elseif 标签都只有一个必选的 test 属性,如表 6-1 所示。

表 6-1 test 属性

属性名称	默 认 值	类 型	描 述
test		Boolean	决定是否显示标签体

例 6-1 条件标签使用示例。

tag_condition.jsp

```
<%@ page language = "java" contentType = "text/html; charset = UTF-8" %>
<%@ taglib prefix = "s" uri = "/struts-tags" %>
<s:if test = "#parameters.str[0] in {'foo'}">
    foo
</s:if>
<s:elseif test = "#parameters.str[0] in {'bar'}">
    bar
</s:elseif>
<s:else>
    others
</s:else>
```

#parameters.str 返回的结果是一个数组,因此需要使用数组索引的方式取得其元素的值。

测试时访问 tag_condition.jsp? str=foo,结果如图 6-2 所示。

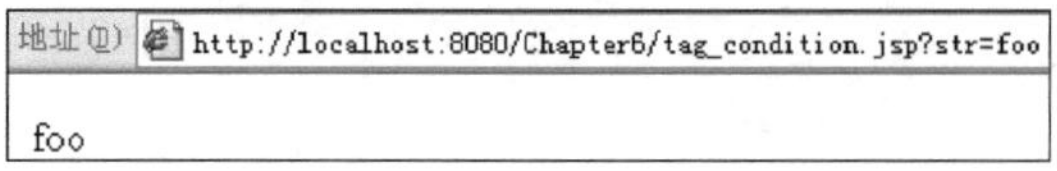

```
地址(D) [图] http://localhost:8080/Chapter6/tag_condition.jsp?str=foo
foo
```

图 6-2 访问 tag_condition.jsp 页面的结果

2．iterator 标签

iterator 标签用于迭代输出，被迭代的值可以是 java.util.Collection、java.util.Iterator 等类型。iterator 标签包含的属性如表 6-2 所示。

表 6-2 iterator 标签包含的属性

属性名称	默认值	类型	描 述
begin	0	Integer	指定迭代开始的值
end		Integer	指定迭代结束的值（包含该值）。默认值是迭代集合（values 属性值）的大小，若 step 属性是负值，则该属性值为 0
status	false	Boolean	设置是否在每次迭代时，将迭代状态实例压入栈中
step	1	Integer	每次迭代时，迭代索引都加上该值。若设置为负值，begin 属性值必须大于 end 属性值
value		String	指定迭代源
var		String	指定访问压入值栈中值的名称

这几个属性都是可选的。如果指定了 status 属性，则将创建一个 IteratorStatus 类的实例，实例名称是 status 属性的值。指定了 status 属性后，就可以访问实例的 count、index、even、odd、first 和 last 等属性。

例 6-2 iterator 标签示例。

tag_iterator.jsp

```
<%@ page language = "java" contentType = "text/html; charset = UTF - 8" %>
<%@ taglib prefix = "s" uri = "/struts - tags" %>
<table border = "1" width = "100">
<s:iterator value = "{1,2,3,4,5}" begin = "2" end = "4" status = "iteratorStatus">
    <s:if test = "! # iteratorStatus.odd">
        <tr bgcolor = "gray">
            <td><s:property /></td>
        </tr>
    </s:if>
    <s:elseif test = "! # iteratorStatus.even">
        <tr bgcolor = "white">
            <td><s:property /></td>
        </tr>
    </s:elseif>
</s:iterator>
</table>
```

访问 tag_iterator.jsp，结果如图 6-3 所示。

图 6-3　访问 tag_iterator.jsp 页面的结果

3．append 标签

append 标签可以将多个迭代器以追加的方式组合在一起。当一个迭代器迭代完后，就对下一个迭代器进行迭代，直到所有的迭代器都迭代结束。append 标签有一个可选的 var 属性，如表 6-3 所示。

表 6-3　var 属性

属性名称	默认值	类型	描　　述
var		String	将组合后的迭代器保存在栈的上下文中，并通过 var 属性值对其引用

例 6-3　append 标签示例。

tag_append.jsp

```
<%@ page language="java" contentType="text/html; charset=UTF-8" %>
<%@ taglib prefix="s" uri="/struts-tags" %>
<s:append var="appendTag">
    <s:param value="{'1-1', '1-2', '1-3'}"></s:param>
    <s:param value="{'2-1', '2-2', '2-3'}"></s:param>
</s:append>
<s:iterator value="#appendTag" status="iteratorStatus">
    <s:property /><s:if test="#iteratorStatus.count == 3"><br></s:if>
</s:iterator>
```

访问 tag_append.jsp，结果如图 6-4 所示。

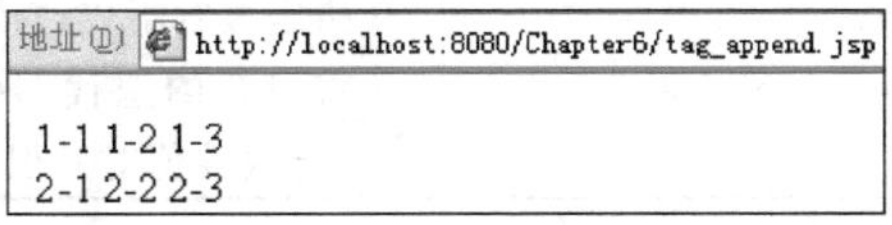

图 6-4　访问 tag_append.jsp 页面的结果

4．merge 标签

merge 标签用于合并多个迭代器。对合并后的迭代器的依次调用可以使得每个被合并的迭代器中的元素有被访问的机会，即每次只访问每个迭代器中的一个元素。merge 标签有一个可选的 var 属性，如表 6-4 所示。

表 6-4 可选的 var 属性

属性名称	默认值	类型	描 述
var		String	将合并后的迭代器保存在栈的上下文中,并通过 var 属性值对其引用

例 6-4 merge 标签示例。

tag_merge.jsp

```
<%@ page language="java" contentType="text/html;charset=UTF-8"%>
<%@ taglib prefix="s" uri="/struts-tags"%>
<s:merge var="mergeTag">
    <s:param value="{'1-1','1-2','1-3'}"></s:param>
    <s:param value="{'2-1','2-2','2-3'}"></s:param>
</s:merge>
<s:iterator value="#mergeTag" status="iteratorStatus">
    <s:property /><s:if test="#iteratorStatus.count % 2 == 0"><br></s:if>
</s:iterator>
```

访问 tag_merge.jsp,结果如图 6-5 所示。

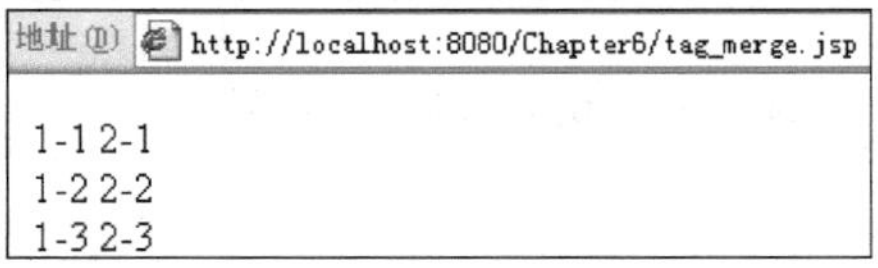

5. generator 标签

generator 标签用于根据 val 属性的值生成一个迭代器。generator 标签包含的属性如表 6-5 所示。

地址(D) http://localhost:8080/Chapter6/tag_merge.jsp
1-1 2-1
1-2 2-2
1-3 2-3

图 6-5 访问 tag_merge.jsp 页面的结果

表 6-5 generator 标签包含的属性

属性名称	必选	类 型	描 述
converter	否	org.apache.struts2.util.IteratorGenerator.Converter	指定一个转换器,用于将解析的字符串转换为对象
count	否	Integer	指定迭代器中最大元素个数
separator	是	String	指定将 val 属性值分解为迭代器中的元素时使用的分隔符
val	是	Integer	指定被解析成迭代器的源
var	否	String	将组合的迭代器保存在 page 上下文中,通过 var 属性值对其引用

例 6-5 generator 标签示例。

tag_generator.jsp

```
<%@ page language="java" contentType="text/html;charset=UTF-8"%>
<%@ taglib prefix="s" uri="/struts-tags"%>
<s:generator val="%{'aaa,bbb,ccc,ddd,eee'}" count="4" separator=",">
    <s:iterator><s:property /></s:iterator>
</s:generator>
```

```
<br />
<s:generator val = "%{'aaa,bbb,ccc,ddd,eee'}" separator = ",">
    <s:iterator><s:property /></s:iterator>
</s:generator>
```

访问 tag_generator.jsp,结果如图 6-6 所示。

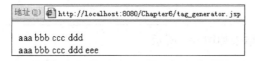

图 6-6　访问 tag_generator.jsp 页面的结果

6. sort 标签

sort 标签用于对列表元素进行排序。包含的属性如表 6-6 所示。

表 6-6　sort 标签包含的属性

属性名称	必选	类型	描述
comparator	是	java.util.Comparator	用于排序的比较器
source	否	String	用于排序的迭代源
var	否	String	将组合的迭代器保存在 page 上下文中,通过 var 属性值对其引用

使用 sort 标签进行排序时,排序规则要由开发人员自己编写,即编写一个实现 Comparator 接口的排序比较器。

例 6-6　sort 标签示例。

(1) 创建排序比较器 SortComparator.java。

SortComparator.java

```java
package example.struts2;
import java.util.Comparator;
public class SortComparator implements Comparator {
    public int compare(Object obj1, Object obj2) {
        return obj1.toString().compareTo(obj2.toString());
    }
}
```

(2) 创建 tag_sort.jsp 页面。

tag_sort.jsp

```
<%@ page language = "java" contentType = "text/html; charset = UTF-8" %>
<%@ taglib prefix = "s" uri = "/struts-tags" %>
<s:bean var = "myComparator" name = "example.struts2.SortComparator"/>
```

```
<s:generator val = "%{'struts2, framework, tag, programming, example'}" separator = ",">
    <s:sort comparator = "#myComparator">
        <s:iterator status = "iteratorStatus">
            <s:property /><s:if test = "!#iteratorStatus.last">、</s:if>
        </s:iterator>
    </s:sort>
</s:generator>
```

访问 tag_sort.jsp,结果如图 6-7 所示。

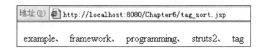

图 6-7　访问 tag_sort.jsp 页面的结果

7. subset 标签

subset 标签用于获取迭代器的子集。包含的属性如表 6-7 所示。

表 6-7　subset 标签包含的属性

属性名称	类型	描述
count	Integer	指定结果子集迭代器中元素的个数
decider	org.apache.struts2.util.SubsetIteratorFilter.Decider	指定一个 decider,用于决定某个特定元素是否应该包含在子集迭代器中
source	String	指定生成结果子集迭代器的源
start	Integer	指定从源中哪个索引值的元素开始生成子集迭代器,索引值从 0 开始取值
var	String	将组合的迭代器保存在 page 上下文中,通过 var 属性值对其引用

这几个属性都是可选项。

在使用 subset 标签时,开发人员可以使用自己开发的实现 Decider 接口的过滤器。

例 6-7　subset 标签示例。

(1) 创建过滤器 SubsetDecider.java。

SubsetDecider.java

```java
package example.struts2;
import org.apache.struts2.util.SubsetIteratorFilter.Decider;
public class SubsetDecider implements Decider {
    public boolean decide(Object obj) throws Exception {
        return obj.toString().length() > 4;
    }
}
```

(2) 创建 tag_subset.jsp 页面。

tag_subset.jsp

```jsp
<%@ page language="java" contentType="text/html; charset=UTF-8" %>
<%@ taglib prefix="s" uri="/struts-tags" %>
<s:bean var="myDecider" name="example.struts2.SubsetDecider"/>
没有设置 decider 属性：<br>
<s:subset source="{'struts2', 'framework', 'tag', 'programming', 'example'}">
    <s:iterator>
        <s:property />
    </s:iterator>
</s:subset>
<br>设置了 decider 属性：<br>
<s:subset source="{'struts2', 'framework', 'tag', 'programming', 'example'}"
                                                    decider="#myDecider">
    <s:iterator>
        <s:property />
    </s:iterator>
</s:subset>
```

访问 tag_subset.jsp，结果如图 6-8 所示。

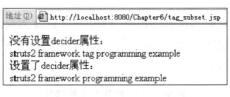

图 6-8　访问 tag_subset.jsp 页面的结果

6.1.2　数据标签

数据标签包含 a、action、bean、date、debug、i18n、include、param、property、push、set、text 和 url 等标签。

1. debug 标签

debug 标签可以输出值栈的内容。该标签会在页面中输出一个[Debug]链接，单击该链接可以查看值栈的内容。debug 标签通常用于 Web 应用开发阶段，方便调试。

debug 标签使用格式如下：

```
<s:debug/>
```

2. param 标签

param 标签可以用来为其他标签提供参数，因此，该标签通常作为其他标签的内嵌标签使用。该标签包含两个可选属性，如表 6-8 所示。

表 6-8　param 标签包含的属性

属性名称	默认值	类型	描述
name		String	用于设置参数的名称
value	栈中指定名称的值	Object	用于设置参数的值表达式

param 标签使用格式如下：

（1）不使用标签体格式。

```
<s:param name = "param_name" value = "param_value" />
```

（2）使用标签体格式。

```
<s:param name = "param_name"> param_value </s:param>
```

3. text 标签

text 标签用于输出国际化消息，该消息必须保存在资源包中，并且资源包的基名和 action 类同名。即必须在 Java 类所在的包中创建一个和 Java 类同名的且扩展名是".properties"的属性文件。

如果在属性文件中没有找到指定的消息，则标签体将作为默认消息。如果没有使用标签体，则将搜索栈，如果有返回值，则输出。如果没有返回值，则输出消息的键(key)。

text 标签包含的属性如表 6-9 所示。

表 6-9　text 标签包含的属性

属性名称	必选	默认值	类型	描述
name	是		String	设置获取资源属性的名称
searchValueStack	否	true	Boolean	如果资源文件中没有找到属性是否栈中搜索
var	否		String	设置访问压入值栈中值的名称

text 标签使用格式如下：

（1）不使用 param 标签。

```
<s:text name = "main.title" />
```

（2）使用 param 标签。

```
<s:text name = "label.welcome">
    <s:param> tsc </s:param>
</s:text>
```

（3）在标签属性中输出资源包中的消息不能使用 text 标签，而采用如下形式：

```
<s:textfield name = "welcome" label = "getText('label.welcome')" />
```

4．i18n 标签

i18n 标签用于获取资源包并放入值栈中。这样 text 标签既可以访问和当前 action 相关联包中的消息，又可以访问其他资源包中的消息。

i18n 标签包含一个 name 属性，如表 6-10 所示。

表 6-10 i18n 标签包含的属性

属性名称	必选	类型	描述
name	是	String	设置资源包文件的基名

i18n 标签使用格式如下：

（1）使用 property 标签。

```
<s:i18n name = "myCustomBundle">
    The i18n value for key main.title in myCustomBundle is:
        <s:property value = "text('main.title')" />
</s:i18n>
```

（2）使用 text 标签。

```
<s:i18n name = "myCustomBundle">
    <s:text name = "main.title"/>
</s:i18n>
```

5．property 标签

property 标签用来获取 value 属性指定的对象属性的值，如果没有指定属性，则默认输出栈顶对象。property 标签包含的属性如表 6-11 所示。

表 6-11 property 标签包含的属性

属性名称	默认值	类型	描述
default		String	设置当 value 属性值为 null 时的默认值
escapeCsv	false	Boolean	设置是否转义输出 CSV
escapeHtml	true	Boolean	设置是否转义输出 HTML
escapeJavaScript	false	Boolean	设置是否转义输出 JavaScript
escapeXml	false	Boolean	设置是否转义输出 XML
value	栈顶对象	Object	设置要输出的值

这些属性是可选的。

property 标签使用格式如下：

```
<s:property value = "myBeanProperty" default = "a default value" />
```

输出 myBeanProperty 属性的值，如果值为 null，则输出"a default value"。

6. set 标签

set 标签可以将一个值赋给指定作用域内的一个变量。如果需要多次访问一个复杂的表达式，则将其赋给一个变量是很有用的，在每次访问复杂表达式的时候，只需访问变量即可。优点表现在以下两方面：

（1）如果复杂表达式计算费时，则可提高性能；

（2）如果复杂表达式难于阅读，则可提高代码的可读性。

set 标签包含的属性如表 6-12 所示。

表 6-12 set 标签包含的属性

属性名称	默认值	类型	描述
scope	action	String	变量的作用域，可选值包括 application、session、request、page、action
value		String	赋给变量的值
var		String	访问压入值栈中值的名称

这三个属性都是可选的。

例 6-8 set 标签示例。

（1）创建 JavaBean 类。

Student.java

```
package example.struts2;
public class Student{
    private String firstName;
    private String lastName;
    public Student(String firstName, String lastName) {
        this.firstName = firstName;
        this.lastName = lastName;
    }
    public String getFirstName() {
        return firstName;
    }
    public void setFirstName(String firstName) {
        this.firstName = firstName;
    }
    public String getLastName() {
        return lastName;
    }
    public void setLastName(String lastName) {
        this.lastName = lastName;
    }
}
```

(2) 创建 Action 类。

SetTagAction.java

```java
package example.struts2;
import com.opensymphony.xwork2.ActionSupport;
public class SetTagAction extends ActionSupport {
    private String firstName;
    private String lastName;
    private Student student;
    public String getFirstName() {
        return firstName;
    }
    public void setFirstName(String firstName) {
        this.firstName = firstName;
    }
    public String getLastName() {
        return lastName;
    }
    public void setLastName(String lastName) {
        this.lastName = lastName;
    }
    public Student getStudent() {
        return student;
    }
    public void setStudent(Student student) {
        this.student = student;
    }
    public String execute() throws Exception {
        student = new Student(firstName, lastName);
        return SUCCESS;
    }
}
```

(3) 编辑 struts.xml 文件，添加 action 定义。

```xml
<action name="setTag" class="example.struts2.SetTagAction">
    <result>/setTagMessage.jsp</result>
</action>
```

(4) 创建 setTagMessage.jsp 文件。

setTagMessage.jsp

```jsp
<%@ page language="java" contentType="text/html; charset=UTF-8" %>
<%@ taglib prefix="s" uri="/struts-tags" %>
<s:set name="firstName" value="student.firstName" />
First Name is:<s:property value="#firstName" /><br>
<s:set name="lastName" value="student.lastName" scope="session" />
Last Name is:<s:property value="#session.lastName" /><br>
```

(5) 创建 studentName.jsp 文件。

studentName.jsp

```
<form action = "setTag.action" method = "post">
    Please input student's name:<br>
    First Name:<input name = "firstName" type = "text"><br>
    Last Name:<input name = "lastName" type = "text"><br>
    <input type = "submit" value = "Submit">
</form>
```

(6) 发布 Web 项目, 启动 Tomcat 服务器, 访问 studentName.jsp, 输入信息提交后的结果如图 6-9 所示。

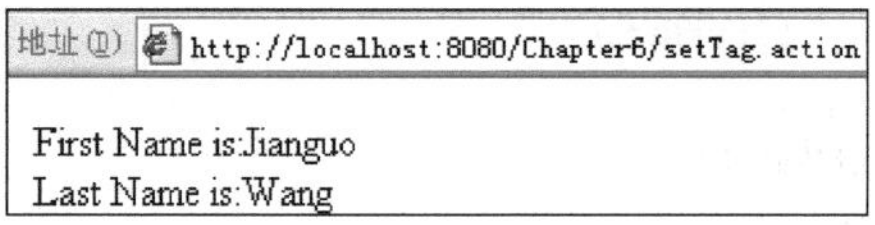

地址(D) http://localhost:8080/Chapter6/setTag.action

First Name is:Jianguo
Last Name is:Wang

图 6-9 set 标签测试的结果

7. push 标签

push 标签用于将值压入值栈, 它包含一个 value 属性, 如表 6-13 所示。

表 6-13 push 标签包含的属性

属 性 名 称	必 选	类 型	描 述
value	是	String	设置压入栈中的值

例 6-9 push 标签示例。

(1) 创建 Action 类。

PushTagAction.java

```
package example.struts2;
import java.util.Map;
import org.apache.struts2.interceptor.SessionAware;
import com.opensymphony.xwork2.ActionSupport;
public class PushTagAction extends ActionSupport implements SessionAware{
    private Map mySession;
    private String firstName;
    private String lastName;
    public String getFirstName() {
        return firstName;
    }
    public void setFirstName(String firstName) {
        this.firstName = firstName;
    }
    public String getLastName() {
        return lastName;
```

```
        }
        public void setLastName(String lastName) {
            this.lastName = lastName;
        }
        public void setSession(Map<String, Object> session) {
            this.mySession = session;
        }
        public String execute() throws Exception {
            Student student = new Student(firstName, lastName);
            mySession.put("student", student);
            return SUCCESS;
        }
    }
```

（2）编辑 struts.xml 文件，添加 action 定义。

```
<action name = "pushTag" class = "example.struts2.PushTagAction">
    <result>/pushTagMessage.jsp</result>
</action>
```

（3）创建 pushTagMessage.jsp 文件。

pushTagMessage.jsp

```
<%@ page language = "java" contentType = "text/html; charset = UTF - 8" %>
<%@ taglib prefix = "s" uri = "/struts - tags" %>
<s:push value = "#session.student">
    First Name is:<s:property value = "firstName" /><br>
    Last Name is:<s:property value = "lastName" /><br>
</s:push>
```

（4）编辑 studentName.jsp 文件，将 form 标签的 action 属性值改为"pushTag.action"。

```
<form action = "pushTag.action" method = "post">
```

（5）发布 Web 项目，启动 Tomcat 服务器，访问 studentName.jsp 进行测试。

8. bean 标签

bean 标签遵循 JavaBean 规范实例化一个类。bean 标签体内可以包含多个 param 标签为 bean 的属性赋值。如果为 bean 标签设置了 var 属性，bean 实例将被放到栈的上下文中。

bean 标签包含的属性如表 6-14 所示。

表 6-14　bean 标签包含的属性

属性名称	必选	类型	描述
name	是	String	用于实例化 bean 的完整实现类名称
var	否	String	用于访问压入值栈中值的名称

例 6-10 bean 标签示例。

(1) 创建 JavaBean。

Person.java

```java
package example.struts2;
public class Person{
    private String firstName;
    private String lastName;
    public String getFirstName() {
        return firstName;
    }
    public void setFirstName(String firstName) {
        this.firstName = firstName;
    }
    public String getLastName() {
        return lastName;
    }
    public void setLastName(String lastName) {
        this.lastName = lastName;
    }
}
```

(2) 创建 JSP 文件。

tag_bean.jsp

```jsp
<%@ page language="java" contentType="text/html; charset=UTF-8" %>
<%@ taglib prefix="s" uri="/struts-tags" %>
<s:bean name="example.struts2.Person" var="person">
    <s:param name="firstName" value="'Jianguo'" />
    <s:param name="lastName" value="'Wang'" />
    Access JavaBean inside tag body: <br>
    First Name is:<s:property value="firstName"/><br>
    Last Name is:<s:property value="lastName"/><br>
</s:bean>
Access JavaBean outside tag body, var attribute must be set : <br>
First Name is:<s:property value="#person.firstName" /><br>
Last Name is:<s:property value="#person.lastName" />
```

(3) 发布 Web 项目，启动 Tomcat 服务器，并访问 tag_bean.jsp，结果如图 6-10 所示。

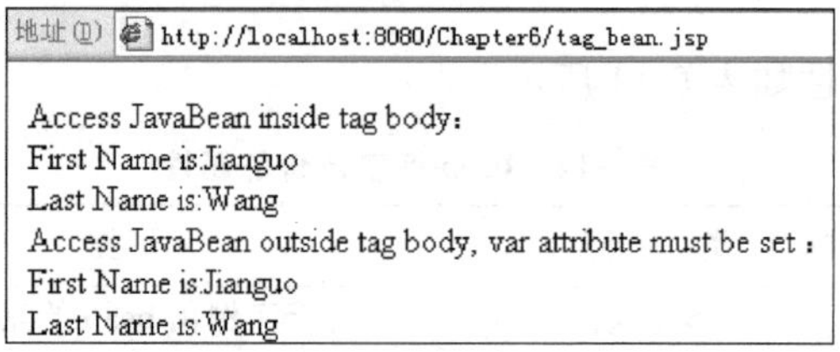

图 6-10 bean 标签测试的结果

9. include 标签

include 标签用于包含 servlet 的输出(servlet 或者 JSP 页面)。include 标签体内可以内嵌 param 标签,用于传递参数。任何提供给被包含页面的参数都不能在输出页面中通过 <s:property...> 标签访问,因为这种方式不会创建值栈。但是可以使用 ${param.ParamName} 进行访问。在 servlet 中可以通过 HttpServletRequest 对象访问。

include 标签只包含一个 value 属性,如表 6-15 所示。

表 6-15 include 标签包含的属性

属性名称	必选	类型	描述
value	是	String	用于设置要包含的 jsp/servlet

例 6-11 include 标签示例。

(1) 创建被包含的 JSP 页面。

include.jsp

```
<%@ page language = "java" contentType = "text/html; charset = UTF-8" %>
The parameter is: ${param.welcome}
```

(2) 创建 tag_include.jsp 页面。

tag_include.jsp

```
<%@ page language = "java" contentType = "text/html; charset = UTF-8" %>
<%@ taglib prefix = "s" uri = "/struts-tags" %>
<s:include value = "include.jsp">
    <s:param name = "welcome">welcome to tsc!</s:param>
</s:include>
```

(3) 发布 Web 项目,启动 Tomcat 服务器,并访问 tag_include.jsp,结果如图 6-11 所示。

图 6-11 include 标签测试的结果

10. url 标签

url 标签用于创建一个 URL。可以在 url 标签体内使用 <param> 标签为 url 标签提供请求参数。如果 param 标签的值是一个数组或者 Iterator,则所有的值都添加到 URL 中。

url 标签包含的属性如表 6-16 所示。

表 6-16 url 标签包含的属性

属性名称	默认值	类型	描述
action		String	设置用于生成 URL 的 action，如果省略，则使用 value 属性的值生成 URL
anchor		String	设置生成的 URL 的锚点(anchor)
encode	true	Boolean	设置是否对参数进行编码
escapeAmp	true	Boolean	设置是否将"&"符号转义为"&amp
forceAddSchemeHostAndPort	false	Boolean	设置是否强制添加协议、主机和端口
includeContext	true	Boolean	设置是否将当前上下文包含在 URL 中
includeParams	none	String	设置是否包含请求参数，可选值 none、get 和 all
method		String	设置 action 使用的方法
namespace		String	设置 action 使用的命名空间
scheme		String	设置协议属性
value		String	设置生成 URL 的目标值，如果省略，则使用 action 属性的值生成 URL
var		String	设置访问压入值栈中值的名称

这些属性都是可选的。

例 6-12 url 标签使用示例。

tag_url.jsp

```
<%@ page language="java" contentType="text/html; charset=UTF-8" %>
<%@ taglib prefix="s" uri="/struts-tags" %>
<b>1. 使用 value 和 anchor 属性：</b>
<s:url value="urlTag" anchor="url" />
<br><b>2. 使用 action 和 forceAddSchemeHostAndPort 属性：</b>
<s:url action="urlTag" forceAddSchemeHostAndPort="true" />
<br><b>3. 同时使用 action 和 value 属性，使用 param 标签传递参数：</b>
<s:url action="urlTagAction" value="urlTagValue" />
<br><b>4. action 和 value 两个属性都省略：</b>
<s:url><s:param name="id" value="'1'" /></s:url>
<br><b>5. 使用 var 属性：</b>
<s:url value="urlTag" var="newURL" />
<s:property value="#newURL"/>
<br><b>6. 使用 scheme 属性：</b>
<s:url value="urlTag" scheme="ftp" />
```

访问 tag_url.jsp，结果如图 6-12 所示。

```
地址(D)  http://localhost:8080/Chapter6/tag_url.jsp

1. 使用 value 和 anchor 属性：   urlTag#url
2. 使用 action 和 forceAddSchemeHostAndPort 属性：   http://localhost:8080/Chapter6/urlTag.action
3. 同时使用 action 和 value 属性，使用 param 标签传递参数：   urlTagValue
4. action 和 value 两个属性都省略：   /Chapter6/tag_url.jsp?id=1
5. 使用 var 属性：   urlTag
6. 使用 scheme 属性：   ftp://localhost/Chapter6/urlTag
```

图 6-12 url 标签测试的结果

11. date 标签

date 标签可以使用不同的方式格式化日期对象，允许开发人员快速方便地格式化日期。不但可以指定自定义格式（如"dd/MM/yyyy hh:mm"），而且可以生成易于阅读的信息（如"in 2 hours, 14 minutes"），或者使用在属性文件中由 struts.date.format 键预先定义的格式。如果没有定义 struts.date.format，则使用默认的 DateFormat.MEDIUM 格式。

date 标签包含的属性如表 6-17 所示。

表 6-17 date 标签包含的属性

属性名称	必选	默认值	类型	描述
format	否		String	设置日期或时间的格式
name	是		String	设置被格式化的日期对象
nice	否	false	Boolean	设置是否输出指定日期和当前日期之间的时差
timezone	否		String	设置格式化日期对象使用的时区
var	否		String	设置访问压入值栈中值的名称

例 6-13 date 标签示例。

tag_date.jsp

```
<%@ page language="java" contentType="text/html; charset=UTF-8" %>
<%@ taglib prefix="s" uri="/struts-tags" %>
<s:bean name="java.util.Date" var="today" />
dd/MM/yyyy:<s:date name="#today" format="dd/MM/yyyy" /><br>
yyyy/MM/dd:<s:date name="#today" format="yyyy/MM/dd" /><br>
yyyy年/MM月/dd日:<s:date name="#today" format="yyyy年/MM月/dd日" /><br>
yyyy/MM/dd hh:mm:<s:date name="#today" format="yyyy/MM/dd hh:mm" /><br>
nice:<s:date name="#today" nice="true" /><br>
original:<s:date name="#today" />
```

访问 tag_date.jsp，结果如图 6-13 所示。

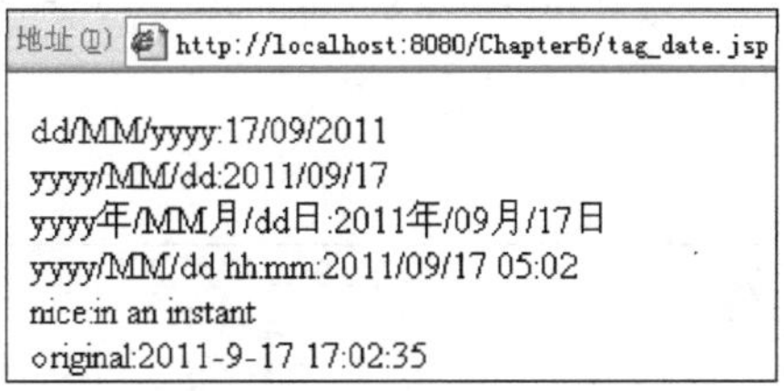

图 6-13 date 标签测试的结果

12. action 标签

action 标签允许开发人员通过在 JSP 页面中指定 action 的名称而直接调用 action。除非指定了 executeResult 参数，否则在 struts.xml 文件中为 action 定义的 result 将被忽略。

在 action 标签内嵌套 param 标签可以为 action 传递参数。action 标签的属性如表 6-18 所示。

表 6-18 action 标签的属性

属性名称	默认值	类型	描述
executeResult	false	Boolean	是否执行或者输出 action 的 result
flush	true	Boolean	action 标签结束时,是否刷新
ignoreContextParams	false	Boolean	当 action 被调用时,是否包含请求参数
name		String	被执行的 action 的名称(不包括扩展名)
namespace	标签所在页面的命名空间	String	被调用 action 的命名空间
var		String	访问被压入值栈中值的名称

其中 name 属性是必选项,其余属性都是可选项。

例 6-14 action 标签示例。

(1) 创建 Action。

ActionTag.java

```java
package example.struts2;
import org.apache.struts2.ServletActionContext;
import com.opensymphony.xwork2.ActionSupport;
public class ActionTag extends ActionSupport {
    private String param;
    public String getParam() {
        return param;
    }
    public void setParam(String param) {
        this.param = param;
    }
    public String execute() throws Exception {
        return SUCCESS;
    }
    public String doDefault() throws Exception {
        ServletActionContext.getRequest().setAttribute("stringByAction",
                    "This is a String put in by the action's doDefault()");
        return SUCCESS;
    }
}
```

(2) 创建 struts.xml 文件。

struts.xml

```xml
<?xml version="1.0" encoding="UTF-8" ?>
<!DOCTYPE struts PUBLIC
    "-//Apache Software Foundation//DTD Struts Configuration 2.0//EN"
    "http://struts.apache.org/dtds/struts-2.0.dtd">
<struts>
    <package name="default" namespace="/" extends="struts-default">
        <action name="actionTag1" class="example.struts2.ActionTag">
```

```xml
            <result>/showMessage.jsp</result>
        </action>
          <action name="actionTag2" class="example.struts2.ActionTag" method="default">
            <result>/showMessage.jsp</result>
        </action>
    </package>
</struts>
```

(3) 创建 showMessage.jsp 文件。

showMessage.jsp

```
<%@ page language="java" contentType="text/html; charset=UTF-8" %>
<%@ taglib prefix="s" uri="/struts-tags" %>
The request parameter is: <s:property value="param"/><br>
String set by doDefault() method: <s:property value="#attr.stringByAction" />
```

(4) 创建 tag_action.jsp 文件。

tag_action.jsp

```
<%@ page language="java" contentType="text/html; charset=UTF-8" %>
<%@ taglib prefix="s" uri="/struts-tags" %>

<B><div>1. Execute result and include it in this page</div></B>
<s:action name="actionTag1" namespace="/" executeResult="true" />
<B><div>2. Execute result and ignore the ContextParams</div></B>
<s:action name="actionTag1" namespace="/" executeResult="true" ignoreContextParams="true"/>
<B><div>3. Invokes method doDefault in action</div></B>
<s:action name="actionTag2!doDefault" namespace="/" executeResult="true" />
<B><div>4. Will not execute result, but retrieve using property tag</div></B>
<s:action name="actionTag2!default" namespace="/" executeResult="false" />
<s:property value="#attr.stringByAction" />
```

(5) 发布 Web 项目，启动 Tomcat 服务器，访问 tag_action.jsp 页面，结果如图 6-14 所示。

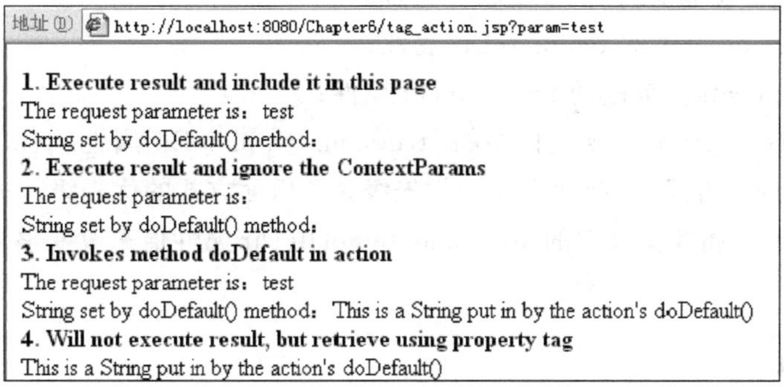

图 6-14　访问 tag_action.jsp 页面的结果

6.2 模板和主题

模板和主题的概念是 Struts 2 框架提供的标签库的核心。在 Web 应用开发中,通过使用模板和主题,开发人员可以进行快速开发,且项目易于维护与升级。

6.2.1 模板

模板是一些(a bit of)代码,通常使用 FreeMarker 编写,可以使用某些标签(HTML 标签)输出。

1. 加载模板

模板加载时,首先搜索应用程序,如果没有,则搜索类路径(classpath)。FreeMarker 是默认的模板引擎,即无论使用哪种格式的视图,都使用 FreeMarker 模板。在内部,JSP、FTL 和 Velocity 标签都使用 FreeMarker 输出。

模板的加载是基于模板目录和主题名称的。struts.properties 属性文件中的 struts.ui.templateDir 属性定义了模板目录(默认值是 template)。如果一个标签使用 Ajax 主题,则如下两个位置将按顺序被搜索:

(1) Web 应用中的/template/ajax/template.ftl;

(2) classpath 中的/template/ajax/template.ftl。

即格式为:/模板目录/主题名称/模板文件名称。

虽然第二种方式(classpath)比较灵活,但是由于性能的原因,开发人员通常会优先使用第一种方式。

2. 选择模板目录

Struts 2 框架提供了几种不同的方式来选择模板的目录:

(1) 使用具体标签中的 templateDir 属性;

(2) 使用 page 范围的 templateDir 属性;

(3) 使用 request 范围的 templateDir 属性;

(4) 使用 session 范围的 templateDir 属性;

(5) 使用 application 范围的 templateDir 属性;

(6) 使用 struts.properties 文件中的 struts.ui.templateDir 属性(默认值是 template)。

在这几种不同的方式中,前面方式的优先级高于后面方式的优先级。另外,通常通过修改 struts.properties 属性文件中的 struts.ui.templateDir 属性值来改变整个 Web 应用的模板目录。

3. 改写模板

Struts 2 框架的核心 Jar 文件(struts2-core-2.2.3.jar)提供的默认模板能够满足大多数的 Web 应用开发。但是为了适应某些 Web 应用开发,需要修改一个模板(在已有的模板上进行修改比开发一个新的模板既快又好)。从核心 Jar 文件中找到想要修改的模板,修改之后,把它保存在 Web 应用中的/template/$theme/$template.ftl 位置。

例如：想要改变 xhtml 主题下的 select 标签的输出效果,在原有模版基础上修改之后,保存在 Web 应用中的/template/xhtml/目录下,模板名称仍是 select.ftl。

6.2.2 主题

主题是一些模板的集合,被打包在一起,来提供公共的功能。

1. 主题分类

Struts 2 框架提供了 4 种主题,分别是 simple、xhtml、css_xhtml 和 ajax 主题。

(1) simple 主题。没有任何特殊性能的最小主题。输出基础的 HTML 元素,通常作为其他主题的基础。例如,textfield 标签输出 HTML 的<input/>标签,该标签没有 label、验证(validation)、错误报告或者其他格式及功能。

(2) xhtml 主题。Struts 2 框架的默认主题。提供了 simple 主题提供的所有功能,并增加了如下一些特性：

- 针对 HTML 标签(form、textfield 和 select 等)提供了标准的两列表格布局;
- 每个 HTML 标签的 label 位置(左边或上边)取决于 labelposition 属性;
- 校验和错误报告信息;
- 在浏览器端使用 100% 的 JavaScript 进行纯 JavaScript 的客户端校验。

如果需要使用不同的布局,不要通过写 HTML 的方式实现,而是创建一个新主题或者使用 simple 主题。

(3) css_xhtml 主题。css_xhtml 主题使用 CSS 样式布局重新实现了 xhtml 主题。提供了 simple 主题提供的所有功能,并增加了如下一些特性：

- 为 HTML 标签(form、textfield 和 select 等)使用<div> 标签,实现标准的两列基于 CSS 的布局;
- 根据 CSS 样式表放置每个 HTML 标签的 label;
- 校验和错误报告信息;
- 在浏览器端使用 100% 的 JavaScript 进行纯 JavaScript 的客户端校验。

(4) Ajax 主题。Ajax 主题使用 Ajax 特性对 xhtml 主题进行了扩展。该主题使用流行的 DOJO AJAX/JavaScript 工具箱。Ajax 特性包括：

- Ajax 客户端校验;
- 支持远程表单提交(和 submit 标签一起使用);
- 高级的 div 模板,提供动态重新加载部分 HTML;
- 高级的 a 模板,提供加载并执行远端的 JavaScript 代码的能力;
- 仅支持 Ajax 的 tabbedPanel 实现;
- 提供 pub-sub 事件模型;
- 提供交互式的自动完成标签。

2. 选择主题

Struts 2 框架能够使用几种不同的方式选择主题,顺序如下:
(1) 使用具体标签中的 theme 属性;
(2) 使用 form 标签的 theme 属性;
(3) 使用 page 范围的 theme 属性;
(4) 使用 request 范围的 theme 属性;
(5) 使用 session 范围的 theme 属性;
(6) 使用 application 范围的 theme 属性;
(7) 使用 struts.properties 文件中的 struts.ui.theme 属性(默认值是 xhtml)。

在这几种不同的方式中,前面方式的优先级高于后面方式的优先级。另外,通常通过修改 struts.properties 属性文件中的 struts.ui.theme 属性值来改变整个 Web 应用的主题。

3. 扩展主题

扩展主题就是在已有主题的基础上创建用户自己的主题。通常有三种方式创建新的主题。

(1) 从无到有创建新主题;

这种方式最难,不是最好的方式,不推荐使用,而是推荐使用 simple 主题作为起点。simple 主题提供了足够的基础,通过扩展和包装基本控制使得创建新的控制变得容易。

(2) 包装一个现有的主题;

xhtml 主题是使用"包装"技术的一个很好的例子。simple 主题输出基本控制,xhtml 主题通过添加一个头部(header)和一个尾部(footer)来"修饰"许多控制。

例 6-15 包装 simple 主题中的 xxx.ftl 模板。

```
<#include "/${parameters.templateDir}/xhtml/controlheader.ftl" />
<#include "/${parameters.templateDir}/simple/xxx.ftl" />
<#include "/${parameters.templateDir}/xhtml/controlfooter.ftl" />
```

通过使用 xhtml 主题下的 controlheader.ftl 和 controlfooter.ftl 模板包装了一个 simple 主题下的 xxx.ftl 模板,创建了一个新的主题。

通过包装技术创建新的主题,需要在 simple 主题提供的基本 HTML 元素基础上做大量的工作。

(3) 扩展一个现有的主题。

通过在存放主题的子目录中包含一个 theme.properties 文件,并且在这个属性文件中使用 parent 元素指定要扩展(继承)的主题。Ajax 主题就是使用了这种技术扩展了 xhtml 主题。

例 6-16 Ajax 主题扩展 xhtml 主题的 theme.properties 文件。

/template/ajax/theme.properties

```
parent = xhtml
```

扩展的主题不需要实现每一个单一的模板。只需要实现要变化的模板即可,其他的模板从父模板中加载。

6.3 UI 标签

和普通标签不一样，UI 标签不提供较多的控制或逻辑，而是集中于使用数据（从 action/值栈或者数据标签）和显示数据。所有的 UI 标签都是由模板和主题驱动的。如果标签中有内容，普通标签就简单地输出标签中的内容，而 UI 标签遵从模板（通常组合在一起成为一个主题）进行实际输出。

UI 标签分为表单标签（Form Tags）、非表单标签（Non-Form UI Tags）和 Ajax 标签（Ajax Tags）三大类。

6.3.1 表单标签的公共属性

所有的表单标签扩展了 UIBean 类。在 UIBean 类中提供了一组公共属性，这些公共属性可以被分为 4 类：和模板相关的属性、和 JavaScript 相关的属性、和 Tooltip 相关的属性以及通用属性。除了这些公共属性外，所有的表单元素标签都有一个特殊的属性：form（${parameters.form}）。form 属性表示用于输出 form 标签的属性，例如 form 的 id。在模板中，可以通过调用 ${parameters.form.id} 获得 form 的 id。

1. 和模板相关的属性

和模板相关的属性的数据类型都是 String。包含的属性如表 6-19 所示。

表 6-19 和模板相关的属性

属性	描述
templateDir	设置模板目录
theme	设置主题名称
template	设置模板名称

2. 和 JavaScript 相关的属性

和 JavaScript 相关的属性都用于 simple 主题，数据类型都是 String 类型。包含的属性如表 6-20 所示。

表 6-20 和 JavaScript 相关的属性

属性	描述
onclick	设置 HTML 的 JavaScript onclick 属性
ondblclick	设置 HTML 的 JavaScript ondblclick 属性
onmousedown	设置 HTML 的 JavaScript onmousedown 属性
onmouseup	设置 HTML 的 JavaScript onmouseup 属性
onmouseover	设置 HTML 的 JavaScript onmouseover 属性
onmouseout	设置 HTML 的 JavaScript onmouseout 属性
onfocus	设置 HTML 的 JavaScript onfocus 属性
onblur	设置 HTML 的 JavaScript onblur 属性
onkeypress	设置 HTML 的 JavaScript onkeypress 属性
onkeyup	设置 HTML 的 JavaScript onkeyup 属性
onkeydown	设置 HTML 的 JavaScript onkeydown 属性
onselect	设置 HTML 的 JavaScript onselect 属性
onchange	设置 HTML 的 JavaScript onchange 属性

3. 和 Tooltip 相关的属性

和 Tooltip 相关的属性的数据类型都是 String 类型。包含的属性如表 6-21 所示。

表 6-21 和 Tooltip 相关的属性

属 性	默 认 值	描 述
tooltip	无	设置该特定组件的工具提示
jsTooltipEnabled	false	设置是否使用 JavaScript 生成工具提示
tooltipIcon	/struts/static/tooltip/tooltip.gif	设置工具提示图标的 URL
tooltipDelay	500	设置显示工具提示的延时(毫秒)

可以为每一个 Form UI 组件(在 xhtml/css_xhtml 或其他扩展它们的主题中)指定工具提示。一旦指定了 Form 组件属性相关的工具提示,就会被应用到所有的它创建的 form UI 组件中,除非在这些创建的 UI 组件中显式地覆盖 tooltip 属性。

例 6-17 textfield 标签使用继承的 tooltipDelay 和 tooltipIconPath 属性。

```
< s:form tooltipDelay = "500" tooltipIconPath = "/myImages/myIcon.gif" .... >
    …
    < s:textfield label = "Customer Name" tooltip = "Enter the customer name" ... />
    …
</s:form >
```

textfield 标签从包含它的 form 标签中继承了 tooltipDelay 和 tooltipIconPath 属性。

例 6-18 textfield 标签使用自己的 tooltipDelay 属性。

```
< s:form tooltipDelay = "500" tooltipIconPath = "/myImages/myIcon.gif" .... >
    …
    < s:textfield label = "Address" tooltip = "Enter your address" tooltipDelay = "5000" />
    …
</s:form >
```

textfield 标签从包含它的 form 标签中继承了 tooltipDelay 和 tooltipIconPath 属性,但是 textfield 标签使用自己的 tooltipDelay 属性覆盖了继承的 tooltipDelay 属性。

4. 通用属性

通用属性如表 6-22 所示。

表 6-22 通用属性

属 性	主 题	数 据 类 型	描 述
accesskey	simple	String	为输出的 HTML 元素设置 accesskey 属性
cssClass	simple	String	设置 HTML 的 class 属性
cssStyle	simple	String	设置 HTML 的 style 属性
cssErrorClass	simple	String	设置 HTML 的 error class 属性
cssErrorStyle	simple	String	设置 HTML 的 error style 属性

续表

属性	主题	数据类型	描述
title	simple	String	设置 HTML 的 title 属性
disabled	simple	String	设置 HTML 的 disabled 属性
key	simple	String	设置该输入字段所表示的属性名称,将自动生成 name、label 和 value 属性的值
label	xhtml	String	设置表单元素的 label 名称
labelPosition	xhtml	String	设置表单元素的 label 的位置(top/left),默认值是 left
requiredposition	xhtml	String	设置表单元素的必填标识符的位置(left/right),默认值是 right
name	simple	String	设置表单元素的字段名称
required	xhtml	Boolean	设置是否在表单元素的 label 上添加"＊"(值为 true 时添加,否则不添加)
tabIndex	simple	String	设置 HTML 的 tabindex 属性
value	simple	Object	设置表单元素的值
id	simple	String	设置 HTML 的 id 属性

一些公共的标签属性可能不被所有或部分的模板所使用。例如,form 标签支持 tabIndex 属性,但是没有一个主题输出它。

在通用属性中,有 value 和 name 两个属性。很多表单标签(除了 form 标签)的 name 和 value 属性之间存在着唯一的关系。name 属性为标签提供名称,反过来,还作为表单提交时的控制属性。在大多数情况下,name 属性映射到一个简单的 JavaBean 属性,例如 "postalCode",在提交时,通过调用 setPostalCode 设值方法设置属性的值。

同样,可以通过调用 JavaBean 访问器(如 getPostalCode 方法)填充一个表单控件。在表达式语言中,可以使用 value 访问 JavaBean 属性。形如"％{postalCode}"的表达式就会调用 getPostalCode 方法。

例 6-19 使用表达式填充一个表单。

```
< s:form action = "updateAddress">
    < s:textfield label = "Postal Code" name = "postalCode" value = "%{postalCode}"/>
    ...
</s:form>
```

然而,由于标签包含了 name 和 value 之间的这种关系,因此,value 属性是可选的。如果没有指定 value 属性,JavaBean 访问器仍被调用。

例 6-20 使用省略 value 属性的表达式填充一个表单。

```
< s:form action = "updateAddress">
    < s:textfield label = "Postal Code" name = "postalCode"/>
    ...
</s:form>
```

虽然大多数属性使用与属性同名的 key(键)和底层模板打交道(${parameters.

label})，但是 value 属性却不是这样。反而，访问它是通过 nameValue 这个键（＄{parameters.nameValue}）。键 nameValue 表明值是由 name 属性生成而不是使用 value 属性明确指定的。

6.3.2 表单标签

表单标签包含 checkbox、checkboxlist、combobox、doubleselect、head、file、form、hidden、inputtransferselect、label、optiontransferselect、optgroup、password、radio、reset、select、submit、textarea、textfield、token 和 updownselect 等标签。

表单标签可以分为两类：form 标签本身和所有其他标签（由单独的表单元素构成）。form 标签的行为和包含在其中的元素的行为不同。

1. form 标签

form 标签输出一个 HTML 输入表单。除了公共属性，还包含如表 6-23 所示的属性。

表 6-23　form 标签的非公共属性

属　　性	默　认　值	数据类型	描　　述
acceptcharset		String	该表单可接受的字符集，值用分号或空格分隔
action	当前 action	String	设置提交的 action 的名称，不用". action"后缀
enctype		String	设置 HTML 表单的 enctype 属性。上传文件时，设置为 multipart/form-data
focusElement		String	设置当页面加载时，具有焦点的元素的 Id
namespace	当前命名空间	String	设置提交的 action 的命名空间
onreset		String	设置 HTML 表单的 onreset 属性
onsubmit		String	设置 HTML 表单的 onsubmit 属性
validate	false	Boolean	设置是否执行客户端/远端校验，只对 xhtml/ajax 主题有效
windowState		String	设置表单提交后显示的窗口状态

这些属性都不是必选项。

form 标签的使用格式为：

```
<s:form>
    …
</s:form>
```

Struts 2 框架决定生成的 HTML 表单中的提交 URL 的顺序/逻辑如下：

(1) 如果没有指定 action 属性，则使用当前请求作为提交 url。

(2) 如果指定了 action，Struts 2 框架会努力获得一个 action 的配置。如果 action 属性值是 struts.xml 文件中定义的有效 action 别名时（命名空间也要匹配），则生成的 url 中会添加". action"后缀。

(3) 如果指定了 action，但其值不是在 struts.xml 文件中定义的 action 别名，Struts 2 框架将使用 action 属性值作为提交 url。

2. submit 标签

submit 标签用于输出一个提交按钮。submit 标签和 form 标签一起使用,实现异步表单的提交。除了公共属性外,还包含如表 6-24 所示的属性。

表 6-24 submit 标签的非公共属性

属性	默认值	数据类型	描述
action		String	设置 action 属性
align		String	设置 HTML 提交按钮的 align 属性
method		String	设置 method 属性
src		String	为 image 类型的提交按钮提供图像路径,对 input 和 button 类型按钮无效
type	input	String	设置提交按钮的类型,有效值是 input、button 和 image

submit 标签的使用格式为:

```
< s:submit … />
```

submit 标签的 type 属性设置不同的值,在客户端浏览器中生成的提交按钮类型也不同。

(1) type 属性值为 input 时,生成的提交按钮的 HTML 代码为:

```
< input type = "submit"…>
```

(2) type 属性值为 button 时,生成的提交按钮的 HTML 代码为:

```
< button type = "submit"…>
```

(3) type 属性值为 image 时,生成的提交按钮的 HTML 代码为:

```
< input type = "image"…>
```

3. reset 标签

reset 标签用于输出表单的重置按钮。reset 标签和 form 标签一起使用,实现表单的重置。除了公共属性外,还包含如表 6-25 所示的属性。

表 6-25 reset 标签的非公共属性

属性	默认值	数据类型	描述
action		String	设置 action 属性
align		String	设置 HTML 重置按钮的 align 属性
method		String	设置 method 属性
type	input	String	设置重置按钮的类型,有效值是 input 和 button

reset 标签的使用格式为：

```
< s:reset ··· />
```

reset 标签的 type 属性设置不同的值，在客户端浏览器中生成的重置按钮类型也不同。
(1) type 属性值为 input 时，生成的重置按钮的 HTML 代码为：

```
< input type = "reset" ··· >
```

(2) type 属性值为 button 时，生成的重置按钮的 HTML 代码为：

```
< button type = "reset" ··· >
```

4. token 标签

token 标签用于防止重复提交表单。如果使用了 TokenInterceptor 或 TokenSessionInterceptor 拦截器，就需要使用 token 标签。

token 标签的使用格式为：

```
< s:token />
```

token 标签通过生成两个隐藏的表单元素来帮助解决表单"双击"提交问题，其 HTML 代码如下：

```
< input type = "hidden" name = "struts.token.name" value = "struts.token" />
< input type = "hidden" name = "struts.token"
                       value = "LJYEHH2H8FZ9X05OPKSSJX1XI1HZOI46" />
```

5. textfield 标签

textfield 标签用于输出一个 HTML 单行文本输入框。除了公共属性外，还包含如表 6-26 所示的一些属性。

表 6-26　textfield 标签的非公共属性

属　　性	默认值	数 据 类 型	描　　述
accesskey		String	为输出的 HTML 元素设置 accesskey 属性
maxlength		Integer	设置 HTML 文本输入框的 maxlength 属性，即能够输入的字符最大个数
readonly	false	Boolean	设置文本输入框的内容是否只读
size		Integer	设置 HTML 文本输入框的 size 属性，即文本输入框的外观大小

textfield 标签的使用格式为：

```
< s:textfield key = "keyName" />
```

textfield 标签效果如图 6-15 所示。

6. password 标签

password 标签用于输出一个 HTML 密码输入框。除了 textfield 标签的属性外,还包含一个可选的 showPassword 属性,如表 6-27 所示。

图 6-15　textfield 标签的效果

表 6-27　password 标签的可选的 showPassword 属性

属性	默认值	数据类型	描述
showPassword	false	Boolean	设置是否显示输入的内容

如果 showPassword 属性值设置为 true,当提交的表单没有通过验证返回该输入页面时,密码输入框的内容还保留。

password 标签的使用格式为:

```
<s:password key = "password" />
```

7. hidden 标签

hidden 标签用于输出一个 HTML 隐藏类型的输入框,使用指定属性的值(存放在值栈中)填入该输入框。除了公共属性外,不包含其他属性。

hidden 标签的使用格式为:

```
<s:hidden name = "propertyName" />
```

8. textarea 标签

textarea 标签用于输出 HTML 多行文本框。除了公共属性外,还包含如表 6-28 所示的一些属性。

表 6-28　textarea 标签的非公共属性

属性	默认值	数据类型	描述
cols		Integer	设置多行文本框的 cols 属性,即列数
readonly	false	Boolean	设置多行文本框的内容是否只读
rows		Integer	设置多行文本框的 rows 属性,即行数
wrap	soft	String	设置多行文本框的 wrap 属性,即是否换行

textarea 标签的使用格式为:

```
<s:textarea label = "Comments" name = "comments" cols = "30" rows = "3"/>
```

wrap 属性有"off"、"soft"和"hard"三个选项。"soft"值表示显示的时候换行,但是提交的时候没有换行符。"hard"值表示显示的时候换行,提交的时候也有换行符。"off"值不允许自动换行,按照用户实际的输入进行显示。

textarea 标签效果如图 6-16 所示。

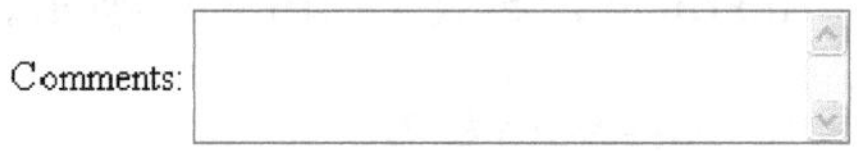

9. file 标签

file 标签用于输出一个 HTML 文件选择框。除了公共属性外,还包含如表 6-29 所示的一些属性。

图 6-16 textarea 标签的效果

表 6-29 file 标签的非公共属性

属性	数据类型	描述
accept	String	设置 HTML 文件输入框的 accept 属性,指示能够接受的文件的 mime 类型
size	Integer	设置 HTML 文件输入框 size 属性,即外观大小

file 标签的使用格式为:

```
<s:file name="uploadFIle" accept="text/html,text/plain" />
```

10. checkbox 标签

checkbox 标签用于输出一个 HTML 复选框,即类型为 checkbox 的 input 元素。除了公共属性,还包含一个 fieldValue 属性,如表 6-30 所示。

表 6-30 checkbox 标签的 fieldValue 属性

属性	默认值	数据类型	描述
fieldValue	true	String	设置复选框的实际值

fieldValue 属性的值是 String 类型,其默认是字符串"true"。

checkbox 标签的使用格式为:

```
<s:checkbox label="checkbox test" name="checkboxField" fieldValue="true"/>
```

11. checkboxlist 标签

checkboxlist 标签用于根据列表(list)创建一系列的复选框。除了公共属性,还包含如表 6-31 所示的一些属性。

表 6-31 checkboxlist 标签的非公共属性

属性	数据类型	描述
list	Map,list,set,array	设置迭代源,必选项。如果 list 是一个 Map(key,value),则 Map 的 key 将作为复选框选项"value"属性的值,Map 的 value 作为选项的内容
listKey	String	设置选项的 value 来源于 list 对象的哪个属性
listValue	String	设置选项的内容来源于 list 对象的哪个属性

checkboxlist 标签的使用格式为：

```
< s:checkboxlist name = "checkboxlist" list = "#{0:'cb1',1:'cb2',2:'cb3'}" label = "Checkboxlist"/>
```

checkboxlist 标签效果如图 6-17 所示。

12. radio 标签

radio 标签用于输出单选按钮，和 checkboxlist 标签具有一样的属性。二者的区别在于 checkboxlist 标签可以多选，radio 标签只能单选。

radio 标签的使用格式为：

```
< s:radio label = "Sex" name = "Sex" list = "{'male','female'}"/>
```

radio 标签效果如图 6-18 所示。

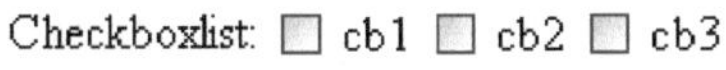

图 6-17　checkboxlist 标签的效果 　　　　图 6-18　radio 标签的效果

13. select 标签

select 标签用于输出一个 HTML 下拉列表框。除了公共属性，还包含如表 6-32 所示的一些属性。

表 6-32　select 标签的非公共属性

属性	默认值	数据类型	描述
emptyOption	false	Boolean	在 header 选项后是否添加空的选项（——）
headerKey		String	设置列表中第一个选项的 key，不能为空
headerValue		String	设置列表中第一个条目的值表达式
list		Map, list, set, array	设置迭代源，必选项。如果 list 是一个 Map（key,value），则 Map 的 key 将作为复选框选项"value"属性的值，Map 的 value 作为选项的内容
listKey		String	设置选项的 value 来源于 list 对象的哪个属性
listValue		String	设置选项的内容来源于 list 对象的哪个属性
multiple	false	Boolean	设置是否允许多选
size		Integer	设置下拉列表框的大小（可以显示选项的个数）

select 标签的使用格式为：

```
< s:select label = "Season" name = "season" headerKey = "-1" headerValue = "Select Season"
    list = "#{'01':'Spring', '02':'Summer', '03':'Autumn', '04':'Winter'}"
    value = "selectedSeason" required = "true"/>
```

select 标签效果如图 6-19 所示。

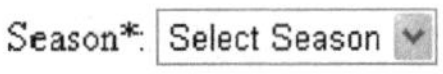

图 6-19　select 标签的效果

单击下拉图标后的效果如图 6-20 所示。

14. combobox 标签

combobox 标签是由一个 HTML 单行文本输入框和下拉列表框组合而成的，提供组合框的功能。用户可以通过下拉列表框将需要的内容放到文本输入框中，或者直接在文本输入框中输入需要的内容。combobox 标签的属性请参考 textfield 标签和 select 标签的属性。

图 6-20　select 标签展开后的效果

combobox 标签的使用格式为：

```
<s:combobox label = "My Favourite Color" name = "myFavouriteColor"
    list = "#{'red':'red','green':'green','blue':'blue'}"
    headerKey = "-1" headerValue = "--- Please Select ---"
    emptyOption = "true" value = "green" />
```

combobox 标签效果如图 6-21 所示。

15. optgroup 标签

optgroup 标签需要内嵌在 select 标签中使用，用于创建一个下拉列表框选项组。optgroup 标签包含的一些属性如表 6-33 所示。

图 6-21　combobox 标签的效果

表 6-33　optgroup 标签包含的属性

属性	数据类型	描述
label	String	设置 label 属性
list	Map，list，set，array	设置迭代源，必选项
listKey	String	设置选项的 value 来源于 list 对象的哪个属性
listValue	String	设置选项的内容来源于 list 对象的哪个属性

optgroup 标签的使用格式为：

```
<s:select label = "My Selection" name = "mySelection"
    list = "%{#{'SUPERMAN':'Superman', 'SPIDERMAN':'spiderman'}}">
    <s:optgroup label = "Adult"
        list = "%{#{'SOUTH_PARK':'South Park'}}" />
    <s:optgroup label = "Japanese"
        list = "%{#{'POKEMON':'pokemon','DIGIMON':'digimon'}}" />
```

optgroup 标签效果如图 6-19 所示。单击下拉图标后的效果如图 6-22 所示。

16. doubleselect 标签

doubleselect 标签用于输出两个关联的 HTML 下拉列表框(select)，其中第二个下拉列表框显示的内容是由第一个下拉列表框中所选内容决定的。

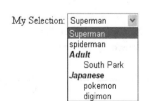

图 6-22　optgroup 标签展开后的效果

除了公共属性外,还包含如下一些属性:doubleList、doubleListKey、doubleListValue、doubleName、doubleValue、doubleId、doubleDisabled、doubleMultiple、doubleSize、doubleHeaderKey、doubleHeaderValue、doubleEmptyOption、doubleCssClass、doubleCssStyle。

这些属性以"double"开始,表示应用于第二个列表框,属性含义请参考 select 标签中对应属性的含义。

doubleselect 标签的使用格式为:

```
<s:doubleselect label = "Doubleselect" name = "menu"
    list = "{'fruit','other'}" doubleName = "dishes"
    doubleList = "top == 'fruit'? {'apple', 'orange'} : {'monkey', 'chicken'}" />
```

doubleselect 标签效果如图 6-23 所示。

单击下拉图标后的效果如图 6-24 所示。

图 6-23　doubleselect 标签的效果　　　图 6-24　doubleselect 标签展开后的效果

17. updownselect 标签

updownselect 标签用于创建一个列表框组件及能够上下移动列表框组件中元素的按钮。当提交表单时,元素按照在列表框组件中安排的顺序(从上到下)进行提交。除了 select 标签具有的属性,还包含如表 6-34 所示的一些属性。

表 6-34　updownselect 标签特有的一些属性

属性	默认值	数据类型	描述
allowMoveDown	true	Boolean	设置是否显示"下移"按钮
allowMoveUp	true	Boolean	设置是否显示"上移"按钮
allowSelectAll	true	Boolean	设置是否显示"全选"按钮
moveDownLabel	v	String	设置"下移"按钮的标题
moveUpLabel	^	String	设置"上移"按钮的标题
selectAllLabel	*	String	设置"全选"按钮的标题

updownselect 标签的使用格式为:

```
<s:updownselect list = "#{'england':'England', 'america':'America', 'germany':'Germany'}"
    headerKey = "-1" headerValue = "--- Select and Move ---"
    name = "countries" emptyOption = "true" />
```

updownselect 标签效果如图 6-25 所示。

18. inputtransferselect 标签

inputtransferselect 标签用于创建一个输入、转移列表框组件,由一个单行文本框、一个列表框组件和在它们之间的用于将文本添加到转移列表框中的按钮组成。当提交表单时,

图 6-25　updownselect 标签的效果

会自动选择列表框中所有的元素。除了公共属性，还包含如表 6-35 所示的一些属性。

表 6-35　inputtransferselect 标签的非公共属性

属　　性	默认值	数据类型	描　　述
addLabel		String	设置"添加"按钮的标题
allowRemoveAll	true	Boolean	设置是否显示"移除全部"按钮
allowUpDown	true	Boolean	设置是否显示"上移下移"按钮
downLabel		String	设置"下移"按钮的标题
leftTitle		String	设置左边文本框组件的标题
multiple	true	Boolean	设置是否可以同时选择列表框中的多个元素
removeAllLabel		String	设置"移除全部"按钮的标题
removeLabel		String	设置"移除"按钮的标题
rightTitle		String	设置右边下拉列表框组件的标题
upLabel		String	设置"上移"按钮的标题

inputtransferselect 标签的使用格式为：

```
<s:inputtransferselect leftTitle = "Left" removeAllLabel = "removeAll"
    label = "Favourite Cartoons Characters"    name = "cartoons"
    list = "{'Popeye', 'He-Man', 'Spiderman'}"/>
```

inputtransferselect 标签效果如图 6-26 所示。

图 6-26　inputtransferselect 标签的效果

19．optiontransferselect 标签

optiontransferselect 标签用于创建一个选项转移列表框组件，由两个列表框和在它们之间的将选项在两个列表框之间移动的按钮组成。当提交表单时，会自动选择列表框中所有的元素。除了公共属性，还包含如表 6-36 所示的一些属性。

除了表 6-36 中列出的属性，optiontransferselect 标签还包含一些以"double"短语开始的属性，这些属性表示应用于第二个列表框，属性含义请参考 select 标签中对应属性的含义。

表 6-36　optiontransferselect 标签的非公共属性

属　　性	默认值	数据类型	描　　述
∧addAllToLeftLabel	<<－－	String	设置"添加所有到左边"按钮的标题
addAllToLeftOnclick		String	设置单击"添加所有到左边"按钮时调用的 JavaScript 函数
addAllToRightLabel	－－>>	String	设置"添加所有到右边"按钮的标题
addAllToRightOnclick		String	设置单击"添加所有到右边"按钮时调用的 JavaScript 函数
addToLeftLabel	<－	String	设置"添加到左边"按钮的标题
addToLeftOnclick		String	设置单击"添加到左边"按钮时调用的 JavaScript 函数
addToRightLabel	－>	String	设置"添加到右边"按钮的标题
addToRightOnclick		String	设置单击"添加到右边"按钮时调用的 JavaScript 函数
allowAddAllToLeft	true	Boolean	设置是否显示"添加所有到左边"按钮
allowAddAllToRight	true	Boolean	设置是否显示"添加所有到右边"按钮
allowAddToLeft	true	Boolean	设置是否显示"添加到左边"按钮
allowAddToRight	true	Boolean	设置是否显示"添加到右边"按钮
allowSelectAll	true	Boolean	设置是否显示"选择所有"按钮
allowUpDownOnLeft	true	Boolean	设置是否显示左侧"上移下移"按钮
allowUpDownOnRight	true	Boolean	设置是否显示右侧"上移下移"按钮
leftDownLabel	v	String	设置左侧"下移"按钮的标题
leftTitle		String	设置左侧列表框的标题
leftUpLabel	^	String	设置左侧"上移"按钮的标题
rightDownLabel	v	String	设置右侧"下移"按钮的标题
rightTitle		String	设置右侧列表框的标题
rightUpLabel	^	String	设置右侧"上移"按钮的标题
selectAllLabel	<*>	String	设置"选择所有"按钮的标题
selectAllOnclick		String	设置单击"选择所有"按钮时调用的 JavaScript 函数
upDownOnLeftOnclick		String	设置单击左侧"上移下移"按钮时调用的 JavaScript 函数
upDownOnRightOnclick		String	设置单击右侧"上移下移"按钮时调用的 JavaScript 函数

optiontransferselect 标签的使用格式为：

```
<s:optiontransferselect
    label = "Optiontransferselect" name = "leftSide" leftTitle = "Left" rightTitle = "Right"
    size = "12" list = "{'Popeye', 'He-Man', 'Spiderman'}" multiple = "true"
    headerKey = "headerKey" headerValue = "--- Please Select ---" emptyOption = "true"
    doubleList = "{'Superman', 'Mickey Mouse', 'Donald Duck'}"
    doubleName = "rightSide" doubleSize = "12"
    doubleHeaderKey = "doubleHeaderKey" doubleHeaderValue = "--- Please Select ---"
    doubleEmptyOption = "true" doubleMultiple = "true" />
```

optiontransferselect 标签效果如图 6-27 所示。

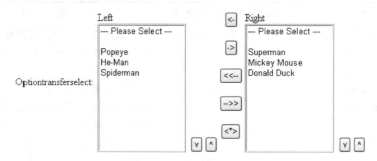

图 6-27　optiontransferselect 标签的效果

20. head 标签

head 标签用于输出一个 HTML 文件的 HEAD 区域的部分内容。

head 标签的使用格式为：

```
<s:head/>
```

21. label 标签

label 标签用于输出一个 HTML 的 LABEL 标签。除了公共属性,还有一个 for 属性,如表 6-37 所示。

表 6-37　label 标签的 for 属性

属　　性	数 据 类 型	描　　述
for	String	设置 HTML 的 for 属性

label 标签的使用格式为：

```
<s:label key="labelText" />
```

6.3.3　非表单标签

非表单标签包含 actionerror、actionmessage、component、div、fielderror 等标签。其中 actionerror、actionmessage、fielderror 三个标签用于输出消息。

1. div 标签

div 标签用于创建一个 HTML 的＜div＞。div 标签可以在 simple、xhtml 及其他主题中使用,但是和 Ajax 主题一起使用效果最好。

div 标签使用格式如下：

```
<s:div>Content to be output!</s:div>
```

2. component 标签

component 标签使用指定的模板输出一个自定义的 UI 构件,并且可以使用 param 标签将附加的对象传递给模板。

(1) 如果使用 Freemarker 模板,则通过使用 $parameters.paramName 或者 $parameters.get('paramName')来检索提供的对象。

(2) 如果使用 JSP 模板,则通过使用 <s:property value="%{parameters.paramName}"/> 或者 <s:property value="%{parameters.get('paramName')}" /> 来检索提供的对象。

目前,自定义 UI 组件可以使用 Velocity、JSP 或者 Freemarker 编写,正确的输出引擎根据文件扩展名进行查找。

另外,如果使用 JSP 模板,则 JSP 模板必须放在 Web 应用程序目录下,而不能放在 classpath 下,这一点与 Freemarker 或 Velocity 模板不一样。

component 标签使用格式如下:

(1) 不使用 param 标签。

```
<s:component template="/myTheme/customTemplate/component.vm"/>
```

(2) 使用 param 标签。

```
<s:component template="/myTheme/customTemplate/component.vm">
    <s:param name="key" value="value"/>
</s:component>
```

例 6-21 component 标签应用示例。

① 在 WebRoot 目录下创建 myTemplate/myTheme 子目录。

② 在 myTheme 子目录下创建模板文件 myJspTemplate.jsp。

myJspTemplate.jsp

```
<%@ page language="java" contentType="text/html; charset=UTF-8" %>
<%@ taglib prefix="s" uri="/struts-tags" %>
This template can only output fruits list!<br>
The fruits contains:
<s:if test="%{parameters.fruits}">
    <s:iterator value="parameters.fruits" status="status">
        <s:property/><s:if test="!#status.last">,</s:if>
    </s:iterator>
</s:if>
```

如果想改变输出效果,只需更改该模板文件即可。

③ 在 Web 应用程序目录下创建测试文件 tag_component.jsp。

tag_component.jsp

```
<%@ page language="java" contentType="text/html; charset=UTF-8" %>
```

```
<%@ taglib prefix = "s" uri = "/struts-tags" %>
<s:component theme = "myTheme" templateDir = "myTemplate"
template = "myJspTemplate.jsp">
    <s:param name = "fruits" value = "{'Apple', 'Banana', 'Grape', 'Pear'}"/>
</s:component>
```

④ 启动 Tomcat 服务器,访问 tag_component.jsp,结果如图 6-28 所示。

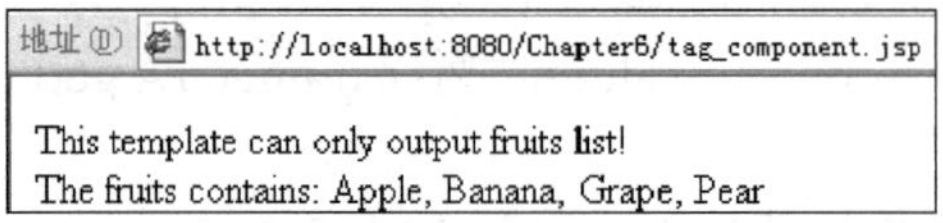

地址(D) http://localhost:8080/Chapter6/tag_component.jsp
This template can only output fruits list!
The fruits contains: Apple, Banana, Grape, Pear

图 6-28　访问 tag_component.jsp 的效果

3. actionerror 标签

如果 action 存在错误消息,则输出;空错误消息(null 或空字符串)不输出。输出消息的具体布局取决于主题,并且 action 错误消息字符串默认是 HTML 转义的。

使用格式如下:

```
<s:actionerror />
```

4. actionmessage 标签

如果 action 存在消息,则输出;空错误消息(null 或空字符串)不输出。输出消息的具体布局取决于主题,并且 action 消息字符串默认是 HTML 转义的。

使用格式如下:

```
<s:actionmessage />
```

5. fielderror 标签

如果 action 字段存在错误消息,则输出。输出消息的具体布局取决于主题,并且 action 字段错误消息字符串默认是 HTML 转义的。

使用格式如下:
(1) 输出 action 中所有字段的错误消息。

```
<s:fielderror />
```

(2) 输出 action 中指定字段的错误消息。

```
<s:fielderror fieldName = "field1" />
```

或者

```
<s:fielderror>
```

```
    <s:param>field1</s:param>
</s:fielderror>
```

第(2)种方式用于输出 action 中 field1 字段的错误消息。

例 6-22 输出消息标签应用示例。

通过 JSP 页面输入姓名,模拟在数据查询,如果输入的姓名为"tsc"则显示"恭喜您,已经查到您的信息!",否则显示"没有您的相关信息!"。

具体操作步骤如下:

① 创建 Action 类。

SearchAction.java

```java
package example.struts2;
import com.opensymphony.xwork2.ActionSupport;
public class SearchAction extends ActionSupport {
    private String stuName;
    public String getStuName() {
        return stuName;
    }
    public void setStuName(String stuName) {
        this.stuName = stuName;
    }
    public String execute() throws Exception {
        if(stuName.equals("tsc")){
            addActionMessage("actionmessage: 恭喜您,已经查到您的信息!");
            return SUCCESS;
        }
        else{
            addActionError("actionerror: 没有您的相关信息!");
            addFieldError("stuName","fielderror: 没有您的相关信息!");
            return ERROR;
        }
    }
}
```

② 编辑 struts.xml。

```xml
<action name="search" class="example.struts2.SearchAction">
    <result name="success">/search.jsp</result>
    <result name="error">/search.jsp</result>
</action>
```

③ 创建 JSP 文件。

search.jsp

```jsp
<%@ page language="java" contentType="text/html; charset=UTF-8" %>
<%@ taglib prefix="s" uri="/struts-tags" %>
```

```
<s:actionmessage/>
<s:actionerror/>
<s:fielderror/>
<s:form action = "search" method = "post">
    <s:textfield name = "stuName" label = "姓名"/>
    <s:submit value = "查询"/>
</s:form>
```

为了能够明显地表示显示的是哪个标签返回的消息,加上了标识信息"actionmessage:"和"actionerror:",在实际使用直接写标签即可。

④ 发布Web项目,启动Tomcat服务器,访问search.jsp,结果如图6-29所示。

当输入"tsc"后,单击"查询"按钮,结果如图6-30所示。

图 6-29　访问 search.jsp 的效果　　　　图 6-30　输入"tsc"的效果

当输入非"tsc"内容后,单击"查询"按钮,结果如图6-31所示。

如果去掉 search.jsp 文件中的＜s:fielderror/＞标签,则在图6-31中将不输出第一条"fielderror:没有您的相关信息!"错误提示信息。

如果 addFieldError 方法中第一个参数名称和 search.jsp 文件中输入文本框的名称不一致,则在图6-31中不输出第二条"fielderror:没有您的相关信息!"错误提示信息。

图 6-31　输入非"tsc"内容的效果

6.3.4　Ajax 标签

Ajax(Asynchronous JavaScript and XML)又称为异步JavaScript和XML,可以实现客户端脚本和服务器之间数据的异步交互,即只更新Web页面中部分内容而不用更新整个Web页面的内容。

Struts 2框架提供了对Ajax技术的支持。在Struts 2框架中使用Ajax标签的步骤如下:

(1) 把和Struts 2框架一起发布的Dojo插件(struts2-dojo-plugin-2.2.3.jar)包含在Web应用的"/WEB-INF/lib"目录下。

(2) 在JSP页面中添加如下语句:

```
<%@ taglib prefix = "sx" uri = "/struts-dojo-tags" %>
```

(3) 在 JSP 页面中包含 head 标签，如下：

```
< sx:head/>
```

配置 head 标签是为了性能或调试目的。

为了方便讲解 Ajax 标签的应用，在 Chapter6 项目中添加 Dojo 插件。这样，支持 Ajax 标签应用的项目目录结构如图 6-32 所示。

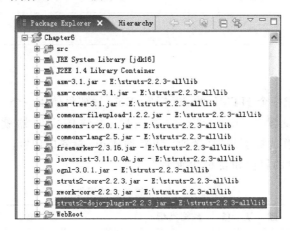

图 6-32　支持 Ajax 标签的项目目录结构

Ajax 标签包含 a、autocompleter、bind、datetimepicker、div、head、submit、tabbedPanel、textarea、tree 和 treenode 等标签，这些标签具有 UI 标签相同的公共属性。

1. head 标签

head 标签输出用于配置 Dojo 的 JavaScript 代码，并且使用所有包含在 Dojo 插件中的标签时也要使用 head 标签。

(1) head 标签包含的属性如表 6-38 所示。

表 6-38　head 标签包含的属性

属性	默认值	数据类型	描述
baseRelativePath	/struts/dojo	String	设置 Dojo 发布目录的相对路径
cache	true	Boolean	设置是否允许浏览器缓存 Struts Dojo 数据
compressed	true	Boolean	设置是否使用压缩版本的 dojo.js 文件
debug	false	Boolean	设置是否使用 Dojo 调试消息
extraLocales		String	设置由 Dojo 加载的区域(locale)名称列表，区域名称之间用逗号分隔
locale		String	设置 Dojo 使用的默认区域
parseContent	false	Boolean	设置在查找 widgets 时是否分析整个文档

为了调试 JavaScript 错误，需 head 标签的 debug 属性设置为 true，这样将会在页面底部显示 Dojo 及 Struts 2 的警告和错误消息。为了提高加载效率，Dojo 核心文件默认是压缩的，但这使得代码难以阅读。因此，为了方便调试 Dojo 和 Struts 2 构件，可以将 compressed

属性设置为 false。但是在项目产品中一定要使用压缩文件,以缩短加载文件的时间。

(2) head 标签的使用格式为:

① 在开发阶段,为了方便排错 JavaScript 问题,建议使用如下配置:

```
< sx:head debug = "true" cache = "false" compressed = "false" />
```

② 在发布的产品中,使用如下配置:

```
< sx:head />
```

2. div 标签

div 标签用于生成一个 HTML 的 div 标签,Dojo 框架将使用一个 XMLHttpRequest 加载标签的内容。当设置了"updateFreq"属性后,内建的定时器就会自动启动,并以"updateFreq"属性值作为刷新周期(毫秒),重新加载 div 标签的内容。

(1) div 标签包含的属性如表 6-39 所示。

表 6-39 div 标签包含的属性

属 性	默 认 值	数据类型	描 述
afterNotifyTopics		String	设置在请求执行成功后将发表的主题列表,各主题之间用逗号分隔
autoStart	true	Boolean	设置是否自动启动计时器
beforeNotifyTopics		String	设置在请求之前将发表的主题列表,各主题之间用逗号分隔
closable	false	Boolean	当 div 放在"tabbedpanel"标签中时,是否显示一个关闭按钮
delay		Integer	设置在获取内容之前等待的时间,以毫秒为单位
errorNotifyTopics		String	设置在请求执行发生错误时将发表的主题列表,各主题之间用逗号分隔
errorText		String	如果在获取内容时有错误,显示给用户的文本信息
executeScripts	false	Boolean	设置是否执行所获取内容中的 JavaScript 代码
formFilter		String	设置过滤表单字段的函数名称
formId		String	设置其字段被序列化并作为参数传递的表单 Id
handler		String	设置处理请求的 JavaScript 函数名称
highlightColor	none	String	设置对指定的元素执行高亮效果的颜色
highlightDuration	2000	Integer	设置高亮效果的持续时间(毫秒)。只有设置"highlightColor"属性后才有效
href		String	设置获取内容的 URL。如果和 Ajax 上下文一起使用,其值必须设置为 url 标签值
indicator		String	设置正在处理请求时要显示的元素 Id
listenTopics		String	设置触发远程调用的主题
loadingText	Loading...	String	设置在获取内容过程中显示的文本信息

续表

属性	默认值	数据类型	描述
notifyTopics		String	设置在请求的前后及发生错误时将发表的主题列表,各主题之间用逗号分隔
parseContent	true	Boolean	设置是否解析返回的 HTML 来查找 Dojo 构件
preload	true	Boolean	设置是否在加载页面时也加载内容
refreshOnShow	false	Boolean	当 div 可见时,就加载其内容。只有嵌套在 tabbedpanel 标签中时才使用该属性
separateScripts	true	Boolean	设置是否为每一个 tag 在单独的域中运行脚本代码
showErrorTransportText	true	Boolean	设置是否显示错误信息
showLoadingText	false	Boolean	设置是否在目标位置显示加载信息
startTimerListenTopics		String	设置启动自动更新计时器的主题
stopTimerListenTopics		String	设置停止自动更新计时器的主题
transport	XMLHTTPTransport	String	设置 Dojo 处理请求使用的传输对象
updateFreq		Integer	设置重新加载内容的频率(毫秒)

(2) div 标签的使用格式为:

```
<sx:div href="%{#url}">Initial Content</sx:div>
```

例 6-23　div 标签示例。

使用 topic 控制计时器,并在每次 div 刷新时,提交表单。

① 创建 Action 类。

DivTagAction.java

```
package example.struts2;
import com.opensymphony.xwork2.ActionSupport;
public class DivTagAction extends ActionSupport {
    private String firstName;
    public String getFirstName() {
        return firstName;
    }
    public void setFirstName(String firstName) {
        this.firstName = firstName;
    }
    public String execute() throws Exception {
        return SUCCESS;
    }
}
```

② 编辑 struts.xml 文件。

```
<action name="divTag" class="example.struts2.DivTagAction">
    <result>/divTagMessage.jsp</result>
</action>
```

③ 创建显示信息页面。

divTagMessage.jsp

```jsp
<%@ page language = "java" contentType = "text/html; charset = UTF-8" %>
<%@ taglib prefix = "s" uri = "/struts-tags" %>
The input data is:<s:property value = "firstName" />
```

④ 创建 tag_Ajax_div.jsp 页面。

tag_Ajax_div.jsp

```jsp
<%@ page language = "java" contentType = "text/html; charset = UTF-8" %>
<%@ taglib prefix = "s" uri = "/struts-tags" %>
<%@ taglib prefix = "sx" uri = "/struts-dojo-tags" %>
<sx:head debug = "true" cache = "false" compressed = "false" />
<sx:div href = "divTag" updateFreq = "5000" autoStart = "false"
    listenTopics = "refresh" startTimerListenTopics = "startTimer"
    stopTimerListenTopics = "stopTimer" highlightColor = "gray" formId = "myForm">
Initial content will be printed here.
</sx:div>
<s:form id = "myForm">
  <s:textfield id = "textInput" name = "firstName"
                        label = "Text to be submited when div reloads"></s:textfield>
</s:form>
<s:submit theme = "simple"
    value = "Start timer" onclick = "dojo.event.topic.publish('startTimer')"/>
<s:submit theme = "simple"
    value = "Stop timer" onclick = "dojo.event.topic.publish('stopTimer')"/>
<s:submit theme = "simple" value = "Refresh" onclick = "dojo.event.topic.publish('refresh')"/>
```

⑤ 发布 Web 项目,启动 Tomcat 服务器,访问 tag_Ajax_div.jsp,结果如图 6-33 所示。

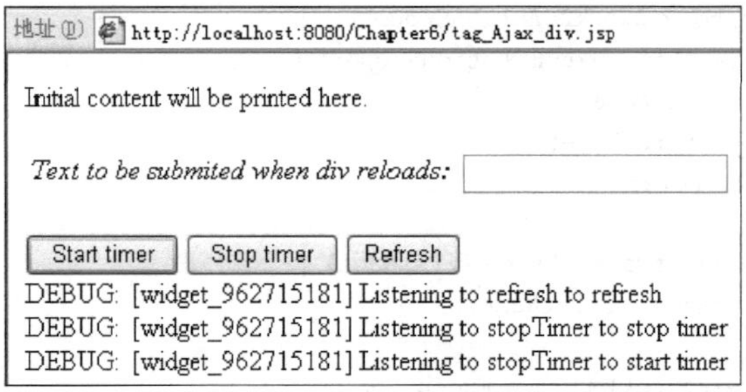

图 6-33　访问 tag_Ajax_div.jsp 的结果

当单击 Start timer 按钮后,就启动 div 标签内部计时器,开始计时,每隔 5 秒刷新 "Initial content will be printed here."区域内容。

3. a 标签

a 标签用于创建一个 HTML 的<a/>元素,即超链接,当单击该链接时,将发送一个异步请求(XMLHttpRequest)。

a 标签包含的属性如表 6-40 所示。

表 6-40　a 标签包含的属性

属　　性	默认值	数 据 类 型	描　　述
ajaxAfterValidation	false	Boolean	设置如果验证成功,是否发送异步请求。该属性只有在"validate"属性的值为 true 时才有效
targets		String	设置将被更新内容的元素 Id 列表,元素之间用逗号分隔
validate	false	Boolean	设置是否进行 Ajax 校验。必须启用"ajaxValidation"拦截器

例 6-24　a 标签示例。

使用 a 标签生成超链接。

① 在例 6-23 的基础上操作,创建 tag_Ajax_a.jsp。

tag_Ajax_a.jsp

```jsp
<%@ page language="java" contentType="text/html; charset=UTF-8"%>
<%@ taglib prefix="s" uri="/struts-tags" %>
<%@ taglib prefix="sx" uri="/struts-dojo-tags" %>
<sx:head debug="true" cache="false" compressed="false" />
<sx:div id="div" cssStyle="border:2px solid red;width:300px">
    This is a div. when below link clicked, the content will be printed in this area!</sx:div>
<br>
<sx:a href="divTag" targets="div">
    Update content of the div!
</sx:a>
<br><br>
<sx:div cssStyle="border:2px solid black;width:300px">
    <s:form id="mform">
        <s:textfield key="firstName"/>
    </s:form>
    <sx:a formId="mform" href="divTag" targets="div">Submit form</sx:a>
</sx:div>
<br>
<sx:div cssStyle="border:2px solid black;width:300px">
    <s:form id="form">
        <s:textfield key="firstName"/>
        <sx:a href="divTag" targets="div">Submit form</sx:a>
    </s:form>
</sx:div>
```

② 启动 Tomcat 服务器,访问 tag_Ajax_a.jsp,结果如图 6-34 所示。

4．autocompleter 标签

autocompleter 标签是一个组合列表框,能够实现文本自动输入到文本输入框。如果使用一个 action 填充 autocompleter 标签,action 的输出结果必须是格式良好的 JSON 字符串。

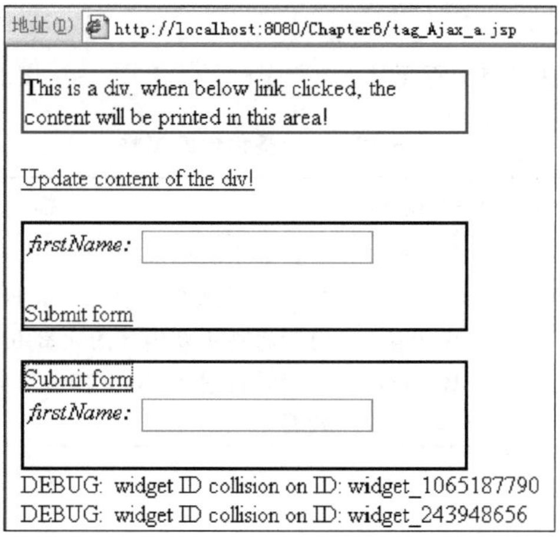

图 6-34 访问 tag_Ajax_a.jsp 的结果

（1）autocompleter 标签包含的属性如表 6-41 所示。

表 6-41 autocompleter 标签包含的属性

属 性	默 认 值	数据类型	描 述
autoComplete	true	Boolean	设置是否在输入文本框中给出提示
dataFieldName	name 属性值	String	设置在返回的 JSON 对象里包含的数据数组中的字段名称
delay	100	Integer	设置搜索之前的延迟时间（毫秒）
dropdownHeight	120	Integer	设置下拉框的高度（像素）
dropdownWidth	同 textbox	Integer	设置下拉框的宽度（像素）
forceValidOption	false	Boolean	设置是否只能选择下拉列表框中的选项
href		String	设置加载选项的 URL
iconPath		String	设置 dropdown 图标的路径
keyName		String	设置被选中的键赋给哪一个字段名
keyValue		String	设置初始的键的值
list		String	设置填充表单的迭代源
listKey		String	设置列表里用来检索选项键的 key
listValue		String	设置列表里用来检索选项值的 value
loadMinimumCount	3	Integer	设置强制加载内容的最小字符个数，即输入多少个字符后开始加载内容
loadOnTextChange	true	Boolean	设置是否每次在文本框中输入一个字符时都重新加载选项
preload	true	Boolean	设置是否加载页面的同时加载选项
resultsLimit	30	Integer	设置显示自动完成选项的个数，设为－1 表示没有限制
searchType	stringstart	String	设置如何执行搜索，有三个可选值：startstring，startword 和 substring
showDownArrow	true	Boolean	设置是否显示向下箭头的按钮
valueNotifyTopics		String	设置当一个值被选中时将发表的主题列表，各主题之间用逗号分隔

其他的属性含义请参考 div 标签中同名属性的含义。

（2）autocompleter 标签的使用格式为：

① 从 action 中获得列表。

```
<sx:autocompleter name="ac" href="%{jsonList}"/>
```

② 使用列表。

```
<sx:autocompleter name="ac" list="{'apple','banana','grape','pear'}" autoComplete="false"/>
```

例 6-25 使用列表的 autocompleter 标签示例。

① 创建 Action 类。

ACTagAction.java

```java
package example.struts2;
import com.opensymphony.xwork2.ActionSupport;
public class ACTagAction extends ActionSupport {
    private String fruitKey;
    public String getFruitKey() {
        return fruitKey;
    }
    public void setFruitKey(String fruitKey) {
        this.fruitKey = fruitKey;
    }
}
```

② 创建 tag_Ajax_autocompleter.jsp。

tag_Ajax_autocompleter.jsp

```jsp
<%@ page language="java" contentType="text/html; charset=UTF-8" %>
<%@ taglib prefix="s" uri="/struts-tags" %>
<%@ taglib prefix="sx" uri="/struts-dojo-tags" %>
<sx:head debug="true" cache="false" compressed="false" />
<s:form action="acSelection">
    <sx:autocompleter name="fruit" list="{'apple','banana','grape','pear'}"/>
    <s:submit/>
</s:form>
```

③ 创建 acTagMessage.jsp。

acTagMessage.jsp

```jsp
<%@ taglib prefix="s" uri="/struts-tags" %>
The fruit you selected is: ${fruitKey}
```

④ 编辑 struts.xml。

```xml
<action name="ac">
    <result>/tag_Ajax_autocompleter.jsp</result>
</action>
<action name="acSelection" class="example.struts2.ACTagAction">
    <result>/acTagMessage.jsp</result>
</action>
```

⑤ 启动 Tomcat 服务器，访问 tag_Ajax_autocompleter.jsp 或者如下网址，结果如图 6-35 所示。

```
http://localhost:8080/Chapter6/ac
```

图 6-35　autocompleter 标签效果

⑥ 单击下拉按钮，结果如图 6-36 所示。

⑦ 选中某一项后单击 Submit 按钮，结果如图 6-37 所示。

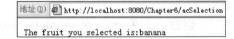

图 6-36　下拉按钮的效果　　　　　图 6-37　单击 Submit 按钮的效果

例 6-26　使用 dataFieldName 属性的 autocompleter 标签示例。

① 创建 Action 类。

ACDivTagAction.java

```java
package example.struts2;
import java.util.ArrayList;
import java.util.List;
import com.opensymphony.xwork2.ActionSupport;
public class ACDivTagAction extends ActionSupport {
    private static List<String> citys = new ArrayList<String>();
    static {
        citys.add("ts");
        citys.add("qhd");
    }
    public List<String> getCitys() {
        return citys;
```

```
        }
}
```

② 创建 tag_Ajax_autocompleter1.jsp。

tag_Ajax_autocompleter1.jsp

```
<%@ page language="java" contentType="text/html; charset=UTF-8" %>
<%@ taglib prefix="sx" uri="/struts-dojo-tags" %>
<sx:head debug="true" cache="false" compressed="false" />
<sx:autocompleter name="city" dataFieldName="city" href="acjson1"/>
```

③ 创建 acTagMessage1.jsp。

acTagMessage1.jsp

```
<%@ taglib prefix="s" uri="/struts-tags" %>
{
    "city":{
        <s:iterator value="citys" status="status">
            '<s:property/>':'<s:property/>'
            <s:if test="!#status.last">,</s:if>
        </s:iterator>
    }
}
```

④ 编辑 struts.xml。

```
<action name="ac1" class="example.struts2.ACDivTagAction">
    <result>/tag_Ajax_autocompleter1.jsp</result>
</action>
<action name="acjson1" class="example.struts2.ACDivTagAction">
    <result>/acTagMessage1.jsp</result>
</action>
```

⑤ 启动 Tomcat 服务器，访问如下网址，结果如图 6-38 所示。

```
http://localhost:8080/Chapter6/ac1
```

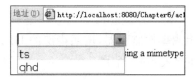

图 6-38　autocompleter 标签效果

例 6-27　服务器的响应是数组的 autocompleter 标签示例。

① 创建 tag_Ajax_autocompleter2.jsp。

tag_Ajax_autocompleter2.jsp

```
<%@ page language = "java" contentType = "text/html; charset = UTF-8" %>
<%@ taglib prefix = "sx" uri = "/struts-dojo-tags" %>
<sx:head debug = "true" cache = "false" compressed = "false" />
<sx:autocompleter name = "city" dataFieldName = "city" href = "acjson2"/>
```

② 创建 acTagMessage2.jsp。

acTagMessage2.jsp

```
<%@ taglib prefix = "s" uri = "/struts-tags" %>
[
    <s:iterator value = "citys" status = "status">
        ['<s:property/>','<s:property/>']
        <s:if test = "!#status.last">,</s:if>
    </s:iterator>
]
```

③ 编辑 struts.xml。

```
<action name = "ac2" class = "example.struts2.ACDivTagAction">
    <result>/tag_Ajax_autocompleter2.jsp</result>
</action>
<action name = "acjson2" class = "example.struts2.ACDivTagAction">
    <result>/acTagMessage2.jsp</result>
</action>
```

④ 启动 Tomcat 服务器,访问如下网址,结果如图 6-39 所示。

```
http://localhost:8080/Chapter6/ac2
```

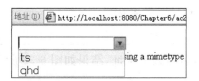

图 6-39 autocompleter 标签效果

5. bind 标签

bind 标签用来为事件源生成事件监听器,按照指定的 URL 生成异步请求并更新目标。

(1) bind 标签包含的属性如表 6-42 所示。

表 6-42 bind 标签包含的属性

属 性	数 据 类 型	描 述
events	String	设置被关联事件名称列表,名称之间用逗号分隔
sources	String	设置被关联元素的 Id 列表,Id 之间用逗号分隔

（2）bind 标签的使用格式为：

① 不关联事件，而是触发远程调用的主题。

```
<sx:bind href="%{#ajaxTest}" listenTopics="/makecall"/>
<s:submit onclick="dojo.event.topic.publish('/makecall')"/>
```

② 关联到提交按钮的 onclick 事件。

```
<sx:bind href="%{#ajaxTest}" sources="button" events="onclick"/>
<s:submit type="submit" key="button"/>
```

③ 提交表单。

```
<sx:bind href="%{#ajaxTest}" sources="chk1" events="onchange" formId="form1"/>
<form id="form1">
    <s:checkbox name="data" label="Hit me" id="chk1"/>
</form>
```

例 6-28 bind 标签示例。

tag_Ajax_bind.jsp

```
<%@ page language="java" contentType="text/html; charset=UTF-8" %>
<%@ taglib prefix="s" uri="/struts-tags" %>
<%@ taglib prefix="sx" uri="/struts-dojo-tags" %>
<sx:head debug="true" cache="false" compressed="false"/>
<sx:div id="div1" cssStyle="border:2px solid red;width:300px">
    This is a div. when below link clicked, the content will be printed in this area!</sx:div>
<sx:bind id="bind1" href="setTag" listenTopics="/makecall" targets="div1"/>
<s:submit align="left" value="Listening to a topic"
                                   onclick="dojo.event.topic.publish('/makecall')"/>
<sx:bind id="bind2" href="divTag" sources="b1" events="onclick" targets="div1"/>
<s:submit align="left" id="b1" value="Attached to event"/>
<sx:bind id="bind3" href="acSelection" sources="tf" targets="div1" events="onkeyup"
                                                                formId="form1"/>
<s:form id="form1">
    <s:textfield id="tf" label="Submit form"></s:textfield>
</s:form>
<sx:bind id="bind4" href="actionTag2" beforeNotifyTopics="/makecall" sources="b3"
                                          events="onclick" targets="div1"/>
<s:submit align="left" id="b3" value="Attached to event and Using beforeNotifyTopics"/>
```

访问 tag_Ajax_bind.jsp 页面的效果如图 6-40 所示。

6. datetimepicker 标签

datetimepicker 标签用于在下拉框容器中输出一个日期/时间选择器。一个单独的 DateTimePicker 构件可以很方便地选择日期/时间，或者以星期、月份、年份为单位增加/减小。

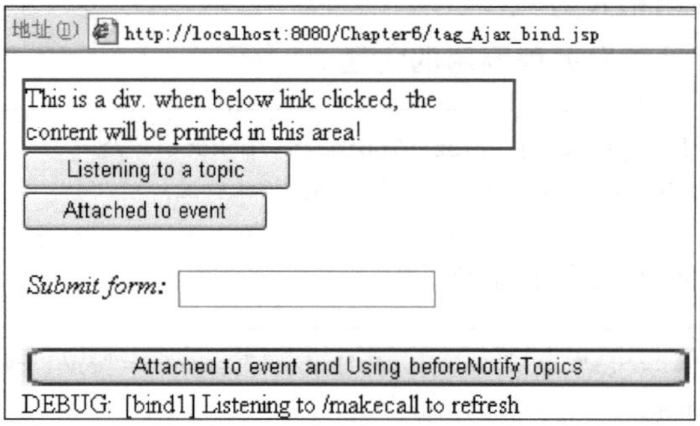

图 6-40　访问 tag_Ajax_bind.jsp 页面效果

datetimepicker 标签能够使用 formatLength（long、short、medium 或 full）或 displayFormat 属性自定义日期/时间显示格式。默认使用当前区域设置。displayFormat 属性支持的日期/时间格式在如下网址中有描述：

http://www.unicode.org/reports/tr35/tr35－4.html#Date_Format_Patterns

日期/时间格式中字母的含义列表如表 6-43 所示。

表 6-43　日期/时间格式中字母的含义

格式	描述
d	一个月中的第几天
D	一年中的第几天
M	月份，用一位或两位表示月份的数字，三位是月份的缩写名称，四位是月份的完整名称
y	年份
h	小时［1～12］
H	小时［0～23］
m	分钟。一位或者两位，不足补零
s	秒。一位或者两位，不足补零

提交给服务器的日期/时间是独立于区域设置的值，值的格式为 yyyy-MM-dd'T'HH：mm：ss，该值使用隐藏字段的 name 属性传递。

（1）datetimepicker 标签包含的属性如表 6-44 所示。

表 6-44　datetimepicker 标签包含的属性

属性	默认值	数据类型	描述
adjustWeeks	false	Boolean	如果值为 true，则日历周数大小随着月份变化。如果值为 false，则使用 42 天的格式
dayWidth	narrow	String	设置如何在标题输出日期的名称，可选值有 narrow、abbr 和 wide
displayFormat		String	设置格式化日期的可视化显示模式，如 dd/MM/yyyy

续表

属性	默认值	数据类型	描述
displayWeeks	6	Integer	设置显示的星期数
endDate	2941-10-12	Date	设置日历中所能显示的最晚日期
formatLength	short	String	设置显示日期/时间的格式类型,可选值有 long、short、medium 和 full
language	浏览器设定的	String	设置显示该构件的语言
startDate	1492-10-12	Date	设置日历中所能显示的最早日期
staticDisplay	false	Boolean	设置是否禁用所有的增量(incremental)控件,只能在当前显示里选择日期
toggleDuration	100	Integer	设置切换的持续时间,单位是毫秒
toggleType	plain	String	设置下拉框的切换类型,可选值有 plain、wipe、explode 和 fade
type	date	String	设置下拉框选择器的类型(日历或者时间),可选值有 date 和 time
weekStartsOn	0	Integer	调整一个星期的第一天,0 表示星期天,6 表示星期六

(2) datetimepicker 标签的使用格式为:

```
<sx:datetimepicker />
```

Datetimepicker 的效果如图 6-41 所示。

图 6-41　datetimepicker 标签效果

例 6-29　datetimepicker 标签示例,选择日期并输出。

① 创建 Action。

DTPickerTagAction.java

```
package example.struts2;
import java.util.Date;
import com.opensymphony.xwork2.ActionSupport;
public class DTPickerTagAction extends ActionSupport {
    private Date selectedDate;
    public Date getSelectedDate() {
```

```
            return selectedDate;
        }
        public void setSelectedDate(Date selectedDate) {
            this.selectedDate = selectedDate;
        }
    }
```

② 创建日期选择文件。

tag_Ajax_dtpicker.jsp

```jsp
<%@ page language="java" contentType="text/html; charset=UTF-8" %>
<%@ taglib prefix="s" uri="/struts-tags" %>
<%@ taglib prefix="sx" uri="/struts-dojo-tags" %>
<sx:head debug="true" cache="false" compressed="false" />
<s:form id="myForm" action="dtpSelection">
    <sx:datetimepicker name="selectedDate" label="Select Date"
                                            displayFormat="MM/dd/yyyy" />
    <s:submit/>
</s:form>
```

③ 编辑 struts.xml。

```xml
<action name="dtpSelection" class="example.struts2.DTPickerTagAction">
    <result>/dtpTagMessage.jsp</result>
</action>
```

④ 创建显示选择的日期文件。

dtpTagMessage.jsp

```jsp
<%@ taglib prefix="s" uri="/struts-tags" %>
The date you selected is: ${selectedDate}
```

⑤ 发布 Web 项目，启动 Tomcat 服务器，访问 tag_Ajax_dtpicker.jsp，结果如图 6-42 所示。选择某一日期并单击 Submit 按钮后，将输出选择的日期。

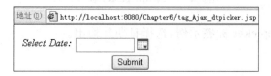

图 6-42　tag_Ajax_dtpicker.jsp 页面效果

7. submit 标签

submit 标签用于输出一个能够实现表单异步提交的按钮。

（1）submit 标签包含的属性如表 6-45 所示，其他属性参考其他标签的同名属性。

表 6-45 submit 标签包含的属性

属性	默认值	数据类型	描述
method		String	设置按钮的 method 属性
src		String	为 image 类型的提交按钮提供图像来源,对 input 和 button 类型的按钮无效
type	input	String	设置提交按钮的类型,可选值有 input,button 和 image

由 type 属性可知,submit 标签共有三种不同类型的按钮,其中:
- input 类型的 submit 标签输出的 HTML 代码为<input type="submit"…>。
- image 类型的 submit 标签输出的 HTML 代码为<input type="image"…>。
- button 类型的 submit 标签输出的 HTML 代码为<button type="submit"…>。

(2) submit 标签的使用格式为:
① 默认类型按钮。

```
<sx:submit />
```

② image 类型按钮。

```
<sx:submit type="image" label="Submit the form" src="submit.gif"/>
```

例 6-30 submit 标签示例。
在例 6-23 的基础上操作,具体操作步骤如下:
① 创建 tag_Ajax_submit.jsp 文件。

tag_Ajax_submit.jsp

```
<%@ page language="java" contentType="text/html; charset=UTF-8" %>
<%@ taglib prefix="s" uri="/struts-tags" %>
<%@ taglib prefix="sx" uri="/struts-dojo-tags" %>
<sx:head debug="true" cache="false" compressed="false" />
<sx:div id="div1" cssStyle="border:2px solid red;width:200px">
    This is a div. </sx:div>
---------------------------------------<br>
<sx:submit id="link1" href="ac" targets="div1" />
<br>------------inside the form---------------
<s:form id="form" action="divTag" namespace="/">
    <s:textfield name="firstName" label="FirstName" />
    <sx:submit targets="div1" />
</s:form>
------------outside the form---------------
<s:form id="form1" action="divTag">
    <s:textfield name="firstName" label="FirstName" />
</s:form>
<sx:submit formId="form1" targets="div1" />
```

② 访问 tag_Ajax_submit.jsp,结果如图 6-43 所示。
③ 单击某一个 Submit 按钮,结果如图 6-44 所示。

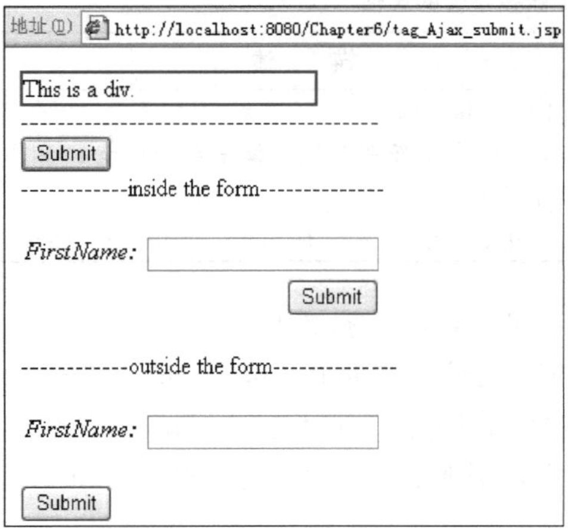

图 6-43　tag_Ajax_submit.jsp 页面效果 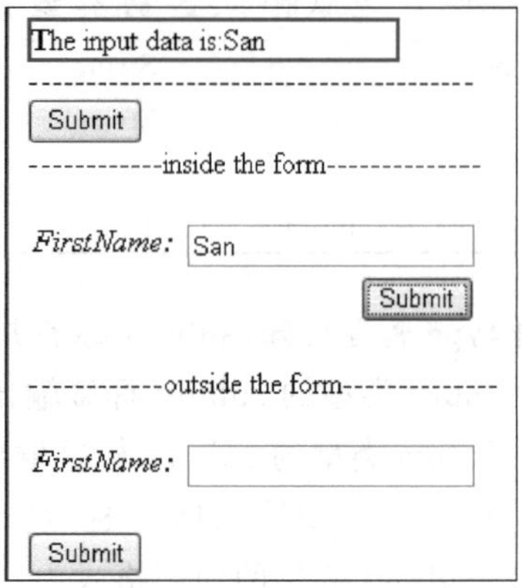　　　图 6-44　单击中间 Submit 按钮的效果

8．tabbedPanel 标签

tabbedpanel 标签是一个 Ajax 组件，每一个 tab 页的内容都可以是本地（静态）的或者是远端（动态，从服务器端获得）的内容，用户每次选择该 tab 页时，其内容被刷新。

tabbedpanel 标签的 id 属性是必选项，并且若想要使用 cookie 特性，必须确保为 tabbedpanel 组件提供唯一的 id。

另外，当 div 标签用在 tabbedpanel 标签内部时，每一个 div 变成了一个 tab。这时 div 标签的一些属性具有特殊的含义，如以下属性。

- refreshOnShow 属性：当 tab 被选择后，将重新加载 div 的内容。
- closable 属性：tab 将有一个关闭按钮。
- preload 属性：在加载页面之后，加载 div 的内容。

tabbedPanel 标签包含的属性如表 6-46 所示，其他属性参考其他标签的同名属性。

表 6-46　tabbedPanel 标签包含的属性

属性	数据类型	描述
afterSelectTabNotifyTopics	String	在被选中的 tab 上单击时，发布的主题列表
beforeSelectTabNotifyTopics	String	在没被选中的 tab 上单击时，发布的主题列表
disabledTabCssClass	String	应用在禁用的 tab 上的 tab 按钮的 Css 类。默认值是 strutsDisabledTab
doLayout	Boolean	如果 doLayout 值为 false，tab 容器的高度等于当前被选 tab 的高度。默认值 false
selectedTab	String	被默认选择的 tab 的 id
useSelectedTabCookie	Boolean	若值为 true，最后选择的 tab 的 id 被保存在 cookie 中。默认值 false

tabbedPanel 标签的 useSelectedTabCookie 属性值为 true 时，并且没有设置 selectedTab 属性，则在输出视图时，将会读取该 cookie，并激活对应的 tab。其中 cookie 的名称是"Struts2TabbedPanel_selectedTab_"＋id。

例 6-31 tabbedPanel 标签示例。

使用本地内容和远端内容及防止选择 tab 的 tabbedPanel 标签示例。

① 创建 tag_Ajax_tabbedPanel.jsp。

tag_Ajax_tabbedPanel.jsp

```jsp
<%@ page language="java" contentType="text/html; charset=UTF-8" %>
<%@ taglib prefix="s" uri="/struts-tags" %>
<%@ taglib prefix="sx" uri="/struts-dojo-tags" %>
<sx:head debug="true" cache="false" compressed="false"/>
----------Utilizing local and remote content------
<sx:tabbedpanel id="tp" cssStyle="width:320px">
    <sx:div id="one" label="Local content" closable="true">
        This tab can be closed!<br/>
        <s:form>
            <s:textfield name="tt" label="Text"/>
        </s:form>
    </sx:div>
    <sx:div id="two" label="Remote content" href="ac">
        This is the remote tab!
    </sx:div>
</sx:tabbedpanel>
<br/>Utilizing notify topics to prevent a tab from being selected
<script type="text/javascript">
    dojo.event.topic.subscribe("/beforeSelect", function(event, tab, tabContainer)
        {event.cancel = true;});
</script>
<sx:tabbedpanel id="tp1" cssStyle="width:330px"
                beforeSelectTabNotifyTopics="/beforeSelect">
    <sx:div id="three" label="Local content" >
        The first tab, and it is the only one we can see.
        <s:form>
            <s:textfield name="tt" label="Text"/>
        </s:form>
    </sx:div>
    <sx:div id="four" label="Remote content" href="divTag" closable="true" >
        Another tab, but it can not be selected for "notify topics" was used.
    </sx:div>
</sx:tabbedpanel>
```

② 访问 tag_Ajax_tabbedPanel.jsp，结果如图 6-45 所示。

9. textarea 标签

textarea 标签用于输出 Dojo 编辑构件。该标签和 UI 标签库中的 textarea 标签具有同样的属性，但可以输出一个功能复杂的文本编辑框。

textarea 标签使用格式为：

```jsp
<sx:textarea />
```

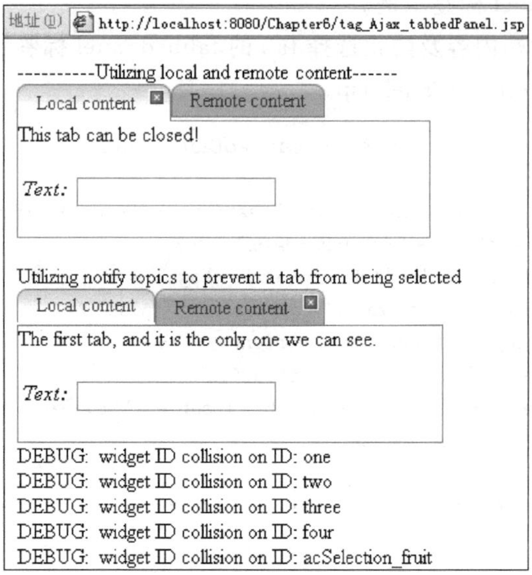

图 6-45　tabbedPanel 标签的效果

textarea 标签的效果如图 6-46 所示。

图 6-46　textarea 标签的效果

例 6-32　textarea 标签示例。

在例 6-23 基础上操作，具体操作步骤如下：

① 创建 tag_Ajax_textarea.jsp。

tag_Ajax_textarea.jsp

```
<%@ page language = "java" contentType = "text/html; charset = UTF - 8" %>
<%@ taglib prefix = "s" uri = "/struts - tags" %>
<%@ taglib prefix = "sx" uri = "/struts - dojo - tags" %>
<sx:head debug = "true" cache = "false" compressed = "false" />
<sx:div id = "div1" cssStyle = "border:2px solid red;width:760px">This is a div.</sx:div>
<s:form id = "form" action = "divTag" namespace = "/">
    <sx:textarea name = "firstName"
            cssStyle = "border:1px solid black;width:760px;height:200px" />
    <sx:submit targets = "div1" />
</s:form>
```

② 访问 tag_Ajax_textarea.jsp，输入内容后提交，结果如图 6-47 所示。

10．tree 标签和 treenode 标签

tree 标签用于在 Ajax 技术支持下输出 tree 构件，而 treenode 标签用于在 Ajax 技术支持下在 tree 构件内输出一个 tree 节点。

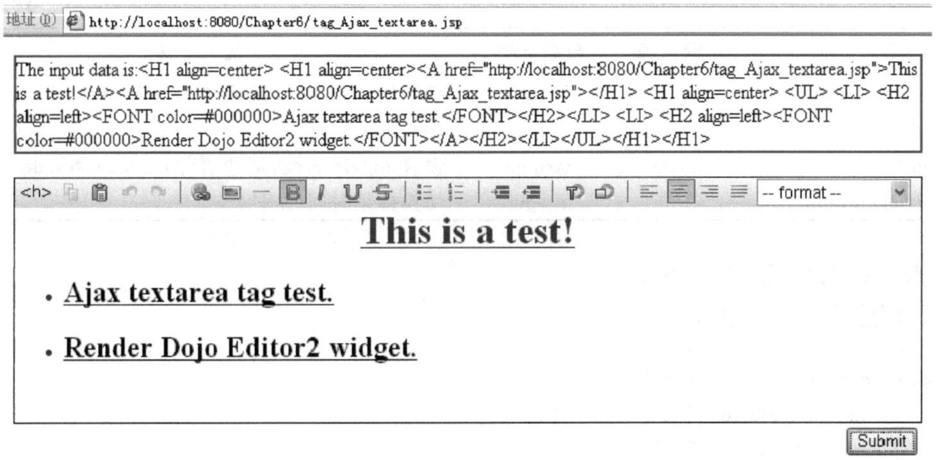

图 6-47　tag_Ajax_textarea.jsp 测试的效果

treenode 标签的属性含义请参考其他 Ajax 标签同名属性的含义。

（1）tree 标签的属性如表 6-47 所示，其他属性参考其他标签的同名属性。

表 6-47　tree 标签的属性

属　　性	默认值	数据类型	描　　述
blankIconSrc		String	设置空白图标的图像来源
childCollectionProperty		String	设置子节点集合属性的属性名称
collapsedNotifyTopics		String	设置一个节点被折叠时发表的主题列表，各主题之间用逗号分隔
expandIconSrcMinus		String	设置节点处于展开（－）状态时图标的图像来源
expandIconSrcPlus		String	设置节点处于收缩（＋）状态时图标的图像来源
expandedNotifyTopics		String	设置一个节点被展开时发表的主题列表，各主题之间用逗号分隔
gridIconSrcC		String	设置子节点下子图标的图像来源
gridIconSrcL		String	设置最后子网格图标的图像来源
gridIconSrcP		String	设置父节点下子图标图像来源
gridIconSrcV		String	设置竖线的图像来源
gridIconSrcX		String	设置单独根节点的网格图像来源
gridIconSrcY		String	设置最后根节点的网格图像来源
href		String	为指定节点设置要加载的子节点列表的 URL
iconHeight	18px	String	设置图标的高度
iconWidth	19px	String	设置图标的宽度
nodeIdProperty		String	设置节点 Id 属性的属性名称
nodeTitleProperty		String	设置节点标题属性的属性名称
rootNode		String	设置根节点属性
selectedNotifyTopics		String	设置一个节点被选择时发表的主题列表，各主题之间用逗号分隔
showGrid	true	Boolean	设置是否显示网格
showRootGrid	true	Boolean	设置 showRootGrid 属性

续表

属 性	默认值	数据类型	描 述
templateCssPath		String	模板的 css 路径,默认值是{contextPath}/struts/tree.css
toggle	fade	String	设置 toggle 属性,可选值有 explode 和 fade
toggleDuration	150	Integer	切换持续时间(毫秒)

(2) tree 标签使用格式为:

① 创建静态树的格式。

```
<sx:tree id = "..." label = "...">
    <sx:treenode id = "..." label = "..." />
    <sx:treenode id = "..." label = "...">
        <sx:treenode id = "..." label = "..." />
    </sx:treenode>
</sx:tree>
```

② 创建动态树的格式。

```
<sx:tree id = "..." rootNode = "..." nodeIdProperty = "..."
         nodeTitleProperty = "..." childCollectionProperty = "..." />
```

(3) tree 标签示例。

例 6-33 创建静态树。

① 创建 tag_Ajax_Static_tree.jsp。

tag_Ajax_Static_tree.jsp

```
<%@ page language = "java" contentType = "text/html; charset = UTF-8" %>
<%@ taglib prefix = "sx" uri = "/struts-dojo-tags" %>
<sx:head debug = "true" cache = "false" compressed = "false" />
<script type = "text/javascript">
dojo.event.topic.subscribe("treeNodeSelected", function(source) {
    alert("The tilte of the selected node is: " + source.node.title);
});
</script>
<sx:tree id = "root" label = "Tree Root" selectedNotifyTopics = "treeNodeSelected">
    <sx:treenode id = "node1" label = "Tree Node 1" />
    <sx:treenode id = "node2" label = "Tree Node 2">
        <sx:treenode id = "node21" label = "Tree Node 21" />
    </sx:treenode>
</sx:tree>
```

② 访问 tag_Ajax_Static_tree.jsp,展开后的结果如图 6-48 所示。

当选择某一个树节点时,会弹出一个对话框,如图 6-49 所示。

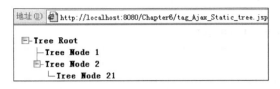

图 6-48 静态树

图 6-49 选择节点后的对话框

例 6-34 创建动态树。

① 创建 Node。

Node.java

```
package example.struts2;
import java.util.ArrayList;
import java.util.List;
public class Node{
    private List<Node> childNodes = new ArrayList<Node>();//子节点集合
    private String nodeId;
    private String nodeTitle;
    public static int counter = 0;
    public Node(String nodeId, String nodeTitle) {
        this.nodeId = nodeId;
        this.nodeTitle = nodeTitle;
    }
    public List<Node> getChildNodes() {
        for(int i = 0; i < 3; i++){                         //同一级元素的个数
            if (counter < 6) {                              //生成的元素的个数
                Node childNode = new Node("childNode" + counter,
                                          "Child Node " + counter);
                childNodes.add(childNode);
                counter++;
            }
        }
        return childNodes;
    }
    public String getNodeId() {
        return nodeId;
    }
    public void setNodeId(String nodeId) {
        this.nodeId = nodeId;
    }
    public String getNodeTitle() {
        return nodeTitle;
    }
    public void setNodeTitle(String nodeTitle) {
        this.nodeTitle = nodeTitle;
    }
}
```

② 创建生成动态树 Action。

DynamicTreeAction.java

```java
package example.struts2;
import com.opensymphony.xwork2.ActionSupport;
public class DynamicTreeAction extends ActionSupport {
    private Node rootNode;
    public Node getRootNode() {
        return rootNode;
    }
    public String execute() {
        Node.counter = 0;
        rootNode = new Node("rootNode", "Root Node");
        return SUCCESS;
    }
}
```

③ 创建选择节点的 Action。

TreeSelectAction.java

```java
package example.struts2;
import com.opensymphony.xwork2.ActionSupport;
public class TreeSelectAction extends ActionSupport {
    private String nodeId;
    public String getNodeId() {
        return nodeId;
    }
    public void setNodeId(String nodeId) {
        this.nodeId = nodeId;
    }
}
```

④ 创建显示动态树的 JSP 文件。

tag_Ajax_Dynamic_tree.jsp

```jsp
<%@ page language="java" contentType="text/html; charset=UTF-8" %>
<%@ taglib prefix="sx" uri="/struts-dojo-tags" %>
<sx:head debug="true" cache="false" compressed="false" />
<sx:tree id="dynaTree" rootNode="rootNode"
    nodeTitleProperty="nodeTitle"
    nodeIdProperty="nodeId"
    childCollectionProperty="childNodes">
</sx:tree>
```

⑤ 创建显示选择节点的 JSP 文件。

itemShow.jsp

```
<%@ page language="java" contentType="text/html; charset=UTF-8" %>
<%@ taglib prefix="s" uri="/struts-tags" %>
Node Id: <s:property value="nodeId"/>
```

⑥ 编辑 struts.xml。

struts.xml

```
<?xml version="1.0" encoding="UTF-8"?>
<!DOCTYPE struts PUBLIC
    "-//Apache Software Foundation//DTD Struts Configuration 2.0//EN"
    "http://struts.apache.org/dtds/struts-2.0.dtd">
<struts>
    <package name="default" namespace="/" extends="struts-default">
        <action name="DynamicTree" class="example.struts2.DynamicTreeAction">
            <result>/tag_Ajax_Dynamic_tree.jsp</result>
        </action>
        <action name="tsa" class="example.struts2.TreeSelectAction">
            <result>itemShow.jsp</result>
        </action>
    </package>
</struts>
```

⑦ 发布 Web 项目，启动 Tomcat 服务器，访问 dynamicTree，展开后的结果如图 6-50 所示。

⑧ 单击叶节点的结果如图 6-51 所示。

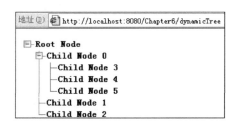

图 6-50　动态树

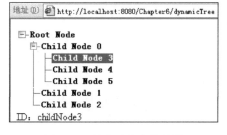

图 6-51　单击叶节点的结果

习题

1. 比较 append 标签和 merge 标签的异同。
2. 写出如下代码的输出结果。

```
<s:action var="myAction" name="actionTagAction1" namespace="/">
    标签内的"myAction"是 null 吗？<s:property value="#myAction == null" /><br>
</s:action>
    标签外的"myAction"是 null 吗？<s:property value="#myAction == null" />
```

3. param 标签使用标签体和不使用标签体的两种格式有何异同？
4. 比较 set 标签和 push 标签的区别。
5. 例 6-10 中的 bean 标签如果不设置 var 属性，结果如何？
6. 修改例 6-11，将 include 标签的 value 属性值改为 action，进行参数传递。
7. 如果 url 标签中的 action 和 value 属性都设置，将使用哪个属性生成 URL？如果两个属性都省略，将如何生成 URL？

第7章 Struts 2框架的国际化

国际化是指应用软件不用做任何修改就可以在不同国家或地区使用，能够支持不同的语言和国家，并按照当地用户的习惯(语言和格式习惯)来显示内容，如日期时间、数字和货币等。当国际化的软件在本地机器上运行时，需要进行本地化，即根据本地机器的语言和地区设置显示相应的内容。

实现国际化的理想方法就是将要显示的内容从程序代码中分离，而不是将其进行硬编码，这样，不用修改应用程序的源码就可以实现国际化，方便应用程序的维护和移植。为了做到这一点，可以将要显示的内容统一保存到一个资源包文件中，当本地化时，再从资源文件中取出符合本地机器的语言和地区设置的内容进行显示。

基于 Struts 2 框架的 Web 应用的国际化也是使用资源文件的方式实现的，并且 Struts 2 框架提供了多种加载国际化资源文件的方法。

7.1 资源文件

在编写国际化程序时，要为不同的国家(地区)和语言编写不同的资源文件，这些资源包同属一个资源系列，共享一个基名。

7.1.1 名称格式

为了使得程序本地化时能够正确加载所需的资源文件，资源文件的名称要符合一定的格式，通常采用 baseName.properties、baseName_language.properties 或 baseName_language_country.properties 的命名格式。

其中 baseName.properties 是缺省的资源文件，是最基本的资源文件，是实现国际化应该提供的资源文件，如果在本地化时找不到合适的资源文件，则将使用缺省的资源文件。

资源文件名称中的 baseName 是资源文件的基名，由用户定义。language 是指语言代码，由两个小写字母组成，它的取值由 ISO-639 定义；例如汉语的代码是 zh，英语的代码是 en。完整的语言代码列表可以参考官方网站：

> http://www.loc.gov/standards/iso639-2/php/English_list.php

country 是指国家和地区代码，由两个大写字母组成，它的取值由 ISO-3166 定义；例如中国的代码是 CN，中国香港地区的代码是 HK。完整的国家和地区代码列表可以参考官方网站：

```
http://www.iso.org/iso/country_codes/iso-3166-1_decoding_table.htm
```

例 7-1 假设资源文件基名为 MyResources，写出使用汉语的资源文件名称。

（1）如果是中国大陆的用户，则资源文件名称可以写为：

```
MyResources_zh_CN.properties。
```

（2）如果中国大陆以外的用户也使用汉语，则资源文件名称可以写为：

```
MyResources_zh.properties。
```

即不指定国家和地区代码。

7.1.2 资源文件的内容

资源文件是一个属性文件，其内容由"key-value"（键-值）对组成，每行一条"键-值"对，又称为一条消息。其格式为：

```
key = value
```

其中 key 是消息的标记，在程序中引用；value 是消息的内容，在程序本地化时要显示的内容。

属性文件中的内容通常是 ASCII 字符，如果不是 ASCII 编码，例如中文，则在本地化时出现乱码，如何解决呢？可以使用 native2ascii 工具进行编码转换，将非 ASCII 编码转换为 Unicode 编码。

native2ascii 命令的使用格式如下：

```
native2ascii -encoding UTF-8 SourceFileName TargetFileName
```

7.2 基于 Struts 2 框架的 Web 应用的国际化体验

例 7-2 使用 struts.custom.i18n.resources 常量实现国际化。

具体操作步骤如下：

（1）创建 Web 项目 Chapter7，并添加 Struts 2 框架的支持。Chapter7 项目目录结构参考 Chapter6 项目的目录结构，web.xml 配置文件的内容同 Chapter6 项目中 web.xml 配置文件的内容。

（2）创建 Action 类。

HelloWorld.java

```
package example.struts2;
import com.opensymphony.xwork2.ActionSupport;
```

```java
public class HelloWorld extends ActionSupport {
    private String message;
    public String getMessage() {
        return message;
    }
    public void setMessage(String message) {
        this.message = message;
    }
    public String execute(){
        if((message == null) || (message.isEmpty())){
            return INPUT;
        }
        else{
            return SUCCESS;
        }
    }
}
```

(3) 创建 struts.xml 配置文件。

struts.xml

```xml
<?xml version="1.0" encoding="UTF-8"?>
<!DOCTYPE struts PUBLIC
    "-//Apache Software Foundation//DTD Struts Configuration 2.0//EN"
    "http://struts.apache.org/dtds/struts-2.0.dtd">
<struts>
    <package name="default" extends="struts-default">
        <action name="helloWorld" class="example.struts2.HelloWorld">
            <result name="success">/inputMessage.jsp</result>
            <result name="input">/inputMessage.jsp</result>
        </action>
    </package>
    <constant name="struts.custom.i18n.resources" value="ApplicationResources" />
</struts>
```

在配置文件中使用 struts.custom.i18n.resources 常量配置资源文件的基名。

(4) 创建 JSP 文件。

inputMessage.jsp

```jsp
<%@ page language="java" contentType="text/html; charset=UTF-8"
    pageEncoding="UTF-8"%>
<%@ taglib prefix="s" uri="/struts-tags" %>
<s:a id="a1" href="helloWorld?request_locale=zh">汉语</s:a>
<s:a id="a2" href="helloWorld?request_locale=en_US">英语</s:a>
<s:a id="a3" href="helloWorld?request_locale=de">其他语言</s:a>
<s:form action="helloWorld">
    <s:textfield name="message" key="info.message"/>
```

```
        <s:submit name = "submit" key = "info.submit"/>
        <s:reset name = "reset" key = "info.reset" />
</s:form>
```

在 a 标签中使用 request_locale 参数,是为了改变当前浏览器的语言环境,模拟不同语言和国家的用户浏览该页面。

(5) 在 src 文件夹下创建三个资源文件,资源文件的基名为 ApplicationResources。

① 支持语言为英语、国家为美国的资源文件。

ApplicationResources_en_US.properties

```
info.message = Please input messages
info.submit = Submit
info.reset = Reset
```

② 支持语言为汉语的资源文件。

ApplicationResources_zh.properties

```
info.message = 请输入内容
info.submit = 提交
info.reset = 重置
```

该文件保存时要采用 UTF-8 编码,并且创建完该文件后,要使用 native2ascii 命令进行转码。

③ 默认的资源文件。

ApplicationResources.properties

```
info.message = Please input messages (Default)
info.submit = Submit
info.reset = Reset
```

ApplicationResources.properties 为默认的资源文件,为了方便说明问题,将其内容和 ApplicationResources_en_US.properties 资源文件的内容相同,为了和该文件区别,加上了"(Default)"。

(6) 发布 Web 项目,启动 Tomcat 服务器,并访问 inputMessage.jsp,结果如图 7-1 所示。

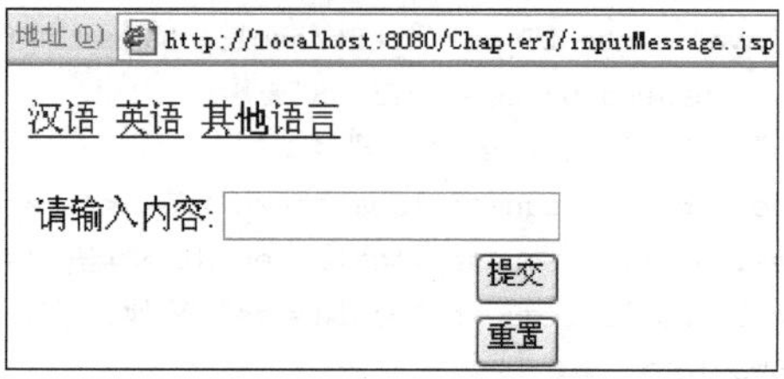

图 7-1　访问 inputMessage.jsp 页面效果

单击"英语"链接,结果如图 7-2 所示。

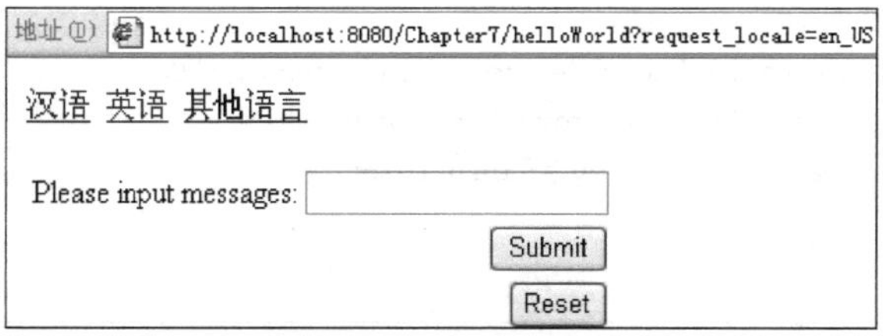

图 7-2 英语语言页面效果

单击"汉语"链接,结果如图 7-1 所示。单击"其他语言"链接,结果如图 7-1 所示。为什么?这是因为单击"其他语言"链接后,要查找 ApplicationResources_de.properties 文件,由于没有提供该文件,因此,Struts 2 框架将使用当前默认的语言(因为本机使用的是中文环境),即汉语进行显示,所以使用了 ApplicationResources_zh.properties 资源文件。如果删除 ApplicationResources_zh.properties 资源文件,则在首次访问 inputMessage.jsp、单击"汉语"链接或者"其他语言"链接时,结果如图 7-3 所示。

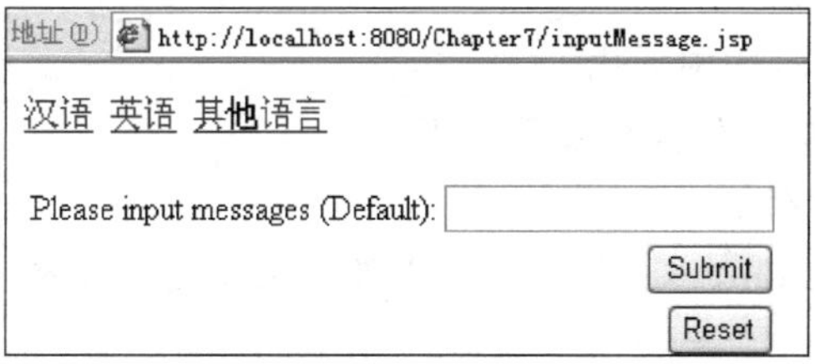

图 7-3 使用了默认资源文件的页面效果

当用户在文本框中输入消息提交后,如何实现消息显示的国际化呢?这可以通过使用 text 标签实现,下面通过一个示例进行讲解。

例 7-3 使用 i18n 标签实现国际化。

(1) 创建资源文件。

MyResources.properties

```
info.message = Please input messages (From MyResources)
info.submit = Submit
info.reset = Reset
```

(2) 修改 inputMessage.jsp 文件,使用 i18n 标签指定资源文件。

inputMessage.jsp

```
<%@ page language = "java" contentType = "text/html; charset = UTF - 8"
                         pageEncoding = "UTF - 8" %>
<%@ taglib prefix = "s" uri = "/struts - tags" %>
<s:a id = "a1" href = "helloWorld?request_locale = zh">汉语</s:a>
<s:a id = "a2" href = "helloWorld?request_locale = en_US">英语</s:a>
<s:a id = "a3" href = "helloWorld?request_locale = de">其他语言</s:a>
```

```
<s:form action = "helloWorld">
    <s:i18n name = "MyResources">
        <s:textfield name = "message" key = "info.message"/>
        <s:submit name = "submit" key = "info.submit"/>
        <s:reset name = "reset" key = "info.reset" />
    </s:i18n>
</s:form>
```

(3) 发布 Web 项目,启动 Tomcat 服务器,访问 inputMessage.jsp,结果如图 7-4 所示。

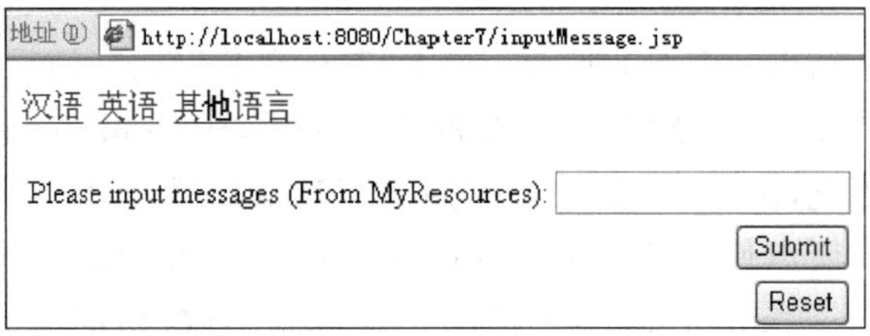

图 7-4 使用 i18n 标签的页面效果

例 7-4 实现消息显示的国际化。

在例 7-2 基础上操作,具体操作步骤如下:

(1) 编辑三个资源文件。

① 在 ApplicationResources.properties 文件中添加如下内容:

```
info.congratulation = congratulations, you have input some messages(Default)!
```

② 在 ApplicationResources_en_US.properties 文件中添加如下内容:

```
info.congratulation = congratulations, you have input some messages!
```

③ 在 ApplicationResources_zh.properties 文件中添加如下内容:

```
info.congratulation = 恭喜,您输入的内容不为空!
```

(2) 创建显示消息的 JSP 文件,使用 text 标签输出国际化信息。

showMessage.jsp

```
<%@ page language = "java" contentType = "text/html; charset = UTF - 8"
                pageEncoding = "UTF - 8" %>
<%@ taglib prefix = "s" uri = "/struts - tags" %>
<s:text name = "info.congratulation"/>
```

(3) 编辑 struts.xml 文件,修改 action 定义中的 name 为"success"的 result,如下:

```
<result name = "success">/showMessage.jsp</result>
```

(4) 发布 Web 项目,启动 Tomcat 服务器,访问 inputMessage.jsp,结果如图 7-1 所示。

① 在图 7-1 所示的界面中输入内容后提交,结果如图 7-5 所示。
② 在图 7-2 所示的界面中输入内容后提交,结果如图 7-6 所示。

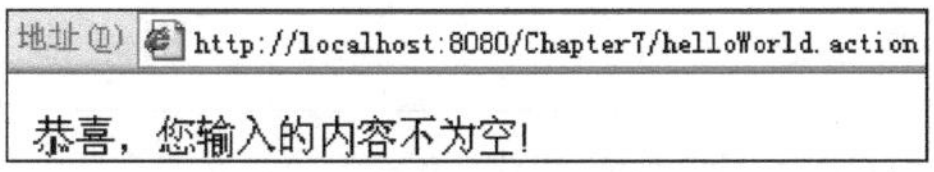

图 7-5　中文环境下的输出 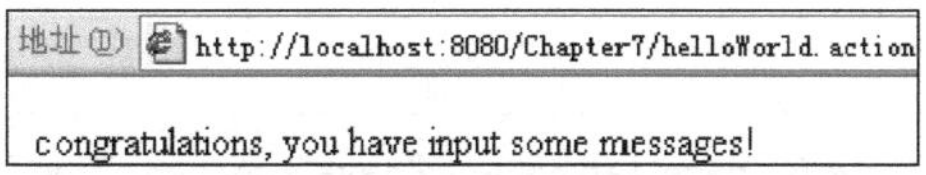　　　　　图 7-6　中文环境下的输出

7.3 如何在资源文件中使用参数

在 Struts 2 框架中,提供了两种在资源文件中设置参数的方法,分别是占位符和 OGNL 表达式。

(1) 资源文件中使用占位符的语法格式:由一对大括号括起来的单个数字(0~9),形如:{0}、{1}、{2}、…。每条消息最多可以包含 10 个不同的占位符,即最多可以接收 10 个不同的参数。

(2) 资源文件中使用 OGNL 表达式的语法格式:以"${"开始,以"}"结束,中间为表达式,形如:${expression}。

下面通过一个示例讲解如何在资源文件中设置参数。

例 7-5　在资源文件中使用参数。

在例 7-4 的基础上操作,当在输入界面的文本框中输入内容并提交后,在显示消息的页面中显示提交的内容。具体操作步骤如下:

(1) 编辑资源文件。

① ApplicationResources.properties 文件内容如下:

```
info.message = Please input messages (Default)
info.submit = Submit
info.reset = Reset
info.congratulation = congratulations, the input message is(Default): {0}!
info.congratulation.ognl = congratulations, the input message is(Default): ${message}!
```

② ApplicationResources_en_US.properties 文件内容如下:

```
info.message = Please input messages
info.submit = Submit
info.reset = Reset
info.congratulation = congratulations, the input message is: {0}!
info.congratulation.ognl = congratulations, the input message is: ${message}!
```

③ ApplicationResources_zh.properties 文件内容如下:

```
info.message = 请输入内容
info.submit = 提交
info.reset = 重置
info.congratulation = 恭喜,您输入的内容是:{0}!
info.congratulation.ognl = 恭喜,您输入的内容是: ${message}!
```

资源文件中的 OGNL 表达式中的参数名称和 Action 类中的属性名称相对应。

（2）编辑消息显示文件。

showMessage.jsp

```
<%@ page language = "java" contentType = "text/html; charset = UTF - 8"
    pageEncoding = "UTF - 8" %>
<%@ taglib prefix = "s" uri = "/struts - tags" %>
-------------- 使用占位符：---------------- <br>
<s:text name = "info.congratulation">
    <s:param>
        <s:property value = "message"/>
    </s:param>
</s:text>
<p>
----------- 使用 OGNL 表达式：------------- <br>
<s:text name = "info.congratulation.ognl"/>
```

在 JSP 文件中通过使用＜s:param＞标签给资源文件中的占位符传递参数。如果有多个不同的占位符，则＜s:param＞标签的顺序和占位符的数字顺序依次对应。即第一个＜s:param＞标签对应{0}，第二个＜s:param＞标签对应{1}，等等，一共可以使用 10 个＜s:param＞标签。

（3）发布 Web 项目，启动 Tomcat 服务器，访问 inputMessage.jsp，输入内容并提交，结果如图 7-7 所示。

在 inputMessage.jsp 文件中（以中文环境为例），label 的内容是"请输入内容"，如果想将 label 的内容改为"请输入姓名"，该如何实现呢？当然可以通过修改资源文件来实现，但是这不是一个好办法。可以通过参数传递的方法来实现，这样就只需修改 JSP 文件就可以达到更新 label 的内容的目的。

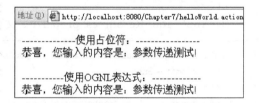

图 7-7　资源文件中使用参数的效果

inputMessage.jsp 文件中的 textfield 标签的 key 属性生成 html 的 label 标签。为了实现参数传递，需使用 textfield 标签的 label 属性通过 getText()方法获取资源文件中的消息。

下面通过示例进行讲解。

例 7-6　使用 getText()方法获取资源文件中的消息。

（1）修改资源文件中 key 为 info.message 的消息，如下：

① ApplicationResources.properties 文件中 info.message 的消息：

```
info.message = Please input {0} (Default)
```

② ApplicationResources_en_US.properties 文件中 info.message 的消息：

```
info.message = Please input {0}
```

③ ApplicationResources_zh.properties 文件中 info.message 的消息：

```
info.message=请输入{0}
```

(2) 修改 inputMessage.jsp 文件，textfield 标签的内容如下：

```
<s:textfield name="message" label="%{getText('info.message',{'姓名'})}"/>
```

getText()方法的第一个参数是资源文件中的 key，第二个参数是一个列表，列表中的元素对应着资源文件中消息的占位符。

(3) 发布 Web 项目，启动 Tomcat 服务器，访问 inputMessage.jsp，结果如图 7-8 所示。

如果将 getText()方法中的"姓名"改为"内容"，访问 inputMessage.jsp 的结果和图 7-1 相同。

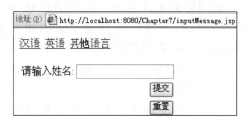

图 7-8 使用 getText()方法传递参数

7.4 访问资源文件中消息的方式

Struts 2 框架提供了几种访问资源文件中消息的方式，分别是：在 Action 中、在 JSP 页面中、在表单标签属性中和在资源文件中。

7.4.1 在 Action 中访问资源文件中的消息

Struts 2 框架的 com.opensymphony.xwork2.TextProvider 接口定义了访问资源文件中消息的 getText()方法，ActionSupport 类实现了该接口。因此，如果要在 Action 中访问资源文件中的消息，只要开发人员编写的 Action 类继承了 ActionSupport 类，就可以在 Action 类中直接使用 getText()方法获取资源文件中的消息。

(1) String getText(String key);

返回资源文件中 key 对应的消息，如果没有找到该 key，则返回 null。

(2) String getText(String key, String defaultValue);

返回资源文件中 key 对应的消息，如果没有找到该 key，则返回方法中提供的默认值(defaultValue 参数指定)。

(3) String getText(String key, String defaultValue, String obj);

返回资源文件中 key 对应的消息，并使用给定的 obj 参数按照 java.text.MessageFormat 格式进行定义。如果没有找到该 key，则返回方法中提供的默认值(defaultValue 参数指定)。

(4) String getText(String key, List<?> args);

返回资源文件中 key 对应的消息，并使用给定的 args 参数(列表)按照 java.text.MessageFormat 格式进行定义。如果没有找到该 key，则返回 null。

(5) String getText(String key, String[] args);

返回资源文件中 key 对应的消息，并使用给定的 args 参数(数组)按照 java.text.

MessageFormat 格式进行定义。如果没有找到该 key，则返回 null。

(6) String getText(String key, String defaultValue, List<?> args);

返回资源文件中 key 对应的消息，并使用给定的 args 参数（列表）按照 java.text.MessageFormat 格式进行定义。如果没有找到该 key，则返回方法中提供的默认值（defaultValue 参数指定）。

(7) String getText(String key, String defaultValue, String[] args);

返回资源文件中 key 对应的消息，并使用给定的 args 参数（数组）按照 java.text.MessageFormat 格式进行定义。如果没有找到该 key，则返回方法中提供的默认值（defaultValue 参数指定）。

(8) String getText(String key, String defaultValue, List<?> args, ValueStack stack);

返回资源文件中 key 对应的消息，并使用给定的 args 参数（列表）按照 java.text.MessageFormat 格式进行定义。如果没有找到该 key，则返回方法中提供的默认值（defaultValue 参数指定）。查找消息时，使用提供的值栈而不是 ActionContext 中的值栈。

(9) String getText(String key, String defaultValue, String[] args, ValueStack stack);

返回资源文件中 key 对应的消息，并使用给定的 args 参数（数组）按照 java.text.MessageFormat 格式进行定义。如果没有找到该 key，则返回方法中提供的默认值（defaultValue 参数指定）。查找消息时，使用提供的值栈而不是 ActionContext 中的值栈。

例 7-7 Action 中输出消息的国际化。

在例 7-4 的基础上操作，具体操作步骤如下：

(1) 编辑资源文件。

① 在 ApplicationResources.properties 文件中添加如下内容：

```
message.null = You did not input messages(Default)!
```

② 在 ApplicationResources_en_US.properties 文件中添加如下内容：

```
message.null = You did not input messages!
```

③ 在 ApplicationResources_zh.properties 文件中添加如下内容：

```
message.null = 您没有输入内容!
```

(2) 修改 Action 类 HelloWorld.java，其 execute 方法中的内容如下：

```java
public String execute(){
    if((message == null) || (message.isEmpty())){
        addActionError(getText("message.null"));
        return INPUT;
    }
    else{
        addActionMessage(getText("info.congratulation", new String[]{message}));
        return SUCCESS;
    }
}
```

（3）编辑 struts.xml，修改 action 定义中的 name 为"success"的 result，如下：

```
<result name="success">/inputMessage.jsp</result>
```

（4）编辑 inputMessage.jsp，添加如下内容：

```
<s:actionmessage/>
<s:actionerror/>
```

（5）发布 Web 项目，启动 Tomcat 服务器，访问 inputMessage.jsp。
① 如果没有输入内容提交，则结果如图 7-9 所示。
② 如果输入内容提交，则结果如图 7-10 所示。

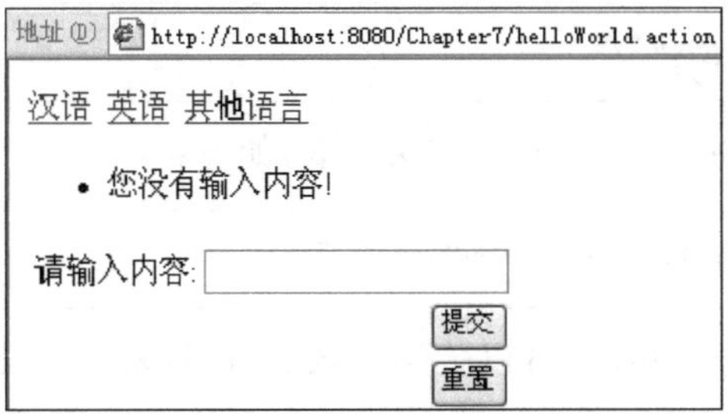

图 7-9　没有输入内容提交后的效果 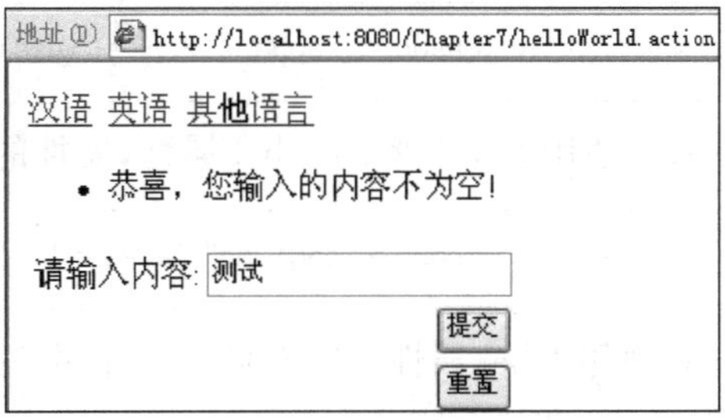　　　　图 7-10　输入内容提交后的效果

7.4.2　在 JSP 页面中访问资源文件中的消息

这里的 JSP 页面是指在 result 中指定的 JSP 页面。在这种类型的 JSP 页面中，使用 text 标签访问资源文件中的消息，标签的 name 属性值是资源文件中的 key。通过在 text 标签中内嵌 param 标签可以向资源文件中的消息传递参数，最多可以内嵌 10 个 param 标签，这种传递参数的方式属于占位符参数方式。

使用 text 标签访问资源文件中消息的格式：
（1）不传递参数。

```
<s:text name="messageKeyName"/>
```

（2）传递参数。
① 使用 Action 类中的属性名称。

```
<s:text name="messageKeyName">
    <s:param>
        <s:property value="actionPropertyName"/>
    </s:param>
    …
</s:text>
```

其中 property 标签的 value 属性值是 Action 类中的属性名称。

② 直接赋值。

```
<s:text name = "messageKeyName">
    <s:param value = "tsc"/>
    …
</s:text>
```

7.4.3 在表单标签中访问资源文件中的消息

可以通过表单标签的 key 或者 label 属性访问资源文件中的消息。使用格式如下：
(1) 使用表单标签的 key 属性，属性的值是资源文件中消息的 key，如下：

```
<s:textfield name = "tagName" key = "messageKeyName"/>
```

(2) 使用表单标签的 label 属性，属性的值通过 getText()方法获取，如下：

```
<s:textfield name = "tagName" label = "%{getText('messageKeyName')}"/>
```

在使用 label 属性访问资源文件中的消息的时候，还可以传递参数。getText()方法请参考 7.4.1 节的介绍。

7.4.4 在资源文件中访问资源文件中的消息

在资源文件中，一条消息的值可以通过使用 OGNL 表达式从另一条消息中获取。OGNL 表达式的格式如下：

```
${getText("messageKeyName")}
```

其中 messageKeyName 是资源文件中消息的 key。

在资源文件中，一条消息的内容从另一条消息中获取有什么好处呢？

假设资源文件中的内容如下：

```
info.zipcode = Zip code
info.zipcode.error = Zip code is invalid.
```

如果想把"Zip code"改为"Post code"，则需要修改两处。如果在资源文件中有多处使用了"Zip code"，则需要修改多处，麻烦并且容易漏改。如果采用 OGNL 表达式的形式，则不会出现这种问题，如下：

```
info.zipcode = Zip code
info.zipcode.error = ${getText("info.zipcode")} is invalid.
```

在获取 info.zipcode.error 的消息时，计算 OGNL 表达式 ${getText("info.zipcode")}的值，并将计算结果"Zip code"和其他字符串连接起来，info.zipcode.error 键的值为"Zip code is invalid."。如果想把"Zip code"改为"Post code"，则只需要修改一处即可。

例 7-8 一条消息的内容从另一条消息中获取的示例。

在例 7-7 的基础上操作,具体操作步骤如下:

(1) 编辑资源文件,修改 message.null 的值。

① ApplicationResources.properties 文件:

```
message.null = ${getText("info.message")}!You did not input messages(Default)!
```

② ApplicationResources_en_US.properties 文件:

```
message.null = ${getText("info.message")}!You did not input messages!
```

③ ApplicationResources_zh.properties 文件:

```
message.null = ${getText("info.message")}!您没有输入内容!
```

(2) 发布 Web 项目,启动 Tomcat 服务器,访问 inputMessage.jsp,并提交,结果如图 7-11 所示。

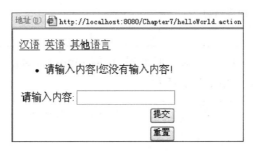

图 7-11 资源文件中使用 OGNL 表达式的效果

这时,只要修改 info.message 键的值,message.null 键的值也就随着发生改变。

7.5 资源文件的分类

通过前面示例的学习,掌握了使用 struts.custom.i18n.resources 常量配置资源文件的方法,这种方法加载的资源文件属于全局资源文件,当前 Web 应用中的所有文件都可以使用。除了使用 struts.custom.i18n.resources 常量配置全局资源文件外,还可以使用 i18n 标签指定资源文件的位置。如果同时使用了 i18n 标签和 struts.custom.i18n.resources 常量,则首先搜索 i18n 标签指定的资源文件,如果没有指定的 key 则搜索全局资源文件。

另外,Struts 2 框架还支持包及 Action 等类型的资源文件。

7.5.1 包资源文件

包资源文件的命名格式为 package_language_country.properties,其中"package"是固定不变的。包资源文件只为所在包中的所有 Action 提供服务,其内容由键-值对组成。包资源文件无需配置,Struts 2 框架就能加载。

如果包资源文件和全局资源文件都存在,则首先搜索包资源文件。如果包资源文件中没有匹配的 key,则搜索全局资源文件。

例 7-9　包资源文件示例。

在例 7-8 基础上操作,具体操作步骤如下:

(1) 在 example.struts2 包中创建资源文件。

① package.properties 文件:

```
info.message = Please input messages(From package,Default)
info.submit = Submit(From package,Default)
```

② package_en_US.properties 文件:

```
info.message = Please input messages (From package)
info.submit = Submit(From package)
```

③ package_zh.properties 文件:

```
info.message = 请输入内容(包)
info.submit = 提交(包)
```

(2) 发布 Web 项目,启动 Tomcat 服务器,访问 inputMessage.jsp,结果同图 7-1,为什么?单击"提交"按钮,结果如图 7-12 所示。

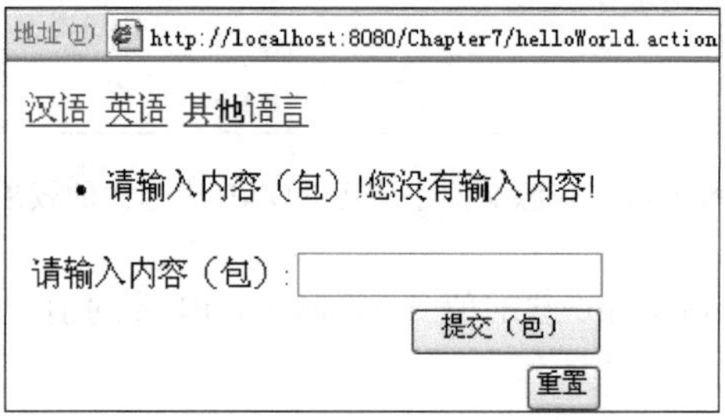

图 7-12　包资源文件生效的效果图

由包资源文件的内容和图 7-12 可以看出,Struts 2 框架首先在包资源文件中搜索 key,然后才在全局资源文件中进行搜索。

7.5.2　Action 资源文件

Action 资源文件的命名格式为 ActionName_language_country.properties,其中"ActionName"是使用该资源文件的 Action 类的具体名称。Action 资源文件只为同名的 Action 类提供服务,其内容由键-值对组成。Action 资源文件无需配置,Struts 2 框架就能加载。

如果 Action 资源文件、包资源文件和全局资源文件都存在,则首先搜索 Action 资源文件。如果 Action 资源文件中没有匹配的 key,则搜索包资源文件和全局资源文件。

例 7-10　Action 资源文件示例。

在例 7-9 基础上操作,具体操作步骤如下:

(1) 在 example.struts2 包中为 HelloWorld Action 创建资源文件。

① HelloWorld.properties 文件：

```
info.message=Please input messages(From HelloWorld Action,Default)
```

② HelloWorld_en_US.properties 文件：

```
info.message=Please input messages (From HelloWorld Action)
```

③ HelloWorld_zh.properties 文件：

```
info.message=请输入内容(HelloWorld Action)
```

(2) 发布 Web 项目，启动 Tomcat 服务器，访问 inputMessage.jsp，并单击"提交"按钮，结果如图 7-13 所示。

图 7-13 Action 资源文件生效的效果图

由 Action 资源文件的内容和图 7-13 可以看出，Struts 2 框架首先在 Action 资源文件中搜索 key，然后才在包资源文件和全局资源文件中进行搜索。

Action 资源文件是否只对具体的 Action 起作用呢？下面通过示例进行讲解。

例 7-11 测试 Action 资源文件的作用范围。

在例 7-10 基础上操作，具体操作步骤如下：

(1) 创建 Action HelloWorld1，内容同 HelloWorld.java。

(2) 编辑 struts.xml，修改 action 定义中的 class 属性，如下：

```
<action name="helloWorld" class="example.struts2.HelloWorld1">
```

(3) 发布 Web 项目，启动 Tomcat 服务器，访问 inputMessage.jsp，并单击"提交"按钮，结果如图 7-12 所示。

7.6 资源文件的加载顺序

Struts 2 框架按照如下顺序搜索资源文件。

1. Action 资源文件

首先在同一包中按如下顺序搜索 Action 资源文件。

(1) 如果存在匹配的 Action 资源文件,则在资源文件中查找 Key。如果该 Action 资源文件中没有找到匹配的 key,或者该 Action 资源文件不存在,则在当前包中搜索和该 Action 实现的接口同名的资源文件。如果和接口同名的资源文件中没有找到匹配的 key,或者该资源文件不存在,则执行第(2)步,进入父一级进行搜索。

假设 SubClassAction 类实现了 ISubClass 接口,如果没有和 SubClassAction 类同名的资源文件(指资源文件的基名)或者该资源文件中没有匹配的 key,则查找和其实现的接口 ISubClass 同名的资源文件。

(2) 进入父一级查找资源文件,即搜索和该 Action 的父类同名的资源文件。其搜索顺序按照第(1)步的顺序进行操作,以此类推。

假设 SubClassAction 类的父类是 ParentClassAction,如果没有和 SubClassAction 类同名的资源文件(指资源文件的基名)或者该资源文件中没有匹配的 key,则查找和其父类 ParentClassAction 同名的资源文件。如果没有和 ParentClassAction 同名的资源文件,或者在该资源文件中没有找到匹配的 key,则查找和 ParentClassAction 类实现的接口同名的资源文件。

Action 资源文件必须和对应的同名 Action 类放在同一目录中。例如:假设 ParentClassAction 类继承了 ActionSupport 类,由于该类在 com.opensymphony.xwork2 包中,因此,以 ActionSupport 命名的资源文件必须放在 com.opensymphony.xwork2 位置才能被加载。

(3) 实现 ModelDriven 接口。如果 Action 类实现了 ModelDriven 接口,则由 getModel()方法获得模型对象后,以模型对象所属的类按上述步骤进行查找资源文件并查找匹配的 key。

2. 包资源文件

(1) 在当前包下查找基名为"package"的包资源文件。

(2) 如果当前包下没有以"package"为基名的包资源文件,或者没有找到匹配的 key,则在其父包中查找以"package"为基名的包资源文件,以此类推。

例如,前面例子中的业务控制器类 HelloWorld.java 在 example.struts2 包中。首先在 example.struts2 包查找包资源文件,如果没有包资源文件或者包资源文件中没有找到匹配的 key,则在父包 example 中查找包资源文件,以此类推。

3. 全局资源文件

全局资源文件通过 struts.custom.i18n.resources 常量设置,对所在 Web 应用中的所有文件都起作用。在执行 Action 的时候,上述第 1、2 两步先被执行,只有在第 1、2 步都没有找到匹配的 key 时,才使用全局资源文件。如果在全局资源文件中也没有找到匹配的 key,则直接输出 key 本身。

4. 局部资源文件

如果用户首先访问的是 JSP 页面,则第 1、2 两步都不会被执行。除了全局资源文件,在 JSP 文件中可以使用 i18n 标签指定所要使用的资源文件,被称为局部资源文件(指其作用范围局部有效)。如果同时使用了 i18n 标签和 struts.custom.i18n.resources 常量设置所要使用的资源文件,则首先搜索 i18n 标签指定的资源文件。如果在该资源文件中没有找到匹配的 key,则搜索全局资源文件。

习题

1. 通过官方网站了解完整的语言代码、国家和地区代码。
2. 在例 7-11 的基础上完成资源文件加载顺序的验证。

第 8 章 Struts 2框架的类型转换

Web 应用是基于 HTTP 协议的,由于 HTTP 协议中没有"数据类型"的概念,因此,所有通过表单输入的数据提交给服务器时都是以字符串的形式存在,即每一个表单项的输入只能是字符串或字符串数组。但是,当服务器端接收到客户端提交的数据后,要使用具体编程语言对数据进行处理(例如在使用 JSP 进行 Web 应用开发时,服务器端要使用 Java 语言),这就必须将接收到的内容转换成正确的数据类型。例如,在图 8-1 所示的添加书目的界面中,单价是浮点型,出版日期是日期型,数量是整型。

由此可见,在 Web 应用开发中,数据的类型转换操作是不可避免的。而在使用传统的 JSP 技术进行 Web 应用开发时,数据类型转换任务需要开发人员编写相应的代码来完成。通常,数据类型转换的任务越大,即用户提交的数据种类越多,所需要的数据类型转换代码也就越多。对于一个大型的 Web 应用来说,这将会变得非常烦琐、乏味,并且难于维护。

图 8-1 添加书目界面

Struts 2 框架提供了比较完善的数据类型转换支持,可以实现用户提交数据的自动类型转换,并且同时还提供了数据类型转换的异常处理功能,这不但提高了开发效率,而且便于 Web 应用的维护。

8.1 Struts 2 框架对类型转换的支持

Struts 2 框架不但能够读写用 OGNL 表达式命名的对象属性,而且还能进行正确的数据类型转换。Struts 2 框架的类型转换对开发人员是透明的。下面是使用 Struts 2 框架类型转换功能的一些提示:

(1) 使用 JavaBean:Struts 2 框架只能创建遵循 JavaBean 规范的对象,JavaBean 需要提供无参构造函数及合适的 getter 和 setter 方法。例如,在添加书目的 Web 应用中,为了使用 Struts 2 框架的类型转换功能,需要创建一个名为 Book 的 JavaBean。

(2) 使用 OGNL 表达式:表单输入元素的命名要符合形如"ObjectName.PropertyName" OGNL 表达式的形式,例如 book.name、book.price、book.date、book.count。这样,Struts 2 框架将会自动创建实际的 book 对象。

(3) Struts 2 框架会调用 getBook().setName()方法为 OGNL 表达式 book.name 赋

值,使用 setBook()方法创建 book 对象。

(4) 如果一个对象实例(如 book)已经存在,将不再被实例化。PrepareInterceptor 拦截器或者 action 的构造函数能够在类型转换之前创建目标对象。

(5) 对于 list 和 map 使用索引符号,如 book[0].name。通常这些 HTML 表单元素在循环中输出。对于 JSP 标签,使用 iterator 标签的状态属性。对于 FreeMarker 标签,使用特殊的属性 ${foo_index}[]。

(6) 对于多选的下拉列表框,不能使用索引符号命名每一个单独的选项,而是使用 book.name 命名元素,Struts 2 框架就会明白应该为每一个选中的选项创建一个新的 book 对象,并设置相应的名称。

8.2 Struts 2 框架内置的类型转换器

Struts 2 框架实现了大多数常见的用于类型转换的转换器,开发人员不用自己编写类型转换代码,就可以完成数据类型转换的任务。Struts 2 框架支持的内置类型转换器包括如下几种。

(1) 布尔类型(boolean/Boolean):在字符串和布尔型数据之间进行转换。
(2) 字符类型(char/Character):在字符串和字符数据之间进行转换。
(3) 数值类型(int/Integer、float/Float、long/Long、double/Double):在字符串和数值型数据之间转换。
(4) 日期类型(dates):为当前请求关联的 Locale 使用 SHORT 格式日期。
(5) 数组类型(arrays):将字符串数组的每个元素转换成其他的数据类型。
(6) 集合类型(collections):如果不能确定对象类型,那么就假定是 String 类型,并创建一个新的 ArrayList。
(7) 字符串类型(String):将 boolean、int、long、double、String 类型的数组或 java.util.Date 类型转换为字符串类型的数据。

8.3 类型转换体验

例 8-1 以添加书目为例,介绍 Struts 2 框架对类型转换的支持。

具体操作步骤如下:

(1) 创建 Web 项目 Chapter8,并添加 Struts 2 框架的支持。Chapter8 项目目录结构参考 Chapter7 项目的目录结构,web.xml 配置文件的内容同 Chapter7 项目中 web.xml 配置文件的内容。

(2) 创建 JavaBean。

Book.java

```
package example.struts2;
import java.util.Date;
```

```java
public class Book {
    private String bookName;
    private double bookPrice;
    private Date bookPublishDate;
    private int bookCount;
    public int getBookCount() {
        return bookCount;
    }
    public void setBookCount(int bookCount) {
        this.bookCount = bookCount;
    }
    public String getBookName() {
        return bookName;
    }
    public void setBookName(String bookName) {
        this.bookName = bookName;
    }
    public double getBookPrice() {
        return bookPrice;
    }
    public void setBookPrice(double bookPrice) {
        this.bookPrice = bookPrice;
    }
    public Date getBookPublishDate() {
        return bookPublishDate;
    }
    public void setBookPublishDate(Date bookPublishDate) {
        this.bookPublishDate = bookPublishDate;
    }
}
```

(3) 创建 Action 类。

BookAction.java

```java
package example.struts2;
import com.opensymphony.xwork2.ActionSupport;
public class BookAction extends ActionSupport {
    private Book book;
    public Book getBook() {
        return book;
    }
    public void setBook(Book book) {
        this.book = book;
    }
    public String execute(){
        if(book == null || book.getBookName().isEmpty()){
            addActionError("请输入完整的书目信息");
            return INPUT;
        }
        else{
            System.out.println("添加的书目信息如下：");
```

```java
            System.out.println("书名: " + book.getBookName());
            System.out.println("单价: " + book.getBookPrice());
            System.out.println("出版日期: " + book.getBookPublishDate());
            System.out.println("数量: " + book.getBookCount());
            return SUCCESS;
        }
    }
}
```

(4) 创建 struts.xml 文件。

struts.xml

```xml
<?xml version="1.0" encoding="UTF-8"?>
<!DOCTYPE struts PUBLIC
    "-//Apache Software Foundation//DTD Struts Configuration 2.0//EN"
    "http://struts.apache.org/dtds/struts-2.0.dtd">
<struts>
    <package name="default" extends="struts-default">
        <action name="addBook" class="example.struts2.BookAction">
            <result name="success">/addBook.jsp</result>
            <result name="input">/addBook.jsp</result>
        </action>
    </package>
</struts>
```

(5) 创建 JSP 页面。

addBook.jsp

```jsp
<%@ page language="java" contentType="text/html; charset=UTF-8" pageEncoding="UTF-8"%>
<%@ taglib prefix="s" uri="/struts-tags" %>
<s:actionerror/>
<s:form action="addBook">
    <s:textfield name="book.bookName" label="书名"/>
    <s:textfield name="book.bookPrice" label="单价"/>
    <s:textfield name="book.bookPublishDate" label="出版日期"/>
    <s:textfield name="book.bookCount" value="1" label="数量"/>
    <s:submit name="submit" value="提交"/>
    <s:reset name="reset" value="重置"/>
</s:form>
<s:if test="book.bookName! = ''">
    添加的书目信息如下: <br>
    书名: <s:property value="book.bookName"/><br>
    单价: <s:property value="book.bookPrice"/><br>
    出版日期: <s:property value="book.bookPublishDate"/><br>
    数量: <s:property value="book.bookCount"/>
</s:if>
```

（6）发布 Web 项目，启动 Tomcat 服务器，访问 addBook.jsp，结果如图 8-2 所示。

① 不输入书名，单击"提交"按钮，结果如图 8-3 所示。

图 8-2　访问 addBook.jsp 的效果 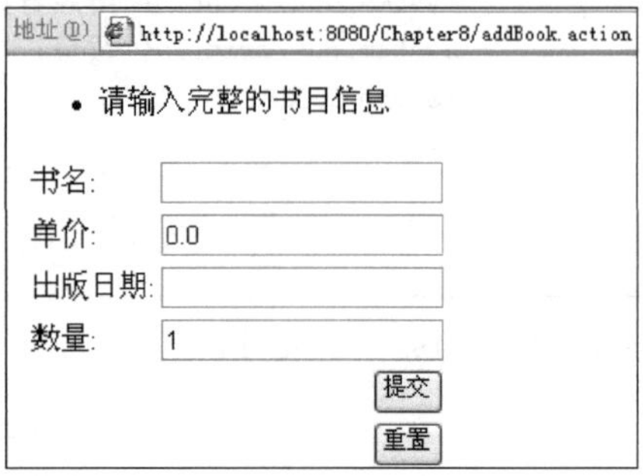　　　　图 8-3　不输入书名提交的效果

② 输入信息后，单击"提交"按钮，结果如图 8-4 所示，控制台的输出如图 8-5 所示。

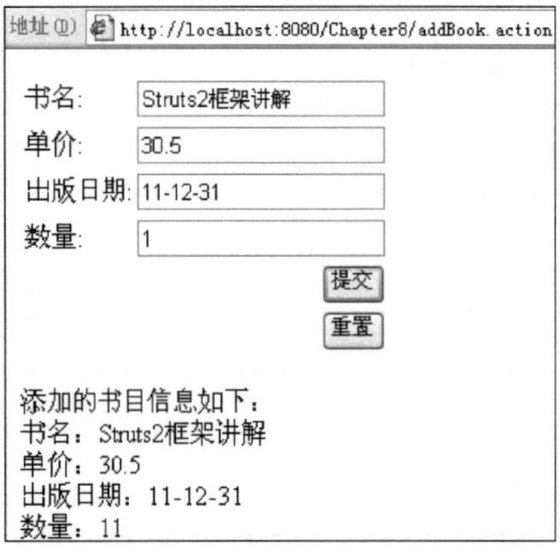

图 8-4　输入信息提交后的效果 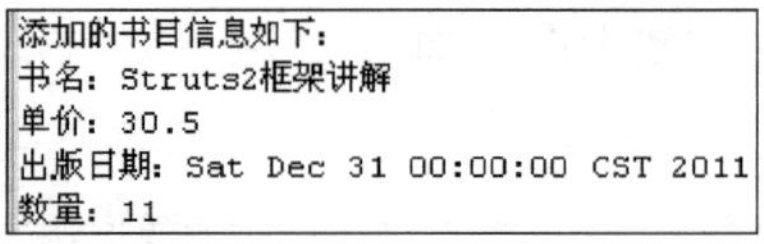　　　　图 8-5　输入信息提交后控制台的输出

由输出结果可以看出，输入的单价、出版日期和数量都被成功地转换为正确的数据类型。

③ 在数量中输入"1,000"后，单击"提交"按钮，结果如图 8-6 所示。

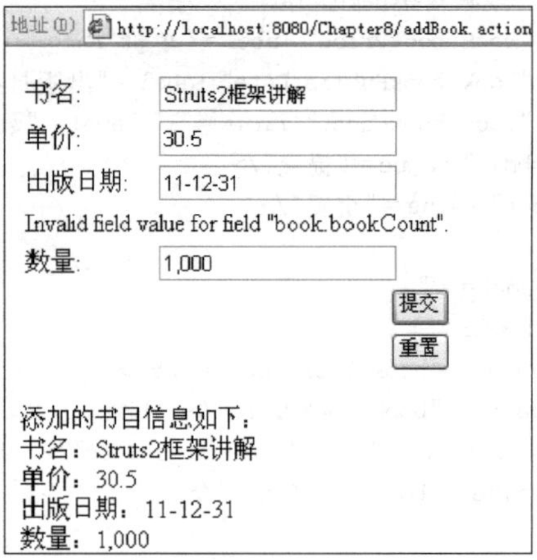

图 8-6　数量输入"1,000"提交后的效果

由图 8-6 可以看出，当输入的数量类型转换失败后，将会出现提示信息。

8.4 处理 List 类型转换

Struts 2 框架提供了对 List 类型转换的支持，如果在 Action 类中不使用泛型定义属性，则必须建立一个属性文件，属性文件的名称形如 ActionName-conversion.properties。属性文件中的内容是键-值对，形如：

```
Element_xxx = 类型的完整类名
```

其中字符串"Element_"是固定形式，xxx 是 Action 或对象中集合类型(List)的属性名称。

例 8-2 以添加书目信息为例，讲解 Struts 2 框架对 List 类型转换的支持。

具体操作步骤如下：

（1）JavaBean 类 Book.java 和例 8-1 中的 JavaBean 类相同。

（2）创建 Action。

BookListAction.java

```java
package example.struts2;
import java.util.ArrayList;
import java.util.Iterator;
import java.util.List;
import com.opensymphony.xwork2.ActionSupport;
public class BookListAction extends ActionSupport {
    private List books;
    public List getBooks() {
        return books;
    }
    public void setBooks(List books) {
        this.books = books;
    }
    public String execute(){
        if(books == null || books.isEmpty()){
            addActionError("请输入完整的书目信息");
            return INPUT;
        }
        else{
            System.out.println("添加的书目信息如下：");
            Iterator<Book> it = books.iterator();
            while(it.hasNext()){
                Book book = it.next();
                System.out.println("书名：" + book.getBookName());
                System.out.println("单价：" + book.getBookPrice());
                System.out.println("出版日期：" + book.getBookPublishDate());
                System.out.println("数量：" + book.getBookCount());
```

```
                System.out.println("------------------");
            }
            return SUCCESS;
        }
    }
}
```

(3) 创建属性文件。

BookListAction-conversion.properties

```
Element_books = example.struts2.Book
```

(4) 编辑 struts.xml 文件,添加内容如下:

```
<action name="addBookList" class="example.struts2.BookListAction">
    <result name="success">/addBookList.jsp</result>
    <result name="input">/addBookList.jsp</result>
</action>
```

(5) 创建 JSP 文件。

addBookList.jsp

```
<%@ page language="java" contentType="text/html; charset=UTF-8"
                                            pageEncoding="UTF-8" %>
<%@ taglib prefix="s" uri="/struts-tags" %>
<s:actionerror/>
<s:form action="addBookList" method="post">
    <s:iterator value="new int[2]" status="status">
        <s:textfield name="%{'books['+#status.index+'].bookName'}" label="书名"/>
        <s:textfield name="%{'books['+#status.index+'].bookPrice'}" label="单价"/>
        <s:textfield name="%{'books['+#status.index+'].bookPublishDate'}"
                                                    label="出版日期"/>
        <s:textfield name="%{'books['+#status.index+'].bookCount'}" value="1"
                                                    label="数量"/>
        <s:label value="--------------------------"></s:label>
    </s:iterator>
    <s:submit name="submit" value="提交"/>
    <s:reset name="reset" value="重置"/>
</s:form>
<s:if test="books!=null">
    添加的书目信息如下:<br>
    <s:iterator value="books">
        书名:<s:property value="bookName"/><br>
        单价:<s:property value="bookPrice"/><br>
        出版日期:<s:property value="bookPublishDate"/><br>
        数量:<s:property value="bookCount"/><br>
```

```
            <s:label value = " -------------------------- "></s:label> <br>
        </s:iterator>
    </s:if>
```

(6) 发布 Web 项目,启动 Tomcat 服务器,访问 addBookList.jsp,结果如图 8-7 所示。输入信息后,单击"提交"按钮,控制台的输出如图 8-8 所示,浏览器输出如图 8-9 所示。

图 8-7　访问 addBookList.jsp 的效果 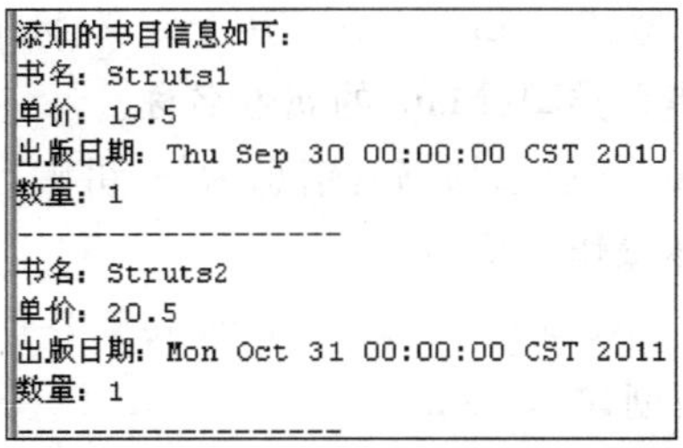　　　图 8-8　输入信息提交后控制台的输出

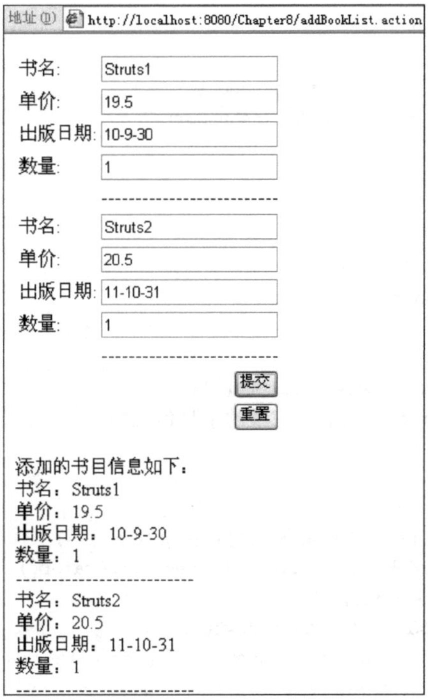

图 8-9　输入信息提交后的效果

8.5 处理 Map 类型转换

Struts 2 框架提供了对 Map 类型转换的支持,如果在 Action 类中不使用泛型定义属性,则必须建立一个属性文件,属性文件的名称形如 ActionName-conversion.properties。属性文件中的内容是键-值对,形如:

```
Key_xxx = 类型的完整类名
Element_xxx = 类型的完整类名
```

其中字符串"Key_"和"Element_"都是固定形式。"Key_xxx"用于定义 Map 类型对象中的 key 的类型,"Element_xxx"用于定义 Map 类型对象中的 value 的类型;"xxx 是 Action 或对象中集合类型(Map)的属性名称。

例 8-3 以添加书目信息为例,讲解 Struts 2 框架对 Map 类型转换的支持。

具体操作步骤如下:

(1) JavaBean 类 Book.java 和例 8-1 中的 JavaBean 类相同。

(2) 创建 Action。

BookMapAction.java

```java
package example.struts2;
import java.util.Iterator;
import java.util.Map;
import com.opensymphony.xwork2.ActionSupport;
public class BookMapAction extends ActionSupport {
    private Map books;
    public Map getBooks() {
        return books;
    }
    public void setBooks(Map books) {
        this.books = books;
    }
    public String execute(){
        if(books == null || books.isEmpty()){
            addActionError("请输入完整的书目信息");
            return INPUT;
        }
        else{
            System.out.println("添加的书目信息如下:");
            Iterator<Book> it = books.values().iterator();
            while(it.hasNext()) {
                Book book = (Book)it.next();
                System.out.println("书名:" + book.getBookName());
                System.out.println("单价:" + book.getBookPrice());
                System.out.println("出版日期:" + book.getBookPublishDate());
                System.out.println("数量:" + book.getBookCount());
              System.out.println("------------------");
            }
```

```
            return SUCCESS;
        }
    }
}
```

(3) 创建属性文件。

BookMapAction-conversion.properties

```
Key_books = java.lang.String
Element_books = example.struts2.Book
```

(4) 编辑 struts.xml 文件,添加内容如下:

```xml
<action name = "addBookMap" class = "example.struts2.BookMapAction">
    <result name = "success">/addBookMap.jsp</result>
    <result name = "input">/addBookMap.jsp</result>
</action>
```

(5) 创建 JSP 文件。

addBookMap.jsp

```jsp
<%@ page language = "java" contentType = "text/html; charset = UTF-8"
                                         pageEncoding = "UTF-8" %>
<%@ taglib prefix = "s" uri = "/struts-tags" %>
<s:actionerror/>
<s:form action = "addBookMap" method = "post">
    <s:iterator value = "new int[2]" status = "status">
        <s:set name = "books" value = "'books.book' + #status.index"/>
        <s:textfield name = "%{#books + '.bookName'}" label = "书名"/>
        <s:textfield name = "%{#books + '.bookPrice'}" label = "单价"/>
        <s:textfield name = "%{#books + '.bookPublishDate'}" label = "出版日期"/>
        <s:textfield name = "%{#books + '.bookCount'}" value = "1" label = "数量"/>
        <s:label value = "--------------------------"></s:label>
    </s:iterator>
    <s:submit name = "submit" value = "提交"/>
    <s:reset name = "reset" value = "重置"/>
</s:form>
<s:if test = "books! = null">
    添加的书目信息如下: <br>
    <s:iterator value = "books">
        书名: <s:property value = "value.bookName"/> <br>
        单价: <s:property value = "value.bookPrice"/> <br>
        出版日期: <s:property value = "value.bookPublishDate"/> <br>
        数量: <s:property value = "value.bookCount"/> <br>
        <s:label value = "--------------------------"></s:label> <br>
    </s:iterator>
</s:if>
```

说明:

① 由于 books 是 Map 类型,因此,表单输入元素的命名要使用 books.key 或者 books['key'] 的形式,本示例中使用了前者形式。

② 由于 books 是 Map 类型,返回的值是键-值对。因此,为了输出对象的属性值,需要使用表达式 value 取得 Map 中的值,即 Book 的实例。对象的属性值通过 value.bookName 的形式获取。

(6) 发布 Web 项目,启动 Tomcat 服务器,访问 addBookMap.jsp,进行测试。

8.6 自定义类型转换器

如果 Struts 2 框架提供的内置类型转换器不能满足 Web 应用开发,开发人员就要自己定制类型转换器。有两种创建自定义类型转换器的方式:①创建基于 OGNL 的类型转换器;②创建基于 Struts 2 框架的类型转换器。

使用自定义类型转换器的步骤如下:

(1) 创建自定义类型转换器。

(2) 注册自定义类型转换器。

根据使用范围,Struts 2 框架提供了两种注册自定义类型转换器的方式:全局注册方式和局部注册方式。

• 全局注册方式。

注册为全局的类型转换器对所有的 Action 都起作用。注册全局的类型转换器需要在 Web 应用的 WEB-INF/classes 目录下创建一个名称为 xwork-conversion.properties 的属性文件。

属性文件由键-值对组成,格式如下:

```
被转换类的完整实现类名称 = 自定义类型转换器的完整实现类名称
```

其中键的名称是被转换类的完整实现类名称,值是自定义类型转换器完整实现类名称。

• 局部注册方式。

注册为局部的类型转换器又称为特定类的类型转换器。注册局部的类型转换器需要在类所在的包中建立一个文件名称形如 ClassName-conversion.properties 的属性文件。

属性文件由键-值对组成,格式如下:

```
propertyName = 自定义类型转换器的完整实现类名称
```

其中 propertyName 是类中被转换的字段名称,值是自定义类型转换器完整实现类名称。

(3) 在 Web 应用中使用自定义类型转换器。

8.6.1 创建基于 OGNL 的类型转换器

创建基于 OGNL 的自定义类型转换器需要实现 TypeConverter 接口中的 convertValue() 方法,声明如下:

```
public Object convertValue(Map<String, Object> context, Object target,
            Member member, String propertyName, Object value, Class toType);
```

该方法用于将给定的值转换为指定的类型。其中各参数的含义如下。

（1）context 参数：用于指定执行转换的上下文。

（2）target 参数：用于指定将被设置的属性所属目标对象。

（3）member 参数：用于指定将被设置的类成员（构造方法、方法或字段）。

（4）propertyName 参数：用于指定将被设置的属性名称。

（5）value 参数：用于指定将被转换的值。

（6）toType 参数：用于指定值被转换到的类型。

由于通过实现 TypeConverter 接口的方式创建自定义类型转换器的工作量非常大，因此，为了简化创建基于 OGNL 的自定义类型转换器的工作，OGNL 又提供了一个 TypeConverter 接口的实现类 DefaultTypeConverter。开发人员可以通过继承 DefaultTypeConverter 类的方式创建自定义类型转换器。

DefaultTypeConverter 类通过调用另外一个 convertValue()方法，对 TypeConverter 接口中的 convertValue()方法进行了实现，如下：

```
public Object convertValue(Map<String, Object> context, Object target, Member member,
            String propertyName, Object value, Class toType) {
        return convertValue(context, value, toType);
}
public Object convertValue(Map<String, Object> context, Object value, Class toType) {
        return convertValue(value, toType);
}
```

由此可见，通过继承 DefaultTypeConverter 类，创建的自定义类型转换器只需要重写具有三个参数的 convertValue()方法即可。

例 8-4 通过继承 DefaultTypeConverter 类，创建自定义类型转换器。

任务描述：添加书目信息时，出版日期按指定格式输入。

在例 8-1 的基础上进行操作（去掉 addBook.jsp 中数量的默认值），具体操作步骤如下：

（1）创建自定义类型转换器。

MyDateConverter.java

```
package example.struts2;
import java.text.ParseException;
import java.text.SimpleDateFormat;
import java.util.Date;
import java.util.Map;
import ognl.DefaultTypeConverter;
import com.opensymphony.xwork2.conversion.TypeConversionException;
public class DateConverter extends DefaultTypeConverter {
    private static SimpleDateFormat sdf = new SimpleDateFormat("yyyy-MM-dd");
```

```java
    public Object convertValue(Map context, Object value, Class toType) {
        if(toType == Date.class){
            System.out.println("类型为 Date.class!");
            try{
                String[] str = (String[])value;
                Date date = sdf.parse(str[0]);
                return date;
            }
            catch(ParseException pe){
                throw new TypeConversionException(pe.getMessage());
            }
        }
        else if(toType == String.class){
            System.out.println("类型为 String.class!");
            return sdf.format(value);
        }
        return null;
    }
}
```

(2) 注册全局的类型转换器。

xwork－conversion.properties

```
java.util.Date = example.struts2.MyDateConverter
```

(3) 发布 Web 项目，启动 Tomcat 服务器，访问 addBook.jsp，输入信息并提交，结果如图 8-10 所示。

图 8-10　输入信息提交后的效果

比较图 8-10 和图 8-4 可知，日期类型转换器生效。

8.6.2　创建基于 Struts 2 框架的类型转换器

Struts 2 框架提供了 StrutsTypeConverter 抽象类，它继承了 DefaultTypeConverter

类,是开发基于 Struts 2 框架的 Web 应用时使用的类型转换器的基类。

StrutsTypeConverter 抽象类的定义如下:

StrutsTypeConverter.java

```java
package org.apache.struts2.util;
import java.util.Map;
import com.opensymphony.xwork2.conversion.impl.DefaultTypeConverter;
public abstract class StrutsTypeConverter extends DefaultTypeConverter {
    public Object convertValue(Map context, Object o, Class toClass) {
        if (toClass.equals(String.class)) {
            return convertToString(context, o);
        } else if (o instanceof String[]) {
            return convertFromString(context, (String[]) o, toClass);
        } else if (o instanceof String) {
            return convertFromString(context, new String[]{(String) o}, toClass);
        } else {
            return performFallbackConversion(context, o, toClass);
        }
    }
    protected Object performFallbackConversion(Map context, Object o, Class toClass) {
        return super.convertValue(context, o, toClass);
    }
    public abstract Object convertFromString(Map context, String[] values, Class toClass);
    public abstract String convertToString(Map context, Object o);
}
```

由此可见,创建继承 StrutsTypeConverter 抽象类的类型转换器需要实现 StrutsTypeConverter 抽象类中的 convertFromString()和 convertToString()两个方法。其中 convertFromString()方法用于将 String 类型转换为对象类型,convertToString()方法用于将对象类型数据转换为 String 类型。

例 8-5 通过继承 StrutsTypeConverter 抽象类,创建自定义类型转换器。

任务描述:添加书目信息时,输入的数量使用逗号分隔。

在例 8-1 的基础上进行操作(去掉 addBook.jsp 中数量的默认值),具体操作步骤如下:

(1) 创建自定义类型转换器。

MyNumberConverter.java

```java
package example.struts2;
import java.util.Map;
import org.apache.struts2.util.StrutsTypeConverter;
public class MyNumberConverter extends StrutsTypeConverter {
    public Object convertFromString(Map context, String[] values, Class toClass) {
        System.out.println("调用 convertFromString()方法!");
        String integerValue = values[0];
        StringBuffer stringValue = new StringBuffer();
        int index = integerValue.indexOf(",", 0);
        if(index == -1){
```

```java
            stringValue.append(integerValue);
        }
        int fromIndex = 0;
        String temp = null;
        while(index > -1){
            temp = integerValue.substring(fromIndex, index);
            stringValue.append(temp);
            fromIndex = index + 1;
            index = integerValue.indexOf(",", fromIndex);
            if(index == -1){
                temp = integerValue.substring(fromIndex);
                stringValue.append(temp);
            }
        }
        return Integer.parseInt(stringValue.toString());
    }
    public String convertToString(Map context, Object obj) {
        System.out.println("调用convertToString()方法!");
        Integer number = (Integer)obj;
        StringBuffer strNum = new StringBuffer(number.toString());
        StringBuffer stringValue = new StringBuffer();
        int numLength = strNum.length();
        int count = numLength / 3;
        int mod = numLength % 3;
        String temp = null;
        if(mod != 0){
            temp = strNum.substring(0, mod);
            stringValue.append(temp);
            stringValue.append(",");
        }
        int i = 0;
        for(i = 0; i < count - 1; i++){
            temp = strNum.substring(i * 3 + mod, (i + 1) * 3 + mod);
            stringValue.append(temp);
            stringValue.append(",");
        }
        temp = strNum.substring(i * 3 + mod, (i + 1) * 3 + mod);
        stringValue.append(temp);
        return stringValue.toString();
    }
}
```

(2) 注册局部的类型转换器。

Book-conversion.properties

```
bookCount = example.struts2.MyNumberConverter
```

(3) 发布 Web 项目,启动 Tomcat 服务器,访问 addBook.jsp,输入信息并提交,结果如图 8-11 所示,控制台输出结果如图 8-12 所示。

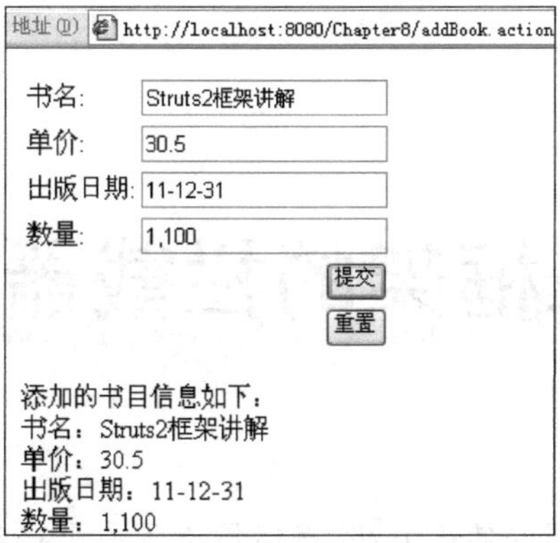

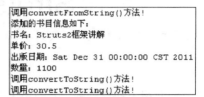

图 8-11　输入信息提交后的效果　　　　图 8-12　控制台输出结果

1. 修改例 8-1，不使用 Struts 2 框架的类型转换，实现输入数据的正确处理。
2. 修改例 8-2 和例 8-3，分别使用 List 泛型、Map 泛型实现。
3. 修改例 8-3 中的 addBookMap.jsp 文件，将表单输入元素的命名改为形如 books['key'] 的形式。
4. 修改例 8-5，使用全局注册的自定义类型转换器转换出版日期和数量。

第9章 Struts 2框架的拦截器

本章首先介绍了拦截器的工作过程和拦截器的使用方法,然后给出了一个自定义拦截器的示例,最后介绍了 Struts 2 框架内置的拦截器和拦截器栈。

9.1 拦截器概述

拦截器是 Struts 2 框架的核心内容之一。Struts 2 框架的许多核心功能都是通过拦截器来实现的,例如避免表单重复提交、数据类型转换、数据校验、文件上传、国际化及权限管理等。Struts 2 框架提供了很多内置的拦截器和拦截器栈,可以满足绝大多数 Web 应用的开发。

Struts 2 框架中的拦截器是面向切面编程(Aspect-Oriented Programming,AOP)设计思想的实现。

9.1.1 AOP

AOP 是一种设计思想,和具体的实现技术无关,只要是符合 AOP 思想的技术实现,都可以看做是 AOP 的实现,Struts 2 框架中的拦截器就是这种思想的具体实现。

AOP 是对面向对象编程(Object Oriented Programming,OOP)的补充。OOP 是当前软件开发的主要模式之一,OOP 借助于类、继承、封装及多态性等概念建立了对象的层次结构,极大提高了代码的复用性。但是,OOP 仍有不足之处,例如在解决事务处理、异常处理、日志及权限管理等方面的问题时,系统中会产生大量的重复代码。这是因为 OOP 思想主要是处理对象间的从上到下的纵向关系,使得事务处理、异常处理等这些技术实现代码分散在各个业务处理实现代码中。OOP 技术如图 9-1 所示。

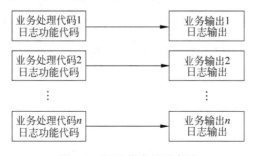

图 9-1 OOP 技术的示意图

AOP 的设计思想是把软件分成两部分,即核心关注点和横切关注点。其中和业务处理操作相关的功能是核心关注点,而与之没有紧密关系的功能则是横切关注点,例如事务处理、日志等功能。横切关注点的一个显著特点是,它们有可能发生在核心关注点的多个位置,并且各个位置的代码都基本相似。AOP 设计思想是处理对象之间的横向关系,它将 OOP 中混在一起的业务处理代码和功能代码分离出来,这样既减少了代码的重复,又降低模块之间的耦合,还增强了系统的可维护性和可读性。AOP 设计思想如图 9-2 所示。

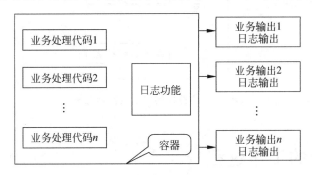

图 9-2　AOP 设计思想的示意图

通过对比图 9-1 和图 9-2,可以看到,在 OOP 中,在写每一个业务处理代码时,都要在其中混有辅助的功能代码,而在 AOP 中,将业务处理代码和辅助的功能代码分离开来,开发人员可以将注意力集中在业务处理代码中。图 9-3 是不使用和使用 AOP 的示意图。

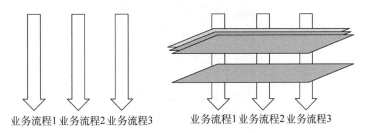

图 9-3　不使用和使用 AOP 的示意图

图 9-3 中左侧是没有使用 AOP 的示意图,在每一个业务流程中,都混有辅助的功能代码,系统越大,代码的冗余就越大。图 9-3 中右侧是使用 AOP 的示意图,将每一个业务流程中的辅助功能代码都分离了出来,灰色横切面是辅助功能代码的实现,如果业务流程需要使用辅助功能代码,只需横穿(即在配置文件中进行配置)横切面即可。

9.1.2　拦截器的作用

Struts 2 框架在调用 action 处理用户请求的前后,通过调用拦截器来完成某些操作,可以实现系统代码之间的解耦,有利于代码的维护和重用。

Struts 2 框架中的拦截器都是可插拔的,这种可插拔性是通过配置文件来完成的。如果需要使用某项功能的时候,只需在配置文件中配置对应的拦截器即可;如果不需要某项功能的时候,只需从配置文件中去除相应的拦截器配置即可。通过这种配置方式,开发者可以按照任务需求来组装拦截器,为指定的 Action 提供所需的功能,而无需修改 Action 代码。因此，开发人员所要做的就是决定一个 Action 需要使用哪些拦截器，然后在配置文件中为这个 Action 配置需要的拦截器。图 9-4 描述了拦截器的可插拔性。

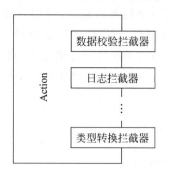

图 9-4　拦截器的可插拔性示意图

9.2　拦截器的工作过程

图 9-5 描述了 Action 和拦截器之间的关系。

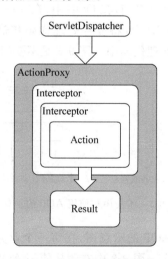

图 9-5　Action 和拦截器的关系

由图 9-5 可以看出，在 Action 的执行前后，都要调用为其配置的拦截器。当 Action 执行完成后，拦截器被调用的顺序和 Action 执行前被调用的顺序相反。

9.3　拦截器的使用方法

拦截器的使用通常有以下几个步骤。
（1）创建拦截器类；
（2）注册拦截器；
（3）使用拦截器。

9.3.1 创建拦截器类

创建拦截器类需要实现 Interceptor(com.opensymphony.xwork2.interceptor.Interceptor) 接口。Interceptor 接口的定义如下：

Interceptor.java

```
package com.opensymphony.xwork2.interceptor;
import com.opensymphony.xwork2.ActionInvocation;
import java.io.Serializable;
public interface Interceptor extends Serializable {
    void destroy();
    void init();
    String intercept(ActionInvocation invocation) throws Exception;
}
```

Interceptor 接口的定义中提供了 init()、destroy() 和 intercept() 三种方法。其中：

(1) init() 方法用于初始化，是在拦截器对象创建之后首先执行的方法。因此，可以在该方法中为拦截器分配所需的资源。

(2) destroy() 方法用于清除分配给拦截器的所有资源，在拦截器对象被销毁之前调用该方法。

(3) intercept() 方法是写拦截器代码的地方，用于拦截用户的请求。就像 action 中的方法一样，intercept() 方法返回一个字符串类型结果码(result)，Struts 2 框架借助于这个结果码将请求转发到其他视图资源。在 intercept() 方法中，ActionInvocation 类型的参数包含了被拦截的 action 的引用，通过调用 ActionInvocation 类型参数的 invoke 方法，将执行 action(如果该拦截器是拦截器栈中的最后一个拦截器)或者下一个拦截器。

通过实现 Interceptor 接口的方式创建拦截器类，开发者需要实现 init()、destroy() 和 intercept() 三个方法。如果在创建的拦截器类中不用实现 init() 和 destroy() 两个方法，则可以通过继承 AbstractInterceptor(com.opensymphony.xwork2.interceptor.AbstractInterceptor) 抽象类的方式创建拦截器类。AbstractInterceptor 是 Struts 2 框架提供的一个实现了 Interceptor 接口的抽象类，该抽象类提供了 init() 和 destroy() 两个方法的空实现。

AbstractInterceptor 抽象类的定义如下：

AbstractInterceptor.java

```
package com.opensymphony.xwork2.interceptor;
import com.opensymphony.xwork2.ActionInvocation;
public abstract class AbstractInterceptor implements Interceptor {
    public void init() {  }
    public void destroy() {  }
    public abstract String intercept(ActionInvocation invocation) throws Exception;
}
```

通过继承 AbstractInterceptor 抽象类的方式创建拦截器类时，只需实现 intercept() 方法即可。

9.3.2 注册拦截器

注册拦截器分为注册拦截器和注册拦截器栈。

1. 注册拦截器

注册拦截器又称为定义拦截器,是通过在配置文件中使用<interceptors>元素和内嵌的<interceptor>子元素在包中实现的。

拦截器的定义格式如下:

```
<package ... >
    <interceptors>
        <interceptor name="interceptorName" class="InterceptorClass"/>
        ...
    </interceptors>
</package>
```

其中:interceptorName 是拦截器注册的名称,在同一个 package 中要唯一。InterceptorClass 是创建的拦截器的完整实现类名称。

2. 注册拦截器栈

另外,在配置文件中,不同的 action 可能需要使用多个相同的拦截器。为了避免代码重复,可以将这些拦截器定义成拦截器栈,当定义 action 时,只需使用定义的拦截器栈。

拦截器栈的定义是通过使用<interceptor-stack>元素和内嵌的<interceptor-ref>子元素在包的<interceptors>元素中实现的。<interceptor-ref>子元素用于指明拦截器栈中所要使用的已经注册的拦截器。拦截器栈中拦截器的顺序很重要,它会影响 action 的行为。

拦截器栈的定义格式如下:

```
<package ... >
    <interceptors>
        <interceptor name="interceptorName" class="InterceptorClass"/>
        ...
        <interceptor-stack name="interceptorStackName">
            <interceptor-ref name="interceptorName"/>
            ...
        </interceptor-stack>
    </interceptors>
</package>
```

其中:interceptorStackName 是拦截器栈注册的名称,在同一个 package 中要唯一。

9.3.3 使用拦截器

使用拦截器是通过在 action 的定义中使用<interceptor-ref …/>元素来实现的。其格式如下：

```
<package ... >
    <action name = "actionName" class = "ActionClass">
        <interceptor-ref name = "interceptorName"/>
        ...
    </action>
</package>
```

<interceptor-ref …/>元素的 name 属性值为已经注册的拦截器或拦截器栈。如果要为拦截器指定参数，则可以在<interceptor-ref …/>元素中内嵌<param …/>子元素来实现。格式如下：

```
<package ...>
    <action name = "actionName" class = "ActionClass">
        <interceptor-ref name = "interceptorName">
            <param name = "paramName">paramValue</param>
            ...
        <interceptor-ref name = "interceptorName">
        ...
    </action>
</package>
```

9.4 自定义拦截器示例

通过自定义拦截器示例学习如何使用拦截器，并对拦截器的工作过程进行演示。

9.4.1 拦截器工作过程示例

例 9-1 单个拦截器示例。

具体操作步骤如下：

（1）创建 Web 项目 Chapter9，并添加 Struts 2 框架的支持。Chapter9 项目目录结构参考 Chapter8 项目的目录结构，web.xml 配置文件的内容同 Chapter8 项目中 web.xml 配置文件的内容。

（2）创建 Action 类。

HelloWorld.java

```java
package example.struts2;
import com.opensymphony.xwork2.ActionSupport;
public class HelloWorld extends ActionSupport{
    private String message;
    public String getMessage() {
        return message;
    }
```

```java
    public void setMessage(String message) {
        this.message = message;
    }
    public String execute(){
        addActionMessage("HelloWorld! Message is: " + message);
        System.out.println("HelloWorld! Message is: " + message);
        return SUCCESS;
    }
}
```

（3）创建拦截器。

MyInterceptor1.java

```java
package example.struts2;
import com.opensymphony.xwork2.ActionInvocation;
import com.opensymphony.xwork2.interceptor.AbstractInterceptor;
public class MyInterceptor1 extends AbstractInterceptor {
    public String intercept(ActionInvocation invocation) throws Exception {
        HelloWorld hw = (HelloWorld)invocation.getAction();
        hw.addActionMessage("Calling MyInterceptor1");
        System.out.println("Calling MyInterceptor1");
        hw.addActionMessage("MyInterceptor1-1:ActionClass name is: " +
                                                hw.getClass().getName());
        System.out.println("MyInterceptor1-1:ActionClass name is: " +
                                                hw.getClass().getName());
        String result = invocation.invoke();
        hw.addActionMessage("MyInterceptor1-2:The result is: " + result);
        System.out.println("MyInterceptor1-2:The result is: " + result);
        return result;
    }
}
```

（4）注册并使用拦截器。

struts.xml

```xml
<?xml version="1.0" encoding="UTF-8" ?>
<!DOCTYPE struts PUBLIC
    "-//Apache Software Foundation//DTD Struts Configuration 2.0//EN"
    "http://struts.apache.org/dtds/struts-2.0.dtd">
<struts>
    <package name="default" extends="struts-default">
        <interceptors>
            <interceptor name="myInterceptor1"
                    class="example.struts2.MyInterceptor1"/>
        </interceptors>
        <action name="helloWorld" class="example.struts2.HelloWorld">
            <result>/inputMessage.jsp</result>
            <interceptor-ref name="defaultStack"/>
```

```
            <interceptor-ref name="myInterceptor1"/>
        </action>
    </package>
</struts>
```

(5) 创建 JSP 页面。

inputMessage.jsp

```
<%@ page language="java" contentType="text/html; charset=UTF-8"
    pageEncoding="UTF-8"%>
<%@ taglib prefix="s" uri="/struts-tags" %>
<s:actionmessage/>
<s:form action="helloWorld">
    <s:textfield name="message" label="Message"/>
    <s:submit name="submit"/>
</s:form>
```

(6) 发布 Web 项目,启动 Tomcat 服务器,访问 inputMessage.jsp。单击 Submit 按钮后,浏览器的输出结果如图 9-6 所示,控制台的输出结果如图 9-7 所示。

图 9-6　浏览器的输出结果

```
Calling MyInterceptor1
MyInterceptor1-1:ActionClass name is: example.struts2.HelloWorld
HelloWorld! Message is:拦截器
MyInterceptor1-2:The result is: success
```

图 9-7　控制台的输出结果

比较图 9-6 和图 9-7 可以看出:在图 9-6 中没有输出"MyInterceptor1-2:The result is: success"信息,而在图 9-7 中输出了这条信息。这说明当 action 执行结束后,根据结果码调用相应的 Web 资源,并将结果码返回给拦截器。

(7) 编辑 Action 类(HelloWorld.java),修改返回语句中的返回值,将 SUCCESS 改为 ERROR,或者其他的在配置文件中没有指定的值。如下:

```
return ERROR;
```

(8) 编辑拦截器类(MyInterceptor1.java),修改返回语句中的返回值,将 result 改为"success"。如下:

```
return "success";
```

（9）重新启动 Tomcat 服务器，访问 inputMessage.jsp。单击 Submit 按钮后，浏览器的输出结果如图 9-8 所示，控制台的输出结果如图 9-9 所示。

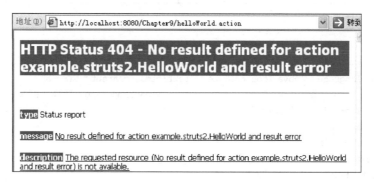

图 9-8　浏览器的输出结果

```
Calling MyInterceptor1
MyInterceptor1-1:ActionClass name is: example.struts2.HelloWorld
HelloWorld! Message is: 拦截器
2011-10-15 11:22:46 com.opensymphony.xwork2.util.logging.jdk.JdkLogger warn
警告: Could not find action or result
No result defined for action example.struts2.HelloWorld and result error
```

图 9-9　控制台的输出结果

比较图 9-8 和图 9-9 可以看出：当 action 执行结束后，根据结果码调用相应的 Web 资源，而不是在拦截器执行结束后，再根据返回值调用相应的 Web 资源。

例 9-2　多个拦截器示例。

在例 9-1 的基础上，使用两个拦截器进行讲解。具体操作步骤如下：

（1）创建拦截器。

MyInterceptor2.java

```java
package example.struts2;
import com.opensymphony.xwork2.ActionInvocation;
import com.opensymphony.xwork2.interceptor.AbstractInterceptor;
public class MyInterceptor2 extends AbstractInterceptor {
    public String intercept(ActionInvocation invocation) throws Exception {
        HelloWorld hw = (HelloWorld)invocation.getAction();
        hw.addActionMessage("Calling MyInterceptor1");
        System.out.println("Calling MyInterceptor2");
        hw.addActionMessage("MyInterceptor2-1:ActionClass name is: " +
                            hw.getClass().getName());
        System.out.println("MyInterceptor2-1:ActionClass name is: " +
                            hw.getClass().getName());
        String result = invocation.invoke();
        hw.addActionMessage("MyInterceptor2-2:The result is: " + result);
        System.out.println("MyInterceptor2-2:The result is: " + result);
```

```
            return result;
        }
}
```

（2）注册并使用拦截器。

struts.xml

```xml
<?xml version = "1.0" encoding = "UTF - 8" ?>
<!DOCTYPE struts PUBLIC
    " - //Apache Software Foundation//DTD Struts Configuration 2.0//EN"
    "http://struts.apache.org/dtds/struts - 2.0.dtd">

<struts>
    <package name = "default" extends = "struts - default">
        <interceptors>
            <interceptor name = "myInterceptor1"
                         class = "example.struts2.MyInterceptor1"/>
            <interceptor name = "myInterceptor2"
                         class = "example.struts2.MyInterceptor2"/>
        </interceptors>
        <action name = "helloWorld" class = "example.struts2.HelloWorld">
            <result>/inputMessage.jsp</result>
            <interceptor - ref name = "defaultStack"/>
            <interceptor - ref name = "myInterceptor1"/>
            <interceptor - ref name = "myInterceptor2"/>
        </action>
    </package>
</struts>
```

（3）重新启动 Tomcat 服务器，访问 inputMessage.jsp。单击 Submit 按钮后，浏览器的输出结果如图 9-10 所示，控制台的输出结果如图 9-11 所示。

图 9-10　浏览器的输出结果

9.4.2　登录示例

在 Web 应用中，一些资源的访问只对登录用户开放，如果用户采用匿名的方式访问这些资源，则系统将会提示用户登录。下面给出具体示例。

```
Calling MyInterceptor1
MyInterceptor1-1:ActionClass name is: example.struts2.HelloWorld
Calling MyInterceptor2
MyInterceptor2-1:ActionClass name is: example.struts2.HelloWorld
HelloWorld! Message is: 拦截器
MyInterceptor2-2:The result is: success
MyInterceptor1-2:The result is: success
```

图 9-11　控制台的输出结果

例 9-3　当用户访问 action 时，首先进行是否已经登录的检查（除登录用的 action 外）。如果用户已经登录，则可以访问这些 action 资源，否则返回到登录页面。

具体操作步骤如下：

(1) 创建 Action 类。

Login.java

```java
package example.struts2;
import java.util.Map;
import org.apache.struts2.interceptor.SessionAware;
import com.opensymphony.xwork2.ActionSupport;
public class Login extends ActionSupport implements SessionAware{
    private String message;
    private Map mySession;
    public void setSession(Map<String, Object> mySession) {
        this.mySession = mySession;
    }
    public String getMessage() {
        return message;
    }
    public void setMessage(String message) {
        this.message = message;
    }
    public String execute(){
        if((message == null) || message.isEmpty()){
            addActionError("请输入信息后提交!");
            return INPUT;
        }
        else{
            addActionMessage("您输入的信息是: " + message);
            mySession.put("loginFlag", "login");
            return SUCCESS;
        }
    }
}
```

Login.java 实现了 SessionAware 接口，当用户输入信息后，将 login 保存到 session 中。Login.java 要实现 SessionAware 接口中的 setSession 方法。

（2）创建拦截器。

LoginStatus.java

```java
package example.struts2;
import java.util.Map;
import com.opensymphony.xwork2.ActionInvocation;
import com.opensymphony.xwork2.ActionSupport;
import com.opensymphony.xwork2.interceptor.AbstractInterceptor;
public class LoginStatus extends AbstractInterceptor {
    public String intercept(ActionInvocation invocation) throws Exception {
        Object action = invocation.getAction();
        if (action instanceof Login){
            return invocation.invoke();
        }
        Map session = invocation.getInvocationContext().getSession();
        String login = (String)session.get("loginFlag");
        if (login != null && login.equals("login")){
            return invocation.invoke();
        }
        else{
            ((ActionSupport) action).addActionMessage("来自拦截器的消息：您还没有登录，请先登录!");
            return "input";
        }
    }
}
```

在该拦截器类中，对访问的action进行了不同的处理。如果访问的action是Login，则调用invoke()方法；如果访问的action不是Login，则要首先判断用于是否曾经输入信息，若是，则调用invoke()方法，否则返回输入页面。

（3）注册并使用拦截器。

struts.xml

```xml
<?xml version = "1.0" encoding = "UTF-8" ?>
<!DOCTYPE struts PUBLIC
    "-//Apache Software Foundation//DTD Struts Configuration 2.0//EN"
    "http://struts.apache.org/dtds/struts-2.0.dtd">
<struts>
    <package name = "default" extends = "struts-default">
        <interceptors>
            <interceptor name = "loginStatus" class = "example.struts2.LoginStatus"/>
        </interceptors>
        <action name = "login" class = "example.struts2.Login">
            <result>/login.jsp</result>
            <result name = "input">/login.jsp</result>
            <interceptor-ref name = "defaultStack"/>
            <interceptor-ref name = "loginStatus"/>
```

```xml
        </action>
        <action name="others">
            <result>/others.jsp</result>
            <result name="input">/login.jsp</result>
            <interceptor-ref name="defaultStack"/>
            <interceptor-ref name="loginStatus"/>
        </action>
    </package>
</struts>
```

由于在创建的拦截器类中，对访问的 action 进行了拦截，为了测试拦截器是否生效，在配置文件中，定义了"login"和"others"两个 action。这样可以观察访问不同的 action 时的输出结果。

（4）创建 JSP 文件。

① 输入页面。

login.jsp

```jsp
<%@page language="java" contentType="text/html; charset=UTF-8" pageEncoding="UTF-8"%>
<%@taglib prefix="s" uri="/struts-tags" %>
<s:actionmessage/>
<s:actionerror/>
<s:form action="login">
    <s:textfield name="message" label="Message"/>
    <s:submit name="submit"/>
</s:form>
<s:a action="others">单击此处，访问 others Action!</s:a>
```

② 访问其他 action 成功后显示的页面。

others.jsp

```jsp
<%@page language="java" contentType="text/html; charset=UTF-8" pageEncoding="UTF-8"%>
<%@taglib prefix="s" uri="/struts-tags" %>
到达此页面，表示您是在输入信息后进行的提交。<br>
<s:a href="login.jsp">返回到输入页面!</s:a>
```

（5）发布 Web 项目，重新启动 Tomcat 服务器，访问 login.jsp，结果如图 9-12 所示。

① 单击图 9-12 中的链接，结果如图 9-13 所示。

图 9-12　访问 login.jsp 的输出结果　　　　图 9-13　单击图 9-12 中链接的输出结果

② 单击图 9-12 中的 Submit 按钮,结果如图 9-14 所示。

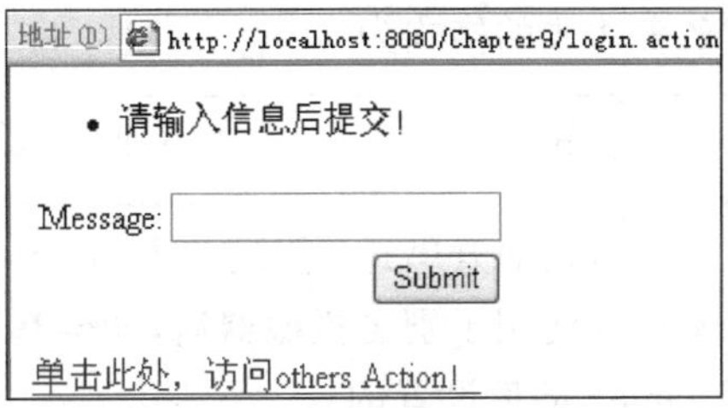

从图 9-13 和图 9-14 可以看出,拦截器对"login"和"others"两个 action 的处理方式是不同的。

③ 输入信息,单击 Submit 按钮,结果如图 9-15 所示。

④ 单击图 9-15 中的链接,结果如图 9-16 所示。

图 9-14 单击图 9-12 中"Submit"按钮的输出结果

图 9-15 输入信息后的输出结果

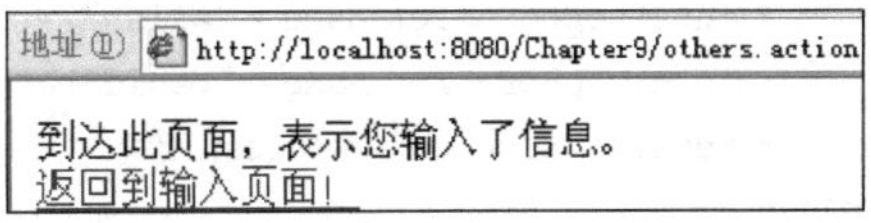

图 9-16 单击图 9-15 中链接的输出结果

这时,用户就可以访问其他 action 了。

9.5 Struts 2 框架的内置拦截器

在 Web 应用开发中使用自定义拦截器时,需要遵循 9.3 节中的三个步骤。为了方便开发,Struts 2 框架提供了很多的内置拦截器,这些拦截器在 struts-default.xml 配置文件中注册,开发人员可以直接使用。也就是说,如果在 Web 应用开发中使用 Struts 2 框架的内置拦截器时,则只需直接使用步骤(3)即可。

Struts 2 框架在 struts-default.xml 配置文件中注册了一些拦截器和拦截器栈,开发者在使用内置的拦截器和拦截器栈时,只需在 struts.xml 配置文件中定义包时,继承 struts-default.xml 文件中定义的 struts-default 抽象包即可,如下:

```
<struts>
    <package name = "packageName" extends = "struts-default">
        ...
    </package>
</struts>
```

这样,在 package 中定义 action 时,就可以使用 Struts 2 框架的内置拦截器和拦截器栈了。

9.5.1 内置的拦截器

1. Alias 拦截器(alias)

别名拦截器,用于在共享相似参数(名称不同)的 action 之间进行名称转换。Action 的别名表达式的格式为：

```
#{ "name1" : "alias1", "name2" : "alias2" }
```

表达式的意思是：假设一个 action 有一个值用于设置名称为 name1 的表达式，那么这个 action 使用了别名拦截器后，就会有一个名称为 alias1 的 setter 方法，并且 alias1 的值是用 name1 的值设置的。

为应用别名拦截器的 action 传递参数时，< param >元素的 name 属性的默认值是 aliases。

例 9-4　Alias 拦截器示例。

```
<action name = "actionName" class = "...">
    <param name = "aliases">#{ 'foo' : 'bar' }</param>
    <interceptor-ref name = "alias"/>
    <interceptor-ref name = "basicStack"/>
    <result>result.jsp</result>
</action>
```

通过使用别名拦截器，foo 参数的值就好像是名称为 bar 的值。

2. ActionAutowiring 拦截器（autowiring）

用于把 action 类自动装配到 Spring beans 中。

3. Chaining 拦截器（chain）

用于将值栈中的每一个对象的属性复制到当前执行的对象中，除了实现 Unchainable 接口的对象。通常是和<result type＝"chain"＞（在前一个 Action 中）一起使用。可选的 includes 和 excludes 参数用于控制如何以及复制哪一个参数。

例 9-5　Chaining 拦截器示例。

```
<action name = "someAction" class = "...">
    <interceptor-ref name = "basicStack"/>
    <result type = "chain">otherAction</result>
</action>
<action name = "otherAction" class = "...">
    <interceptor-ref name = "chain"/>
    <interceptor-ref name = "basicStack"/>
    <result>result.jsp</result>
</action>
```

4. StrutsConversionError 拦截器（conversionError）

用于将 ActionContext 中的转换错误信息添加到字段错误中（假设 action 实现了 ValidationAware 接口）。另外，任何包含校验错误的字段都将原值进行了保存，这样，之后的对该值的请求就可以得到原值。例如：如果提交的字符串"abc"不能转换为整型，却又想再次显示原字符串"abc"而不是整型值（例如 0），这一点则非常重要。

例 9-6 conversionError 拦截器示例。

```
<action name = "someAction" class = "...">
    <interceptor-ref name = "params"/>
    <interceptor-ref name = "conversionError"/>
    <result>result.jsp</result>
</action>
```

5. Cookie 拦截器(cookie)

基于 cookie 的名称/值,设置 stack/action 中的值。

Cookie 拦截器有以下两个必选参数。

(1) cookiesName:用于指明注入 action 中的 cookies 的名称;如果需要多个 cookies 的名称,则用逗号分隔。如果需要所有 cookies 的名称,可以使用"*"。

(2) cookiesValue:如果 cookies 的名称匹配 cookiesName 属性,并且值也匹配,则 cookies 的值被注入到 Struts 2 框架的 action 中。

如果 cookiesName 为空,则将认为没有 cookie 被注入 Struts 2 框架的 action 中。

如果 cookiesValue 为空,则认为所有的和 cookiesName 参数指定的 cookie 名称(不考虑它们的值)匹配的 cookies 的名称都将被注入 Struts 2 框架的 action 中。

例 9-7 cookie 拦截器示例。

(1) 将名称为 cookie1 或 cookie2 的 cookies 的值"cookie1value"或"cookie2value"注入 Struts 的 action 中。

```
<action name = "someAction" class = "...">
    <interceptor-ref name = "cookie">
        <param name = "cookiesName">cookie1, cookie2</param>
        <param name = "cookiesValue">cookie1value, cookie2value</param>
    </interceptor-ref>
</action>
```

(2) 将名称为 cookie1 或 cookie2 的 cookies 的值(无论何值)注入 Struts 2 框架的 action 中。

```
<action name = "someAction" class = "......">
    <interceptor-ref name = "cookie">
        <param name = "cookiesName">cookie1, cookie2</param>
        <param name = "cookiesValue">*</param>
    <interceptor-ref>
</action>
```

6. ClearSession 拦截器(clearSession)

用于清除 HttpSession 对象。

7. CreateSession 拦截器(createSession)

用于自动创建一个 HttpSession 对象,对于某些需要 HttpSession 对象才正常工作的拦截器(如 TokenInterceptor)非常有用。

例 9-8 CreateSession 拦截器示例。

```
<action name="someAction" class="......">
    <interceptor-ref name="createSession"/>
    <interceptor-ref name="defaultStack"/>
    <result name="input">input_with_token_tag.jsp</result>
</action>
```

当在 JSP 模板中使用了 token 标签时,需使用 CreateSession 拦截器。

8. Debugging 拦截器(debugging)

用于提供多个不同的调试窗口来观察页面中的内部数据。该拦截器只有在 struts.properties 文件中的 devMode 属性设为 true 时,才被激活。

当为 action 应用了 Debugging 拦截器后,其使用格式如下:

```
http://localhost:8080/debugging.action?debug=xml
```

debug 参数有 4 个可选值,分别如下。

(1) xml:将参数、上下文、session 和值栈转储为 XML 文档。

(2) console:弹出一个 OGNL 控制台窗口,允许用户基于值栈测试 OGNL 表达式。

(3) command:测试 OGNL 表达式并返回字符串结果码,只在 OGNL 控制台中使用。

(4) browser:显示在对象参数(默认值为♯context)中指定的对象字段的值。

9. ExecuteAndWait 拦截器(execAndWait)

在后台执行 Action,同时立即发送给用户一个等待页面,这对于执行时间很长的 action 很有用,可以防止 HTTP 请求超时。

ExecuteAndWait 拦截器有以下三个可选的参数。

(1) threadPriority:指定线程的优先级,默认值是 Thread.NORM_PRIORITY。

(2) delay:显示等待页面的初始延迟时间(单位毫秒),默认值是不等待,即返回 wait 结果码就显示等待页面。

(3) delaySleepInterval:和 delay 参数一起使用。用于设置检查后台进程是否执行完成的时间间隔,默认值是 100 毫秒。

例 9-9 ExecuteAndWait 拦截器示例。

(1) 没有使用参数的 ExecuteAndWait 拦截器配置。

```
<action name="someAction" class="…">
    <interceptor-ref name="completeStack"/>
    <interceptor-ref name="execAndWait"/>
    <result name="wait">wait.jsp</result>
    <result>success.jsp</result>
</action>
```

（2）使用参数的 ExecuteAndWait 拦截器配置。

```
<action name="someAction" class="...">
    <interceptor-ref name="completeStack"/>
    <interceptor-ref name="execAndWait">
        <param name="delay">2000<param>
        <param name="delaySleepInterval">100<param>
    <interceptor-ref>
    <result name="wait">wait.jsp</result>
    <result>success.jsp</result>
</action>
```

10．ExceptionMapping 拦截器（exception）

ExceptionMapping 拦截器具有异常处理特性的核心功能。异常处理允许开发人员将异常映射到结果码，就像 action 返回结果码一样，而不是抛出一个无法预料的异常。建议将 ExceptionMapping 拦截器设置为拦截器栈中的第一个拦截器，这样能够确保捕捉任何异常，即使是由其他拦截器抛出的异常。

ExceptionMapping 拦截器有以下三个可选参数。

（1）logEnabled：设置异常是否被记录到日志中（true 或 false）。

（2）logLevel：设置使用哪种日志级别（trace，debug，info，warn，error，fatal），默认值是 debug。

（3）logCategory：设置记录日志的类，默认使用 com.opensymphony.xwork2.interceptor.ExceptionMappingInterceptor 类。

例 9-10　ExceptionMapping 拦截器示例。

```
<package name="default" extends="struts-default">
    <global-exception-mappings>
        <exception-mapping exception="java.lang.Exception" result="error"/>
    </global-exception-mappings>
    <action name="someAction">
        <interceptor-ref name="exception"/>
        <interceptor-ref name="basicStack"/>
        <exception-mapping exception="example.struts2.CustomException"
                                                result="customError"/>
        <result name="customError">customError.jsp</result>
        <result name="success">success.jsp</result>
    </action>
</package>
```

11．FileUpload 拦截器（fileUpload）

FileUpload 拦截器基于 MultiPartRequestWrapper，能够自动应用于任何包含文件的请求。添加如下的参数，其中[File Name]是 HTML 表单中上传文件表单元素的指定名称：

- [File Name]：实际的文件，File 类型。

- [File Name]ContentType：文件的内容类型，字符串类型。
- [File Name]FileName：上传文件的实际名称(不是 HTML 表单元素的名称)，字符串类型。

只要通过 Action 中提供的和上面三个参数对应的 setters 方法就可以访问这些文件，例如 setDocument(File document)，setDocumentContentType(String contentType)等，其中 document 是 HTML 表单元素的名称。

如果 Action 实现了 ValidationAware 接口,该拦截器将会增加几个字段错误。这些错误消息是基于 struts-messages.properties 属性文件中的几个 i18n 值，该属性文件是所有的 i18n 请求都要处理的一个默认的 i18n 文件。开发人员可以改写这些键所对应的消息：

- struts.messages.error.uploading：当文件上传失败时显示的通用错误消息。
- struts.messages.error.file.too.large：当上传的文件太大时显示的错误消息。
- struts.messages.error.content.type.not.allowed：当上传的文件不是指定的内容类型时显示的错误信息。
- struts.messages.error.file.extension.not.allowed：当上传的文件不是指定的扩展名时显示的错误信息。

FileUpload 拦截器有以下三个可选参数。

- maximumSize：拦截器允许 action 中处理的上传文件的最大字节数，和 struts.properties 文件中设置允许最大上传文件字节数的属性不同。默认值是 2MB。
- allowedTypes：拦截器允许 action 中处理的上传文件的内容类型列表(用逗号分隔)，如果没有指定，则允许上传任意类型的文件。
- allowedExtensions：拦截器允许 action 中处理的上传文件的扩展名列表(用逗号分隔)，如果没有指定，则允许上传任意扩展名的文件。

例 9-11 FileUpload 拦截器示例。

(1) 创建 JSP 文件。

```
<s:form action="uploadFile" method="post" enctype="multipart/form-data">
    <s:file name="upload" label="File"/>
    <s:submit/>
</s:form>
```

(2) 创建 Action。

```
public UploadAction extends ActionSupport {
    private File file;
    private String contentType;
    private String filename;
    public void setUpload(File file) {
        this.file = file;
    }
    public void setUploadContentType(String contentType) {
        this.contentType = contentType;
    }
```

```
        public void setUploadFileName(String filename) {
            this.filename = filename;
        }
        public String execute() {
            …
            return SUCCESS;
        }
    }
```

(3) 配置拦截器。

```
<action name = "upload" class = "…">
    <interceptor-ref name = "fileUpload">
        <param name = "allowedTypes">image/png,image/gif,image/jpeg</param>
    </interceptor-ref>
    <interceptor-ref name = "basicStack"/>
    <result>result.jsp</result>
</action>
```

12. I18n 拦截器（i18n）

将 session 中指定 locale 设置为当前 action 请求的 locale。另外，该拦截器将会查找一个具体的 HTTP 请求参数，并将 locale 设置为参数提供的值。对要求多语言（国际化）支持的应用非常有用。

I18n 拦截器有以下三个可选参数。

- parameterName：HTTP 请求参数的名称，该参数规定了切换到的 locale，并保存到 session 中。参数的默认值是 request_locale。
- requestOnlyParameterName：HTTP 请求参数的名称，该参数仅规定了当前请求切换到的 locale，不保存到 session 中。参数的默认值是 request_only_locale。
- attributeName：session 中保存 locale 的键的名称，参数的默认值是 WW_TRANS_I18N_LOCALE。

例 9-12 FileUpload 拦截器示例。

```
<action name = "someAction" class = "…">
    <interceptor-ref name = "i18n"/>
    <interceptor-ref name = "basicStack"/>
    <result>result.jsp</result>
</action>
```

13. Logging 拦截器（logger）

用于记录并输出一个 action 执行的开始和结束的信息，在 INTO 级别进行记录。只有英文信息，没有国际化的信息。

例 9-13 Logging 拦截器示例。

(1) 在 action 执行前后输出信息。

```
<action name="someAction" class="...">
    <interceptor-ref name="completeStack"/>
    <interceptor-ref name="logger"/>
    <result>result.jsp</result>
</action>
```

(2) 输出拦截器开始执行前后的信息。

```
<action name="someAction" class="...">
    <interceptor-ref name="logger"/>
    <interceptor-ref name="completeStack"/>
    <result>result.jsp</result>
</action>
```

该示例说明拦截器配置的顺序非常重要。

14. ModelDriven 拦截器（modelDriven）

监视实现 ModelDriven 接口的 action，并将 action 的非空模型（getModel 的结果）添加到值栈中。另外，如果模型能够使用参数，则必须在 StaticParameters 和 Parameters 两个拦截器前使用 ModelDriven 拦截器。

ModelDriven 拦截器有如下一个可选参数。

refreshModelBeforeResult：如果在 action 执行之后 result 执行之前刷新值栈中的模型，则将该参数设置为 true，这将导致至少调用两次 getModel() 方法。

例 9-14 ModelDriven 拦截器示例。

```
<action name="someAction" class="...">
    <interceptor-ref name="modelDriven"/>
    <interceptor-ref name="basicStack"/>
    <result>result.jsp</result>
</action>
```

15. ScopedModelDriven 拦截器（scopedModelDriven）

支持作用域范围的模型驱动的 action，该拦截器只有在 Action 实现了 ScopedModelDriven 接口时才被激活，如果检测到，将会从配置的作用域范围检索模型类，然后提供给 Action。

ScopedModelDriven 拦截器有如下三个参数。

- className：模型类的名称，默认值是 getModel() 方法返回的对象的类名称。
- name：存储或检索作用域范围内的实例时使用的 key。默认值是模型类的名称。
- scope：存储和检索模型的作用域范围，默认值是"request"。

例 9-15 ScopedModelDriven 拦截器示例。

（1）基本用法。

```
<action name = "someAction" class = "...">
    <interceptor - ref name = "scopedModelDriven"/>
    <interceptor - ref name = "basicStack"/>
    <result>result.jsp</result>
</action>
```

（2）使用参数的用法。

```
<action name = "upload" class = "...">
    <interceptor - ref name = "scopedModelDriven"/>
        <param name = "scope">session</param>
    </interceptor - ref>
    <interceptor - ref name = "basicStack"/>
    <result>result.jsp</result>
</action>
```

16. Parameters 拦截器（params）

将请求参数设置到 Action 中，即设置值栈中所有的参数。该拦截器有两个可选参数。
- ordered：如果要使用 top-down 属性，则将该参数值设置为 true。
- excludeParams：通过设置该参数，可以强制拦截器忽略匹配的参数。参数的值是逗号分隔的正则表达式列表。Struts 2 框架定义的拦截器栈已经忽略了一些参数：
  - dojo\..*：任何名称中包含"dojo."的参数。
  - ^struts\..*：任何名称以"struts."开始的参数。

例 9-16 Parameters 拦截器示例。

```
<action name = "someAction" class = "...">
    <interceptor - ref name = "params"/>
    <result>result.jsp</result>
</action>
```

17. ActionMappingParameters 拦截器（actionMappingParams）

根据 Action 的映射为当前请求设置值栈中的所有参数。ActionMappingParameters 拦截器和 Parameters 拦截器相似，只是参数的来源不同。前者参数来源于 ActionMapping，后者参数来源于 Action 上下文的 getParameters()方法。

18. Prepare 拦截器（prepare）

如果 Action 实现了 Preparable 接口，则可以调用它的 prepare 方法。如果想在 Action 类中实际的 execute()方法执行之前执行一些逻辑，则可以使用 Prepare 拦截器。

Prepare 拦截器有一个可选参数。

alwaysInvokePrepare：默认值是 true。如果值为 true，则 prepare 方法总被激活，否则

不被激活。

例 9-17 Prepare 拦截器示例。

```
<action name = "someAction" class = "...">
    <interceptor-ref name = "params"/>
    <interceptor-ref name = "prepare"/>
    <interceptor-ref name = "basicStack"/>
    <result>result.jsp</result>
</action>
```

调用两次 params 拦截器，一次在 Prepare 拦截器之前，一次在 Prepare 拦截器之后。开发人员通过使用 Prepare 拦截器预加载数据，为第二次参数设置做准备。

19. StaticParameters 拦截器(staticParams)

用于把 action 配置文件中定义的静态参数（即 struts.xml 文件中＜action＞标签的直接子标签＜param＞设置的参数）填充到 action 中。如果 action 实现了 Parameterizable 接口，静态参数的映射也将被直接传递给 action。

例 9-18 StaticParameters 拦截器示例。

```
<action name = "someAction" class = "...">
    <interceptor-ref name = "staticParams"/>
    <result>result.jsp</result>
</action>
```

20. Scope 拦截器(scope)

用于将 Action 的状态保存到 Session 和 Application 中的一种简单机制。Scope 拦截器的可选参数如下：

- session：绑定到 session 作用域中的 action 属性列表。
- application：绑定到 application 作用域中的 action 属性列表。
- key：session/application 属性 key 前缀，可选值有：
    - CLASS：基于 action 的命名空间和 action 类生成一个唯一的 key 前缀。该值为默认值。
    - ACTION：基于 action 的命名空间和 action 名称生成一个唯一的 key 前缀。
    - 任何其他的值作为 key 前缀。
- type：可选值有：
    - start：所有 session 作用域的属性都被重置为默认值。
    - end：在 action 运行后，清除 session 作用域的属性。
    - 任何其他抛出 IllegalArgumentException 异常的值。
- sessionReset：如果设置了参数名称（默认值是"session.reset"），则所有的 session 值都被重置为 action 的默认值或者 application 作用域的值。和 type="start"的作用一样。

- reset：布尔类型，默认值 false，如果设置为 true，和设置所有的 session 值都被重置为 action 的默认值或者 application 作用域的值具有一样的效果。
- autoCreateSession：布尔类型，如果需要自动创建 session，则设置为 true。

例 9-19　Scope 拦截器示例。

（1）使用 session 和 autoCreateSession 参数。

```
<action name = "someAction" class = "……">
    <interceptor-ref name = "basicStack"/>
    <interceptor-ref name = "scope">
        <param name = "session">filter,orderBy</param>
        <param name = "autoCreateSession">true</param>
    </interceptor-ref>
    <result>result.jsp</result>
</action>
```

session 参数给出了 action 属性列表(filter 和 orderBy)，由 Scope 拦截器在 session 作用域内自动管理这些属性。

（2）如果在一个 action(ActionA)中有一个 getter 方法，在另一个 action(ActionB)中有一个 setter 方法，Scope 拦截器可用于从 ActionA 传递任意对象给 ActionB。同时还要设置一个 key 参数。

```
<action name = "scopea" class = "example.struts2.ScopeActionA">
    <result name = "success" type = "dispatcher">/jsp/test.jsp</result>
    <interceptor-ref name = "basicStack"/>
    <interceptor-ref name = "scope">
        <param name = "key">funky</param>
        <param name = "session">person</param>
        <param name = "autoCreateSession">true</param>
    </interceptor-ref>
</action>
<action name = "scopeb" class = "example.struts2.ScopeActionB">
    <result name = "success" type = "dispatcher">/jsp/test.jsp</result>
    <interceptor-ref name = "scope">
        <param name = "key">funky</param>
        <param name = "session">person</param>
        <param name = "autoCreateSession">true</param>
    </interceptor-ref>
    <interceptor-ref name = "basicStack"/>
</action>
```

注意：在 ScopeActionA 中需要一个 getPerson()方法，在 ScopeActionB 中需要一个 setPerson(Person person)方法，并且确保指定 key，如果没有 key，Scope 拦截器将存储变量，但是不会在另一个 Action 中设置它们。

21. ServletConfig 拦截器（servletConfig）

基于 action 实现的接口，设置 action 的属性。例如，如果 action 实现了 ParameterAware 接

口，则 action 上下文的参数 map 将会被设置给 action 的属性。

支持的接口有：ServletContextAware、ServletRequestAware、ServletResponseAware、ParameterAware、RequestAware、SessionAware、ApplicationAware、PrincipalAware。

例 9-20　ServletConfig 拦截器示例。

```
<action name = "someAction" class = "......">
    <interceptor-ref name = "servletConfig"/>
    <interceptor-ref name = "basicStack"/>
    <result>result.jsp</result>
</action>
```

22. Timer 拦截器（timer）

记录 action 执行的总时间（包括内嵌的拦截器和视图），单位为毫秒。为了使该拦截器正常工作，日志框架级别必须设置为至少 INFO 级别。

Timer 拦截器的两个可选参数：
- logLevel：设置使用的日志级别，可选值有 trace、debug、info、warn、error 和 fatal。默认值是 info。
- logCategory：如果不提供，则使用默认的 com.opensymphony.xwork2.interceptor.TimerInterceptor。

例 9-21　Timer 拦截器示例。

（1）仅记录 action 的执行时间。

```
<action name = "someAction" class = "…">
    <interceptor-ref name = "completeStack"/>
    <interceptor-ref name = "timer"/>
    <result>result.jsp</result>
</action>
```

（2）记录 action 和拦截器的执行时间。

```
<action name = "someAction" class = "…">
    <interceptor-ref name = "timer"/>
    <interceptor-ref name = "completeStack"/>
    <result>result.jsp</result>
</action>
```

23. Token 拦截器（token）

确保每个请求只有一个 token 被处理。可以确保后退按钮和双击操作不会引起不必要的负面影响。例如，开发人员可以使用该拦截器防止粗心用户进行在线存储时，可能出现的双击按钮情况。使用该拦截器时，要在表单中使用 token 标签，任何没有提供 token 的请求（使用 token 标签）将被当作一个包含无效 token 的请求处理。

Token 拦截器生成的 action 错误消息使用 struts.messages.invalid.token 进行国

际化。

例 9-22 Token 拦截器示例。

（1）创建 JSP 文件。

```
<s:form action="someAction">
    <s:token/>
    <s:textfield name="message" label="message"/>
    <s:submit/>
</s:form>
```

（2）使用 Token 拦截器。

```
<action name="someAction" class="...">
    <interceptor-ref name="token"/>
    <interceptor-ref name="basicStack"/>
    <result>result.jsp</result>
</action>
```

（3）如果对 action 中的某一个方法（如 someMethod）不进行 token 的有效性检查，则使用 excludeMethods 参数。如下：

```
<action name="someAction" class="...">
    <interceptor-ref name="token">
        <param name="excludeMethods">someMethod</param>
    </interceptor-ref>
    <interceptor-ref name="basicStack"/>
    <result>result.jsp</result>
</action>
```

24．TokenSessionStore 拦截器（tokenSession）

TokenSessionStore 拦截器是 Token 拦截器的升级，为处理无效的 token 提供了高级的逻辑处理。和 Token 拦截器不同，TokenSessionStore 拦截器在使用同一个 session 的多请求事件中，提供智能的故障处理。也就是说，TokenSessionStore 拦截器只处理一次提交，后续的提交将被忽略，这样 action 就能够进行正确处理请求。其使用方式和 Token 拦截器的使用方式一样。

例 9-23 TokenSessionStore 拦截器示例。

```
<action name="someAction" class="...">
    <interceptor-ref name="tokenSession"/>
    <interceptor-ref name="basicStack"/>
    <result>result.jsp</result>
</action>
```

25．AnnotationValidation 拦截器（validation）

使用标准校验框架校验 Action，校验规则在 XML 格式的配置文件中定义，配置文件的名称形如 ActionClass-validation.xml。该校验器通常作为拦截器栈中最后一个或者倒数第二个拦截器使用。

如果被调用的方法名称在 excludeMethods 参数中指定，则该拦截器什么也不做。例如，excludeMethods 参数的值设置为"input，back"。则 foo!input.action 和 foo!back.action 这两个请求将被 AnnotationValidation 拦截器忽略。该拦截器通常和 workflow 拦截器一起使用。

AnnotationValidation 拦截器的可选参数有：

- alwaysInvokeValidate：默认值是 true。如果值为 true，validate()方法总被调用，否则不被调用。
- programmatic：默认值是 true。如果值为 true，并且 action 是可校验的，就会调用 validate()方法，以及任何以"validate"字符串开始的方法。
- declarative：默认值是 true。基于配置文件（xml 格式）或注解执行校验。

例 9-24　AnnotationValidation 拦截器示例。

（1）基本应用。

```
<action name="someAction" class="...">
    <interceptor-ref name="params"/>
    <interceptor-ref name="validation"/>
    <interceptor-ref name="workflow"/>
    <result>result.jsp</result>
</action>
```

（2）不校验 Action 类中的某个方法（someMethod）。

```
<action name="someAction" class="...">
    <interceptor-ref name="params"/>
    <interceptor-ref name="validation">
        <param name="excludeMethods">someMethod</param>
    </interceptor-ref>
    <interceptor-ref name="workflow"/>
    <result>result.jsp</result>
</action>
```

（3）只校验 Action 类中注解的方法。

```
<action name="someAction" class="...">
    <interceptor-ref name="params"/>
    <interceptor-ref name="validation">
        <param name="validateAnnotatedMethodOnly">true</param>
    </interceptor-ref>
    <interceptor-ref name="workflow"/>
    <result>result.jsp</result>
</action>
```

26. DefaultWorkflow 拦截器（workflow）

DefaultWorkflow 拦截器不进行任何校验操作，其作用是确保在拦截器链继续执行前没有校验错误。

如果被调用的方法名称在 excludeMethods 参数中指定,则该拦截器什么也不做。例如,excludeMethods 参数的值设置为"input,back"。则 foo!input.action 和 foo!back.action 这两个请求将被 DefaultWorkflow 拦截器忽略。

DefaultWorkflow 拦截器的可选参数如下:

inputResultName:默认值是"input"。当发现 action / field 错误时,决定返回的结果码名称。

例 9-25 DefaultWorkflow 拦截器示例。

(1) 基本应用。

```
<action name = "someAction" class = "...">
    <interceptor-ref name = "params"/>
    <interceptor-ref name = "validation"/>
    <interceptor-ref name = "workflow"/>
    <result>result.jsp</result>
</action>
```

(2) workflow 不处理 Action 类中的 someMethod1 和 someMethod2 方法。

```
<action name = "someAction" class = "...">
    <interceptor-ref name = "params"/>
    <interceptor-ref name = "validation"/>
    <interceptor-ref name = "workflow">
        <param name = "excludeMethods">someMethod1,someMethod2</param>
    </interceptor-ref name = "workflow">
    <result>result.jsp</result>
</action>
```

(3) 只处理 Action 类中的 someWorkflowMethod 方法,并且返回的结果码为"error"。

```
<action name = "someAction" class = "...">
    <interceptor-ref name = "params"/>
    <interceptor-ref name = "validation"/>
    <interceptor-ref name = "workflow">
        <param name = "inputResultName">error</param>
        <param name = "excludeMethods">*</param>
        <param name = "includeMethods">someWorkflowMethod</param>
    </interceptor-ref>
    <result>result.jsp</result>
</action>
```

27. MessageStore 拦截器(store)

将实现 ValidationAware 接口的 action 的消息/错误和字段错误存储到 HTTP Session 中,这样,在后面执行过程中还可以访问这些内容。

MessageStore 拦截器的可选参数有:

- allowRequestParameterSwitch：是否允许使用请求参数转换拦截器的操作模式。
- requestParameterSwitch：指示拦截器所在操作模式的请求参数。
- operationMode：拦截器可使用的操作模式，可选值有："STORE"，"RETRIEVE"，"AUTOMATIC"和"NONE"。默认值是"NONE"。

操作模式可选值的含义如下：

- "STORE"模式：拦截器将实现 ValidationAware 接口的 action 的消息/错误和字段错误存储到 HTTP Session 中。
- "RETRIEVE"模式：拦截器将检索保存的 action 的消息/错误和字段错误存，并放回到实现 ValidationAware 接口的 action 中。
- "AUTOMATIC"模式：拦截器将总是检索保存的 action 的消息/错误和字段保存，并放回到实现 ValidationAware 接口的 action 中，并且在 Action 执行后，如果 Result 是一个 ServletRedirectResult 实例，action 的消息/错误和字段错误自动存储到 HTTP Session 中。
- "NONE"模式：拦截器将不做任何事，为默认模式。

例 9-26 MessageStore 拦截器示例。

```
<action name="applicationFailed" class="...">
    <interceptor-ref name="store">
        <param name="operationMode">RETRIEVE</param>
    </interceptor-ref>
    <result>applicationFailed.jsp</result>
</action>
```

28. Checkbox 拦截器（checkbox）

由于一个 checkbox 附带一个同名的具有"_checkbox"前缀的隐藏属性的 input，这样，如果复选框没有选中就被提交，则将"false"值插入到参数中，就好像提交了一个"false"值一样，这和不使用 Checkbox 拦截器是不同的。如果不使用 Checkbox 拦截器，当复选框没有选中而被提交时，得到的将是一个 null 值。

Checkbox 拦截器的可选参数如下：

setUncheckedValue：为没有选中的复选框设置一个值，覆盖默认值"false"。

例 9-27 Checkbox 拦截器示例。

```
<action name="someAction" class="...">
    <interceptor-ref name="checkbox"/>
    <interceptor-ref name="params"/>
    <result>result.jsp</result>
</action>
```

29. ProfilingActivation 拦截器（profiling）

当 devMode 设置为 true 时，可以通过请求参数设置是否允许性能监视。

ProfilingActivation 拦截器的可选参数如下：

profilingKey：改变激活参数的名称。

例 9-28　ProfilingActivation 拦截器示例。

（1）为了能够进行性能测试，首先要确保 action 应用了 ProfilingActivation 拦截器。

```
<action name = "someAction" class = "...">
    <interceptor-ref name = "profiling">
        <param name = "profilingKey">profilingKey</param>
    </interceptor-ref>
    <result>result.jsp</result>
</action>
```

（2）通过参数激活。

```
http://host:port/context/namespace/someAction.action?profilingKey = true
```

30. Roles 拦截器（roles）

用于保证只有当用户拥有正确角色的时候才可以执行 action。Roles 拦截器可选的参数有：

- allowedRoles：逗号分隔的允许的角色列表。
- disallowedRoles：逗号分隔的不允许的角色列表。

例 9-29　Roles 拦截器示例。

```
<action name = "someAction" class = "...">
    <interceptor-ref name = "completeStack"/>
    <interceptor-ref name = "roles">
        <param name = "allowedRoles">admin,member</param>
    </interceptor-ref>
    <result>result.jsp</result>
</action>
```

只允许 admin 和 members 角色执行 someAction。

31. JSONValidation 拦截器（jsonValidation）

将校验及 action 错误序列化成 JSON。该拦截器不做任何校验工作，因此，必须放在栈中"validation"拦截器的后面。

JSONValidation 拦截器可选的参数有：

validationFailedStatus：如果设置了该参数，则在校验失败后，其值将被用作响应的状态。

例 9-30　JSONValidation 拦截器示例。

```
<action name = "someAction" class = "...">
    <interceptor-ref name = "basicStack"/>
    <interceptor-ref name = "validation">
        <param name = "excludeMethods">input,back,cancel</param>
```

```xml
        </interceptor-ref>
        <interceptor-ref name="jsonValidation"/>
        <interceptor-ref name="workflow"/>
</action>
```

32. AnnotationWorkflow 拦截器(annotationWorkflow)

调用 action 中任何被注解的方法,特别地,该拦截器支持@Before、@BeforeResult 和 @After 注解。可以使用同一注解标记多个方法,但是它们的执行顺序不能得到保障。

33. Multiselect 拦截器(multiselect)

Select 和 checkboxlist 等有多个选项值的元素能够附带一个同名的具有"__multiselect_"前缀的隐藏属性的 input。这样,当如果没有选项选中就被提交时,还能接收到一个值(空字符串数组),这和不使用 Multiselect 拦截器是不同的。如果不使用 Multiselect 拦截器,当没有选项选中就被提交时,则没有值。

9.5.2 内置的拦截器栈

除了内置拦截器,Struts 2 框架还提供了一些内置拦截器栈,拦截器栈中拦截器的顺序很重要。

1. 基本拦截器栈 basicStack

定义如下:

```xml
<interceptor-stack name="basicStack">
    <interceptor-ref name="exception"/>
    <interceptor-ref name="servletConfig"/>
    <interceptor-ref name="prepare"/>
    <interceptor-ref name="checkbox"/>
    <interceptor-ref name="multiselect"/>
    <interceptor-ref name="actionMappingParams"/>
    <interceptor-ref name="params">
        <param name="excludeParams">dojo\..*,^struts\..*</param>
    </interceptor-ref>
    <interceptor-ref name="conversionError"/>
</interceptor-stack>
```

excludeParams 参数设置了 params 拦截器强制忽略与正则表达式"dojo\..*"和"^struts\..*"匹配的参数,对任何名称中包含"dojo."的参数及任何名称以"struts."开始的参数不进行处理。

2. 校验工作流拦截器栈 validationWorkflowStack

定义如下:

```xml
<interceptor-stack name="validationWorkflowStack">
    <interceptor-ref name="basicStack"/>
    <interceptor-ref name="validation"/>
    <interceptor-ref name="workflow"/>
</interceptor-stack>
```

一个拦截器栈中可以使用已经定义的拦截器栈。校验工作流拦截器栈是在 basicStack 的基础上增加了 "validation" 和 "workflow" 两个拦截器。

3. JSON 校验拦截器栈 jsonValidationWorkflowStack

定义如下：

```xml
<interceptor-stack name="jsonValidationWorkflowStack">
    <interceptor-ref name="basicStack"/>
    <interceptor-ref name="validation">
        <param name="excludeMethods">input,back,cancel</param>
    </interceptor-ref>
    <interceptor-ref name="jsonValidation"/>
    <interceptor-ref name="workflow"/>
</interceptor-stack>
```

JSON 校验拦截器栈在 basicStack 的基础上增加了 validation、jsonValidation 和 workflow 三个拦截器。

excludeMethods 参数指定了 validation 拦截器忽略对 input、back 和 cancel 方法的请求，即对 input、back 和 cancel 方法不进行校验。

4. 文件上传拦截器栈 fileUploadStack

定义如下：

```xml
<interceptor-stack name="fileUploadStack">
    <interceptor-ref name="fileUpload"/>
    <interceptor-ref name="basicStack"/>
</interceptor-stack>
```

用于对上传的文件进行拦截。

5. 模型驱动拦截器栈 modelDrivenStack

定义如下：

```xml
<interceptor-stack name="modelDrivenStack">
    <interceptor-ref name="modelDriven"/>
    <interceptor-ref name="basicStack"/>
</interceptor-stack>
```

6. action 链拦截器栈 chainStack

定义如下：

```
<interceptor-stack name="chainStack">
    <interceptor-ref name="chain"/>
    <interceptor-ref name="basicStack"/>
</interceptor-stack>
```

7. 国际化拦截器栈 i18nStack

定义如下：

```
<interceptor-stack name="i18nStack">
    <interceptor-ref name="i18n"/>
    <interceptor-ref name="basicStack"/>
</interceptor-stack>
```

8. paramsPrepareParamsStack 拦截器栈

定义如下：

```
<interceptor-stack name="paramsPrepareParamsStack">
    <interceptor-ref name="exception"/>
    <interceptor-ref name="alias"/>
    <interceptor-ref name="i18n"/>
    <interceptor-ref name="checkbox"/>
    <interceptor-ref name="multiselect"/>
    <interceptor-ref name="params">
        <param name="excludeParams">dojo\..*,^struts\..*</param>
    </interceptor-ref>
    <interceptor-ref name="servletConfig"/>
    <interceptor-ref name="prepare"/>
    <interceptor-ref name="chain"/>
    <interceptor-ref name="modelDriven"/>
    <interceptor-ref name="fileUpload"/>
    <interceptor-ref name="staticParams"/>
    <interceptor-ref name="actionMappingParams"/>
    <interceptor-ref name="params">
        <param name="excludeParams">dojo\..*,^struts\..*</param>
    </interceptor-ref>
    <interceptor-ref name="conversionError"/>
    <interceptor-ref name="validation">
        <param name="excludeMethods">input,back,cancel,browse</param>
    </interceptor-ref>
    <interceptor-ref name="workflow">
        <param name="excludeMethods">input,back,cancel,browse</param>
    </interceptor-ref>
</interceptor-stack>
```

除了在 prepare 拦截器之前包含了一个额外的 params 拦截器外，该拦截器栈和 defaultStack 拦截器栈其实是一样的。

9. 默认拦截器栈 defaultStack

定义如下：

```
<interceptor-stack name="defaultStack">
    <interceptor-ref name="exception"/>
    <interceptor-ref name="alias"/>
    <interceptor-ref name="servletConfig"/>
    <interceptor-ref name="i18n"/>
    <interceptor-ref name="prepare"/>
    <interceptor-ref name="chain"/>
    <interceptor-ref name="debugging"/>
    <interceptor-ref name="scopedModelDriven"/>
    <interceptor-ref name="modelDriven"/>
    <interceptor-ref name="fileUpload"/>
    <interceptor-ref name="checkbox"/>
    <interceptor-ref name="multiselect"/>
    <interceptor-ref name="staticParams"/>
    <interceptor-ref name="actionMappingParams"/>
    <interceptor-ref name="params">
        <param name="excludeParams">dojo\..*,^struts\..*</param>
    </interceptor-ref>
    <interceptor-ref name="conversionError"/>
    <interceptor-ref name="validation">
        <param name="excludeMethods">input,back,cancel,browse</param>
    </interceptor-ref>
    <interceptor-ref name="workflow">
        <param name="excludeMethods">input,back,cancel,browse</param>
    </interceptor-ref>
</interceptor-stack>
```

defaultStack 包含了所有常见的拦截器，它在 struts-default.xml 配置文件中被设置成了 struts-default 包的默认拦截器栈。因此，开发人员可以通过继承 struts-default 包的方式，直接使用 defaultStack 拦截器栈中的常用拦截器。

10. completeStack 拦截器栈

定义如下：

```
<interceptor-stack name="completeStack">
    <interceptor-ref name="defaultStack"/>
</interceptor-stack>
```

该拦截器栈是 defaultStack 栈的另一个名称，是为了应用中的向后兼容。

11. 执行等待拦截器栈 executeAndWaitStack

定义如下：

```xml
<interceptor-stack name="executeAndWaitStack">
    <interceptor-ref name="execAndWait">
        <param name="excludeMethods">input,back,cancel</param>
    </interceptor-ref>
    <interceptor-ref name="defaultStack"/>
    <interceptor-ref name="execAndWait">
        <param name="excludeMethods">input,back,cancel</param>
    </interceptor-ref>
</interceptor-stack>
```

execAndWait 拦截器应该总是最后一个拦截器。

9.5.3 内置拦截器的配置

1. 覆盖拦截器栈中的拦截器参数的两种方法

(1) 方法一,代码如下:

```xml
<action name="someAction" class="…">
    <interceptor-ref name="basicStack"/>
    <interceptor-ref name="validation">
        <param name="excludeMethods">myValidationExcudeMethod</param>
    </interceptor-ref>
    <interceptor-ref name="jsonValidation"/>
    <interceptor-ref name="workflow"/>
</action>
```

将 jsonValidationWorkflowStack 拦截器栈中的拦截器复制,然后修改 validation 拦截器的 excludeMethods 参数值为 myValidationExcudeMethod。

(2) 方法二,代码如下:

```xml
<action name="someAction" class="…">
    <interceptor-ref name="jsonValidationWorkflowStack">
        <param name="validation.excludeMethods">myValidationExcludeMethod</param>
    </interceptor-ref>
</action>
```

使用 interceptor-ref 元素指向一个存在的拦截器栈,如本例中的 jsonValidationWorkflowStack 拦截器栈,然后覆盖 validation 拦截器的 excludeMethods 参数。在 param 标签中,name 属性值的格式形如 interceptorName.parameterName。其中:interceptorName 表示参数要被覆盖的拦截器名称,parameterName 表示要被覆盖的参数本身。

2. 参数覆盖不具有继承性

参数覆盖在拦截器中不具有继承性,也就是说对覆盖参数的最后设置将被使用。例如,parentStack 拦截器栈中的 postPrepareParameterFilter 拦截器的参数 defaultBlock 被覆盖,如下：

```
<interceptor-stack name="parentStack">
    <interceptor-ref name="postPrepareParameterFilter">
        <param name="defaultBlock">true</param>
    </interceptor-ref>
</interceptor-stack>
```

在一个 action 中覆盖 postPrepareParameterFilter 拦截器中的 allowed 参数，如下：

```
<package name="childPackage" namespace="/child" extends="parentPackage">
    <action name="someAction" class="…">
        <interceptor-ref name="parentStack">
            <param name="postPrepareParameterFilter.allowed">myObject.name</param>
        </interceptor-ref>
    </action>
</package>
```

只有 postPrepareParameterFilter 拦截器的 allowed 参数被覆盖，其他参数（如 defaultBlock）的值是 null。

习题

1. 使用实现 interceptor 接口的方式改写例 9-1。
2. 使用拦截器栈的方式改写例 9-2。
3. 修改例 9-3，完善其功能，当用户输入指定的用户名（如：tsc）和密码（如：tsc）后才可以访问其他 action。

第10章 Struts 2框架的输入校验

本章首先介绍了输入校验的分类以及如何实现输入校验的国际化,然后介绍了Struts 2框架的内置校验器,最后介绍了自定义校验器的实现。

10.1 输入校验概述

在第8章介绍了数据的类型转换,通过使用基于Struts 2框架的类型转换功能,可以将通过表单提交的数据转换到正确的类型。虽然输入数据能够进行正确的类型转换,但是却不能保证这些输入数据具有合理性,即不一定符合业务逻辑要求。例如在第8章例8-1的添加书目示例中,出版日期字段的输入内容是"2099-09-09"时,虽然能够正确转换到日期类型,但是和当前时间比,这个时间却不合理。

为了解决Web应用中输入数据的合理性问题,开发人员需要提供额外的输入校验代码对输入数据进行合理性检查。为了减轻开发人员数据校验的负担,Struts 2框架提供了一个校验框架,该框架极大简化了数据的输入验证工作,有利于Web应用的开发与维护。

Struts 2框架的输入校验是通过XML配置文件或者注解来实现的,当然在Action中进行手动(编码)校验也是允许的,并且也可以手动校验和XML配置文件或者注解方式的校验一起使用。

数据校验依赖于validation和workflow两个拦截器(这两个拦截器都包含在默认拦截器栈defaultStack中)。validation拦截器进行校验并创建具体字段错误的列表。workflow拦截器检查存在的校验错误:如果有,就返回"input"(默认值)结果码,返回到包含校验错误的表单页面。

在基于Struts 2框架的Web应用中,输入校验既可以在服务器端完成,又可以在客户端完成。

1. 服务器端输入校验

在服务器端进行输入校验有两种主要方法。

(1) 基于硬编码的方法:Struts 2框架可以通过在业务控制器类中重写validate()方法来完成对输入数据的校验。另外,Struts 2框架还允许开发人员使用形如validateXxx()形式的校验方法对输入数据进行校验。其中Xxx表示action中不同方法的名称。

(2) 基于配置文件的方法:采用硬编码的方法,代码的耦合度高。因此,Struts 2框架还提供了一种基于配置文件的输入校验方法,这种校验方法是把校验规则保存在特定的 XML 格式配置文件中,实现了输入校验和 Action 类的分离,降低了代码的耦合性。

2. 客户端输入校验

除了服务器端的输入校验,Struts 2 框架还提供了客户端的输入校验方法。

使用 JavaScript 进行客户端校验,可以完成对输入数据的初步校验。这种校验方式不用和服务器进行交互,速度快,但是使用客户端输入校验具有书写复杂、不易维护的缺点。因此,Struts 2 框架为开发人员提供了一种基于框架的客户端输入校验方法,只需进行简单的设置,Struts 2 框架就可以自动生成客户端校验用的 JavaScript 脚本。

3. Ajax 校验

Struts 2 框架只为一部分输入校验器提供了客户端校验,而使用 Ajax 校验,服务器端使用的所有校验器,可以不需重新加载页面,就能显示校验错误消息。Ajax 校验由一个服务器端和一个客户端组成,其中服务器端包含在 Struts 2 框架的核心 Jar 包中(一个拦截器和一些实用的 JavaScript 文件),客户端包含在 Dojo 插件中。

为了便于讲解本章内容,首先创建 Web 项目 Chapter10,并添加 Struts 2 框架的支持。Chapter10 项目目录结构参考 Chapter9 项目的目录结构,web.xml 配置文件的内容同 Chapter9 项目中 web.xml 配置文件的内容。

10.2 服务器端输入校验

服务器端输入校验包含两种方式:硬编码方式和配置文件方式。

10.2.1 使用编码进行输入校验

例 10-1 使用 validate()方法进行输入校验。

任务描述:以第 8 章的添加书目为例进行讲解,对出版日期和数量进行输入校验。

具体操作步骤如下:

(1) 创建 addBook.jsp 文件,内容同第 8 章例 8-1 中 addBook.jsp 文件的内容。

(2) 创建 Book.java 文件,内容同第 8 章例 8-1 中 Book.java 文件的内容。

(3) 创建 Action 类。

BookAction.java

```java
package example.struts2;
import java.util.Date;
import com.opensymphony.xwork2.ActionSupport;
public class BookAction extends ActionSupport {
    private Book book;
    public Book getBook() {
        return book;
    }
```

```java
    public void setBook(Book book) {
        this.book = book;
    }
    public String execute(){
        return SUCCESS;
    }
    public void validate() {
        if (book.getBookName().isEmpty()){
            addFieldError("book.bookName", "书名不能为空!");
        }
        if (book.getBookPrice() <= 0){
            addFieldError("book.bookPrice", "单价不应小于等于0!");
        }
        if (book.getBookPublishDate() == null ||
                        book.getBookPublishDate().after(new Date())){
            addFieldError("book.bookPublishDate",
                            "出版日期不应为空并且应在今天之前!");
        }
        if (book.getBookCount() == null || book.getBookCount() <= 0){
            addFieldError("book.bookCount","数量不应为空并且要大于0!");
        }
    }
}
```

(4) 创建 struts.xml 文件。

struts.xml

```xml
<?xml version="1.0" encoding="UTF-8"?>
<!DOCTYPE struts PUBLIC
    "-//Apache Software Foundation//DTD Struts Configuration 2.0//EN"
    "http://struts.apache.org/dtds/struts-2.0.dtd">
<struts>
    <package name="default" extends="struts-default">
        <action name="addBook" class="example.struts2.BookAction">
            <result>/addBook.jsp</result>
            <result name="input">/addBook.jsp</result>
        </action>
    </package>
</struts>
```

(5) 发布 Web 项目，启动 Tomcat 服务器，访问 addBook.jsp。不输入内容，提交后的结果如图 10-1 所示。

(6) 将图 10-1 中的数量改为 1.1，然后单击"提交"按钮，结果如图 10-2 所示。

在该示例中进行了数据类型转换和输入校验，由图中的结果可以看出，Struts 2 框架首先执行数据类型转换，然后再进行输入校验。

使用 validate() 方法进行输入校验时，会对当前 Action 中的所有的方法有效。由于 Struts 2 框架可以在同一个 Action 中使用的不同方法来处理不同的请求，因此，如果只想对

Action 中的某个方法进行输入校验,则可以使用 validateXxx()方法来实现,其中 Xxx 是将 Action 中方法名称为 xxx()的首字母大写。

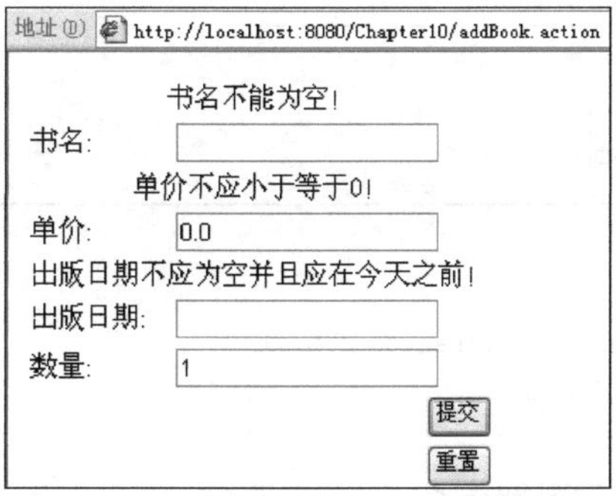

图 10-1　不输入内容提交后的效果 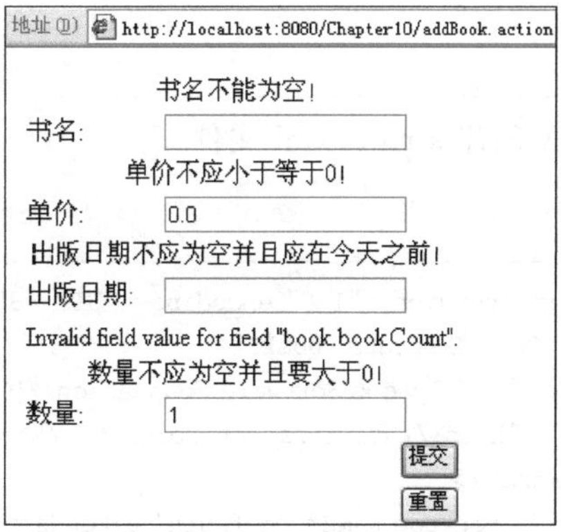　　　　图 10-2　数量为浮点数提交后的效果

例 10-2　使用 validateXxx()方法对添加的书目信息进行输入校验。

(1) 创建 Action。

BookMethodAction.java

```java
package example.struts2;
import java.util.Date;
import com.opensymphony.xwork2.ActionSupport;
public class BookMethodAction extends ActionSupport {
    private Book book;
    public Book getBook() {
        return book;
    }
    public void setBook(Book book) {
        this.book = book;
    }
    public String add(){
        return SUCCESS;
    }
    public void validateAdd() {
        if (book.getBookName().isEmpty()){
            addFieldError("book.bookName","书名不能为空!");
        }
        if (book.getBookPrice() <= 0){
            addFieldError("book.bookPrice","单价不应小于等于 0!");
        }
        if (book.getBookPublishDate() == null ||
                            book.getBookPublishDate().after(new Date())){
            addFieldError("book.bookPublishDate",
                    "出版日期不应为空并且应在今天之前!");
        }
        if (book.getBookCount() == null || book.getBookCount() <= 0){
```

```
                    addFieldError("book.bookCount","数量不应为空并且要大于0!");
                }
        }
}
```

(2) 创建 struts.xml 文件。

struts.xml

```xml
<?xml version = "1.0" encoding = "UTF - 8" ?>
<!DOCTYPE struts PUBLIC
    " - //Apache Software Foundation//DTD Struts Configuration 2.0//EN"
    "http://struts.apache.org/dtds/struts - 2.0.dtd">
<struts>
    <package name = "default" extends = "struts - default">
        <action name = "addBook" method = "add"
                            class = "example.struts2.BookMethodAction">
            <result>/addBook.jsp</result>
            <result name = "input">/addBook.jsp</result>
        </action>
    </package>
</struts>
```

(3) addBook.jsp 和 Book.java 文件的内容参考例 10-1 中同名文件的内容。

(4) 发布 Web 项目,启动 Tomcat 服务器,访问 addBook.jsp。不输入内容,提交后的结果如图 10-1 所示。

10.2.2 使用配置文件进行输入校验

在使用硬编码方式进行输入校验时,输入校验代码和业务处理代码被混在一起,因此,这种校验方式的代码耦合度高。除了使用硬编码进行输入校验的方式外,Struts 2 框架还提供了一种基于配置文件的输入校验方式。在这种校验方式中,校验规则被保存在特定的 XML 格式配置文件中,从而实现了输入校验代码和业务处理代码的分离,降低了代码间的耦合性。

1. 校验规则文件的结构

校验规则文件的结构是由 xwork-validator-1.0.2.dtd 文件定义的,内容如下:

xwork - validator - 1.0.2.dtd

```
<?xml version = "1.0" encoding = "UTF - 8"?>
<!--
    XWork Validators DTD.
    Used the following DOCTYPE.
    <!DOCTYPE validators PUBLIC
            " - //OpenSymphony Group//XWork Validator 1.0.2//EN"
            "http://www.opensymphony.com/xwork/xwork - validator - 1.0.2.dtd">
```

```
    -->
<!ELEMENT validators (field|validator)+>
<!ELEMENT field (field-validator+)>
<!ATTLIST field
    name CDATA #REQUIRED
>
<!ELEMENT field-validator (param*, message)>
<!ATTLIST field-validator
    type CDATA #REQUIRED
    short-circuit (true|false) "false"
>
<!ELEMENT validator (param*, message)>
<!ATTLIST validator
    type CDATA #REQUIRED
    short-circuit (true|false) "false"
>
<!ELEMENT param (#PCDATA)>
<!ATTLIST param
    name CDATA #REQUIRED
>
<!ELEMENT message (#PCDATA)>
<!ATTLIST message
    key CDATA #IMPLIED
>
```

（1）文件中"<!--"和"-->"标记符之间的内容表明，一个校验规则文件开始的内容是如下内容：

```
<!DOCTYPE validators PUBLIC
    "-//OpenSymphony Group//XWork Validator 1.0.2//EN"
    "http://www.opensymphony.com/xwork/xwork-validator-1.0.2.dtd">
```

（2）validators 是校验规则文件的根元素。validators 元素可以包含一个或多个 field 或 validator 子元素。

（3）field 子元素是基于字段定义输入校验的验证规则。field 子元素有一个必选的 name 属性，用于设置要验证的表单字段的名称。

另外，field 子元素可以包含一个或多个 field-validator 子元素，用于设置校验器。

（4）field-validator 子元素有 type 和 short-circuit 两个属性。type 是必选属性，用于设置验证器的名称。如果为同一个表单字段配置了多个验证器，当某个验证器验证失败后，其后的验证器是否还执行，可以使用 short-circuit 属性进行设置，默认值为 false。如果 short-circuit 属性值设置为 true，则其后的验证器不执行。

另外，field-validator 子元素可以包含 0 个、1 个或多个 param 子元素以及 1 个 message 子元素。

（5）validator 子元素是基于验证器定义输入校验的验证规则。validator 子元素有 type 和 short-circuit 两个属性，其中 type 是必选属性，用于设置验证器的名称，short-circuit 属性用于设置当某个验证器验证失败后，其后的验证器是否还执行，默认值为 false。如果 short-circuit 属性值设置为 true,则其后的验证器不执行。

另外,validator 子元素包含 0 个、1 个或多个 param 子元素以及一个 message 子元素。

(6) param 子元素为验证器传递参数。param 子元素有一个必选的 name 属性,用于设置参数的名称,参数的值嵌套在 param 元素内。

(7) message 子元素设置当验证器验证失败时显示的错误消息。message 子元素有一个可选的 key 属性,用于指定错误消息在国际化资源文件中的键(key)。

2. 校验规则文件的名称

校验规则文件有两种命名方式,其名称分别是 ActionName-validation.xml 和 ActionName-actionAlias-validation.xml,其中 ActionName 是 Action 类的名称。校验规则文件和对应的 Action 保存在相同的目录下。

actionAlias 是指在 Struts 2 框架的配置文件中给定的 action 的名称。通常,action 的 name 属性值和 method 名称匹配,但是也可以不同。

在这两种命名方式中,ActionName-validation.xml 中的校验规则是在调用 Action 类中的任何方法时都会起作用,而 ActionName-actionAlias-validation.xml 中的校验规则仅在 actionAlias 和 action 的 name 属性值相同时才起作用。如果同时配置了两个校验规则文件,则在 ActionName-validation.xml 文件中只放所有 method 都使用的验证规则。

3. 基于字段的校验规则

字段校验规则使用<field>子元素定义,其格式如下:

```
<validators>
    <field name = "Form's fieldName to be validated">
        <field-validator type = "validatorName">
            <param name = "paramName">paramValue</param>
            <message>messages when validation error occurred</message>
        </field-validator>
        ...
    </field>
    ...
</validators>
```

例 10-3 使用基于字段的校验规则对添加的书目信息进行输入校验。

具体操作步骤如下:

(1) 创建 Action 类。

BookXMLAction.java

```
package example.struts2;
import com.opensymphony.xwork2.ActionSupport;
public class BookXMLAction extends ActionSupport {
    private Book book;
    public Book getBook() {
        return book;
```

```java
    }
    public void setBook(Book book) {
        this.book = book;
    }
    public String execute(){
        return SUCCESS;
    }
}
```

(2) 创建校验规则文件,和 BookXMLAction.java 文件保存在相同的位置。

BookXMLAction-validation.xml

```xml
<?xml version="1.0" encoding="UTF-8"?>
<!DOCTYPE validators PUBLIC
        "-//OpenSymphony Group//XWork Validator 1.0.2//EN"
        "http://www.opensymphony.com/xwork/xwork-validator-1.0.2.dtd">
<validators>
    <field name="book.bookName">
        <field-validator type="requiredstring">
            <message>必须输入书名!</message>
        </field-validator>
    </field>
    <field name="book.bookPrice">
        <field-validator type="required">
            <message>必须输入单价!</message>
        </field-validator>
        <field-validator type="double">
            <param name="minExclusive">0</param>
            <message>单价应该大于${minExclusive}!</message>
        </field-validator>
    </field>
    <field name="book.bookPublishDate">
        <field-validator type="required">
            <message>必须输入出版日期!</message>
        </field-validator>
        <field-validator type="date">
            <param name="min">2009-01-01</param>
            <param name="max">2011-12-31</param>
            <message>出版时间应在${min}至${max}之间!</message>
        </field-validator>
    </field>
    <field name="book.bookCount">
        <field-validator type="required">
            <message>必须输入数量!</message>
        </field-validator>
        <field-validator type="int">
            <param name="min">1</param>
            <message>数量要大于0!</message>
```

```
        </field-validator>
    </field>
</validators>
```

(3) 编辑 struts.xml,内容如下:

struts.xml

```
<?xml version = "1.0" encoding = "UTF-8" ?>
<!DOCTYPE struts PUBLIC
    " - //Apache Software Foundation//DTD Struts Configuration 2.0//EN"
    "http://struts.apache.org/dtds/struts-2.0.dtd">
<struts>
    <package name = "default" extends = "struts-default">
        <action name = "addBook" class = "example.struts2.BookXMLAction">
            <result>/addBook.jsp</result>
            <result name = "input">/addBook.jsp</result>
        </action>
    </package>
</struts>
```

(4) addBook.jsp 和 Book.java 文件的内容参考例 10-1 中同名文件的内容。

(5) 发布 Web 项目,启动 Tomcat 服务器,访问 addBook.jsp。不输入内容,提交后的结果如图 10-3 所示。

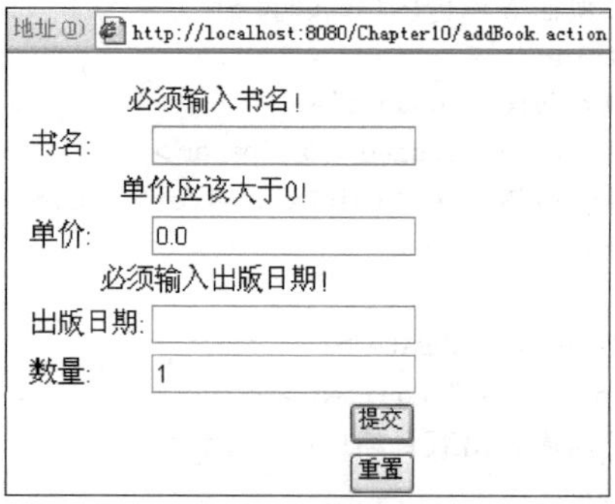

图 10-3　不输入内容提交的结果

4. 基于 validator 校验器的校验规则

校验器校验规则使用<validator>子元素定义,格式如下:

```
<validators>
    <validator type = 'validatorName'>
        <param name = "fieldName">Form's fieldName to be validated</param>
        <param name = "paramName">paramValue</param>
        ...
        <message>messages when validation error occurred</message>
```

```
        </validator>
        ...
</validators>
```

例 10-4 使用基于校验器的校验规则对添加的书目信息进行输入校验。

在例 10-3 的基础上进行操作,具体操作步骤如下:

(1) 修改 BookXMLAction-validation.xml 文件,内容如下:

BookXMLAction – validation.xml

```xml
<?xml version="1.0" encoding="UTF-8"?>
<!DOCTYPE validators PUBLIC
        "-//OpenSymphony Group//XWork Validator 1.0.2//EN"
        "http://www.opensymphony.com/xwork/xwork-validator-1.0.2.dtd">
<validators>
    <validator type='requiredstring'>
        <param name="fieldName">book.bookName</param>
        <message>必须输入姓名!</message>
    </validator>
    <validator type='double'>
        <param name="fieldName">book.bookPrice</param>
        <param name="minExclusive">0</param>
        <message>单价应该大于${minExclusive}!</message>
    </validator>
    <validator type='required'>
        <param name="fieldName">book.bookPublishDate</param>
        <message>必须输入出版日期!</message>
    </validator>
    <validator type='date'>
        <param name="fieldName">book.bookPublishDate</param>
        <param name="min">2009-01-01</param>
        <param name="max">2011-12-31</param>
        <message>出版时间应在${min}至${max}之间!</message>
    </validator>
    <validator type='required'>
        <param name="fieldName">book.bookCount</param>
        <message>必须输入数量!</message>
    </validator>
    <validator type='int'>
        <param name="fieldName">book.bookCount</param>
        <message>数量要大于0!</message>
    </validator>
</validators>
```

(2) addBook.jsp、Book.java 和 struts.xml 文件的内容参考例 10-3 中同名文件的内容。

(3) 发布 Web 项目,并启动 Tomcat 服务器,访问 addStudent.jsp 文件进行测试,结果如图 10-3 所示。

在同一个校验规则文件中，可以混合使用＜field＞子元素和＜validator＞子元素定义校验规则。

当使用字段校验器时，基于字段校验器的语法总是优于基于 Validator 校验器的语法，因为前者更容易根据字段组织字段校验器，特别是当一个字段需要有多个字段校验器时更显得方便。

5. 校验规则的搜索顺序

(1) 校验规则文件的搜索顺序。

由上面的内容可知，校验规则文件的名称有两种命名方式，ActionName-validation.xml 和 ActionName-actionAlias-validation.xml。如果同一个 Action 同时存在两种命名形式的校验规则文件，那么哪一个校验规则文件起作用呢？

Struts 2 框架将首先搜索 ActionName-validation.xml 文件，然后搜索 ActionName-alias-validation.xml 文件。因此，如果两个校验规则文件中的校验规则有重复时，则校验规则将重复执行。

另外，如果在定义某业务控制器类(如 BookAction)时，继承了 BookParentAction 类，同时实现了 IBookAction 接口。这时校验规则文件的搜索顺序如下：

① IBookAction-validation.xml；
② IBookAction-alias-validation.xml；
③ BookParentAction-validation.xml；
④ BookParentAction-alias-validation.xml；
⑤ BookAction-validation.xml；
⑥ BookAction-alias-validation.xml。

也就是说接口的校验规则文件优先于实现类的校验规则文件被搜索，父类的校验规则文件优先于子类的校验规则文件被搜索。如果 Action 类名称相同的两种形式的校验规则文件都存在时，则别名校验规则文件被后搜索。

(2) 校验规则文件中校验器的搜索顺序。

在同一校验规则文件中，既可以定义基于字段的校验规则，又可以定义基于 validator 校验器的校验规则。那么在同一个校验规则文件中，Struts 2 框架将按照如下的顺序进行校验：

① 使用＜validator＞定义的校验器优先于使用＜field＞定义的校验器；
② 如果都是使用＜validator＞定义的校验器，则按照校验器的定义顺序进行校验；
③ 如果都是使用＜field＞定义的校验器，则按照校验器的定义顺序进行校验；
④ 如果某个短路的校验器校验失败，则相同表单字段的后续校验器就不再执行校验。

10.3 客户端输入校验

在 10.2 节介绍了 Struts 2 框架的服务器端输入校验，也就是说输入数据的校验都是在服务器端完成的，这时数据已经被传送到了服务器。如果数据校验出错，则将错误信息返回给用户，等待用户输入正确的数据。这将造成资源浪费，并且如果网速很慢，等待时间会较长。除了服务器端输入校验，Struts 2 框架还支持使用客户端输入校验。

在基于 Struts 2 框架的客户端输入校验中，所有的输入校验工作都在客户端完成。另外，基于 Struts 2 框架的客户端输入校验并不是在所有的主题下都可以使用。

例 10-5 基于 Struts 2 框架的客户端输入校验示例。

以添加书目信息为例进行讲解，在例 10-3 或例 10-4 的基础上进行操作，具体操作步骤如下：

(1) 创建 JSP 文件。

addBookClient.jsp

```jsp
<%@ page language="java" contentType="text/html; charset=UTF-8" %>
<%@ taglib prefix="s" uri="/struts-tags" %>
<s:head/>
<s:form action="addBookClient" validate="true" theme="css_xhtml">
    <s:textfield name="book.bookName" label="书名"/>
    <s:textfield name="book.bookPrice" label="单价"/>
    <s:textfield name="book.bookPublishDate" label="出版日期"/>
    <s:textfield name="book.bookCount" value="1" label="数量"/>
    <s:submit name="submit" value="提交" align="left"/>
    <s:reset name="reset" value="重置" align="left"/>
</s:form>
<s:if test="book.bookName!=''">
    添加的书目信息如下：<br>
    书名：<s:property value="book.bookName"/><br>
    单价：<s:property value="book.bookPrice"/><br>
    出版日期：<s:property value="book.bookPublishDate"/><br>
    数量：<s:property value="book.bookCount"/>
</s:if>
```

要将<s:form>设置为"css_xhtml"主题，并且将 validate 属性设置为"true"。

(2) 编辑 struts.xml。

struts.xml

```xml
<?xml version="1.0" encoding="UTF-8"?>
<!DOCTYPE struts PUBLIC
    "-//Apache Software Foundation//DTD Struts Configuration 2.0//EN"
    "http://struts.apache.org/dtds/struts-2.0.dtd">
<struts>
    <package name="default" extends="struts-default">
        <action name="client">
            <result>/addBookClient.jsp</result>
        </action>
        <action name="addBookClient" class="example.struts2.BookXMLAction">
            <result>/addBookClient.jsp</result>
            <result name="input">/addBookClient.jsp</result>
        </action>
```

```
        </package>
</struts>
```

在配置文件中定义了 client。

（3）发布 web 项目，启动 Tomcat 服务器，访问如下网址，结果如图 10-4 所示。

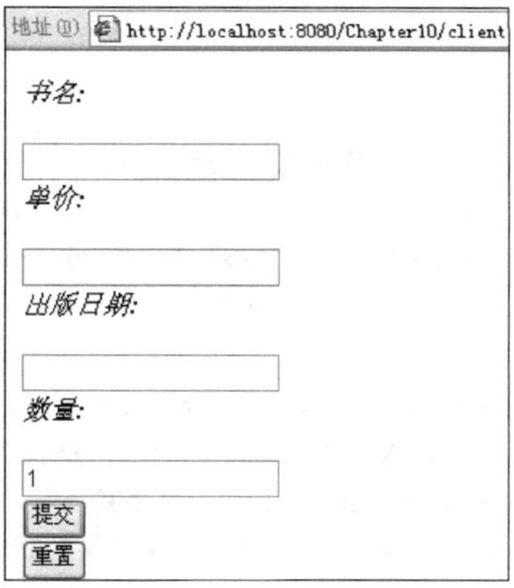

图 10-4　输入页面效果

```
http://localhost:8080/Chapter10/client
```

如果直接访问 addBookClient.jsp，则会报错，如图 10-5 所示。

图 10-5　访问 addBookClient.jsp 的效果

（4）查看图 10-4 页面的源码，部分源码如下：

```html
<script src="/Chapter10/struts/utils.js" type="text/javascript"></script>
<script type="text/javascript" src="/Chapter10/struts/css_xhtml/validation.js"></script>
<form id="addBookClient" name="addBookClient"
        onsubmit="return validateForm_addBookClient();"
        action="/Chapter10/addBookClient.action" method="post"
        onreset="clearErrorMessages(this);clearErrorLabels(this);">
<script type="text/javascript">
    function validateForm_addBookClient() {
        form = document.getElementById("addBookClient");
        clearErrorMessages(form);
        clearErrorLabels(form);
        var errors = false;
        var continueValidation = true;
```

```
                ...
                return !errors;
        }
</script>
```

由于将 validate 属性设置为 true,因此,Struts 2 框架生成了校验用的 JavaScript 函数 validateForm_addBookClient(),并且在＜form＞表单元素中增加了一个 onsubmit 属性,在提交表单的时候先调用该属性的值,进行客户端校验,客户端校验通过后,再调用 action 属性的值。

(5) 单击"提交"按钮,结果如图 10-6 所示。

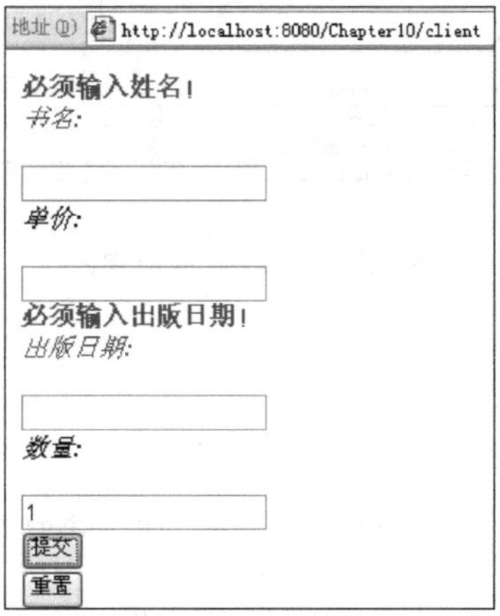

图 10-6　提交后的效果

在图 10-6 中的 url 并没有发生变化,说明使用的是客户端校验。如果使用的是服务器端校验,网址会发生变化。

10.4　Ajax 校验

Struts 2 框架只为一部分校验器提供了客户端校验(使用 JavaScript)。使用 Ajax 校验,服务器端使用的所有校验器,可以不用重新加载页面,就能显示校验错误消息。Ajax 校验包含一个服务器端和一个客户端,其中服务器端包含在 Struts 2 框架的核心 Jar 文件中(一个拦截器和一些实用的 JavaScript 文件),客户端包含在 Dojo 插件中。

Ajax 校验流程图如图 10-7 所示。

使用 Ajax 校验需要安装 Dojo 插件。另外,Ajax 校验是由 jsonValidation 拦截器执行的,为了执行 Ajax 校验,需要把 jsonValidation 拦截器包含在 jsonValidationWorkflowStack 拦截器栈中。

例 10-6　Ajax 校验示例。

以添加书目信息为例进行讲解,在例 10-3 或例 10-4 的基础上进行操作,具体操作步骤如下:

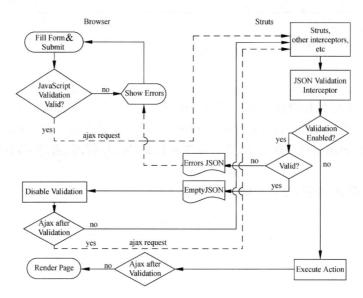

图 10-7　Ajax 校验流程图

（1）创建 JSP 文件。

addBookAjax.jsp

```
<%@ page language="java" contentType="text/html; charset=UTF-8"%>
<%@ taglib prefix="s" uri="/struts-tags"%>
<%@ taglib prefix="sx" uri="/struts-dojo-tags"%>
<sx:head debug="true" cache="false" compressed="false"/>
<s:form action="addBookAjax">
    <s:textfield name="book.bookName" label="书名"/>
    <s:textfield name="book.bookPrice" label="单价"/>
    <s:textfield name="book.bookPublishDate" label="出版日期"/>
    <s:textfield name="book.bookCount" value="1" label="数量"/>
    <sx:submit validate="true"/>
</s:form>
<s:if test="book.bookName! = ''">
    添加的书目信息如下：<br>
    书名：<s:property value="book.bookName"/><br>
    单价：<s:property value="book.bookPrice"/><br>
    出版日期：<s:property value="book.bookPublishDate"/><br>
    数量：<s:property value="book.bookCount"/>
</s:if>
```

为了执行 Ajax 校验，必须将<sx:submit>标签的 validate 属性值设置为 true。

（2）编辑 struts.xml 文件。

struts.xml

```
<?xml version="1.0" encoding="UTF-8"?>
<!DOCTYPE struts PUBLIC
```

```
            " - //Apache Software Foundation//DTD Struts Configuration 2.0//EN"
            "http://struts.apache.org/dtds/struts - 2.0.dtd">
<struts>
    <package name = "default" extends = "struts - default">
        <action name = "addBookAjax" class = "example.struts2.BookXMLAction">
            <interceptor - ref name = "jsonValidationWorkflowStack"/>
            <result>/addBookAjax.jsp</result>
            <result name = "input">/addBookAjax.jsp</result>
        </action>
    </package>
</struts>
```

为 addBookAjax 配置了 jsonValidationWorkflowStack 拦截器栈。

（3）发布 Web 项目，并启动 Tomcat 服务器，访问 addBookAjax.jsp，不输入信息，单击"添加"按钮，结果如图 10-8 所示。

（4）输入信息，提交后的结果如图 10-9 所示。

图 10-8　不输入信息提交后的效果　　　　图 10-9　日期不正确提交后的效果

比较例 10-6 和例 10-5，可以发现，在客户端输入校验中，对日期不能正确校验，而 Ajax 校验则能够对日期正确校验。

（5）输入正确信息，提交后的结果如图 10-10 所示。

图 10-10　输入正确信息后提交的效果

当 Ajax 校验通过后,数据将交给 action 进行处理。

10.5 输入校验的国际化

在上面所有的输入校验示例中,都是将校验错误提示信息进行了硬编码,这样不适合 Web 应用的国际化。为了国际化校验信息,将这些校验信息保存到一个属性文件中,在显示校验信息的位置使用"key"属性。下面对服务器端的两种输入校验方式(硬编码方式和配置文件方式)中的校验错误信息进行国际化的方法进行介绍。

1. 对使用编码方式进行输入校验中的校验错误信息的国际化

例 10-7 对使用编码方式进行输入校验中的校验错误信息的国际化。

在例 10-1 的基础上进行操作,具体操作步骤如下:

(1) 创建国际化资源文件。

ApplicationResources_zh.properties

```
bookNameError = 书名不能为空!
bookPriceError = 单价不应小于等于 0!
bookPublishDateError = 出版日期不应为空并且应在今天之前!
bookCountError = 数量不应为空并且要大于 0!
```

将 Action 中的校验错误信息保存在资源文件中。

(2) 编辑 struts.xml 文件。

struts.xml

```
<?xml version = "1.0" encoding = "UTF-8" ?>
<!DOCTYPE struts PUBLIC
    "-//Apache Software Foundation//DTD Struts Configuration 2.0//EN"
    "http://struts.apache.org/dtds/struts-2.0.dtd">
<struts>
    <package name = "default" extends = "struts-default">
        <action name = "addBook" class = "example.struts2.BookAction">
            <result>/addBook.jsp</result>
            <result name = "input">/addBook.jsp</result>
        </action>
    </package>
    <constant name = "struts.custom.i18n.resources" value = "ApplicationResources"/>
</struts>
```

在该配置文件中配置了国际化资源文件的基名 ApplicationResources。

(3) 编辑 Action 类。

BookAction.java

```
package example.struts2;
import java.util.Date;
```

```java
import com.opensymphony.xwork2.ActionSupport;
public class BookAction extends ActionSupport {
    private Book book;
    public Book getBook() {
        return book;
    }
    public void setBook(Book book) {
        this.book = book;
    }
    public void validate() {
        if (book.getBookName().isEmpty()){
            addFieldError("book.bookName", getText("bookNameError"));
        }
        if (book.getBookPrice() <= 0){
            addFieldError("book.bookPrice", getText("bookPriceError"));
        }
        if (book.getBookPublishDate() == null ||
                              book.getBookPublishDate().after(new Date())){
            addFieldError("book.bookPublishDate", getText("bookPublishDateError"));
        }
        if (book.getBookCount() == null || book.getBookCount() <= 0){
            addFieldError("book.bookCount", getText("bookCountError"));
        }
    }
}
```

使用 getText() 方法获取资源文件中的国际化资源消息。

（4）发布 Web 项目，启动 Tomcat 服务器，访问 addBook.jsp。不输入内容，提交后的结果如图 10-1 所示。

2. 对使用配置文件方式进行输入校验中的校验错误信息的国际化

在使用配置文件方式进行输入校验中，是通过使用＜message＞元素显示校验错误信息的。在进行国际化时，通过使用＜message＞元素的 key 属性来获取国际化信息。

例 10-8 对使用配置文件方式进行输入校验中的校验错误信息的国际化。

在例 10-3 的基础上进行操作，具体操作步骤如下：

（1）编辑国际化资源文件。

ApplicationResources_zh.properties

```
bookNameError = 书名不能为空！
bookPriceRequiredError = 必须输入单价！
bookPriceError = 单价应该大于 ${minExclusive}！
bookPublishDateRequiredError = 必须输入出版日期！
bookPublishDateError = 出版时间应在 ${min}至 ${max}之间！
bookCountRequiredError = 必须输入数量！
bookCountError = 数量要大于 0！
```

(2) 编辑 BookXMLAction-validation.xml 文件。

BookXMLAction - validation.xml

```xml
<?xml version = "1.0" encoding = "UTF - 8"?>
<!DOCTYPE validators PUBLIC
        " - //OpenSymphony Group//XWork Validator 1.0.2//EN"
        "http://www.opensymphony.com/xwork/xwork - validator - 1.0.2.dtd">
<validators>
    <field name = "book.bookName">
        <field - validator type = "requiredstring">
            <message key = "bookNameError" />
        </field - validator>
    </field>
    <field name = "book.bookPrice">
        <field - validator type = "required">
            <message key = "bookPriceRequiredError" />
        </field - validator>
        <field - validator type = "double">
            <param name = "minExclusive">0</param>
            <message key = "bookPriceError" />
        </field - validator>
    </field>
    <field name = "book.bookPublishDate">
        <field - validator type = "required">
            <message key = "bookPublishDateRequiredError" />
        </field - validator>
        <field - validator type = "date">
            <param name = "min">2009 - 01 - 01</param>
            <param name = "max">2011 - 12 - 31</param>
            <message key = "bookPublishDateError" />
        </field - validator>
    </field>
    <field name = "book.bookCount">
        <field - validator type = "required">
            <message key = "bookCountRequiredError" />
        </field - validator>
        <field - validator type = "int">
            <message key = "bookCountError" />
        </field - validator>
    </field>
</validators>
```

(3) 发布 Web 项目，启动 Tomcat 服务器，访问 addBook.jsp。不输入内容，提交后的结果如图 10-3 所示。

10.6　Struts 2 框架的内置校验器

内置校验器都有一个 fieldName 参数，用于设置校验器要验证的表单字段的名称。如果使用 validator 校验语法，则必须设置 fieldName 参数，否则不需要。在介绍 Struts 2 框架的内置校验器时,只介绍其他参数。

10.6.1 类型转换校验器

类型转换校验器(conversion)用于检查该表单字段是否存在类型转换错误,可应用于 field 校验和 validator 校验。该校验器可以在输出的默认类型转换错误消息的基础上输出一条自定义的类型转换错误消息。conversion 校验器的参数 repopulateField 用于设置当出现类型转换错误返回 INPUT 结果码映射的视图时,是否用原值自动填充表单字段。如果需要自动填充,则将 repopulateField 参数值设置为 true。

例 10-9 conversion 校验器示例。

(1) 基于 field 的校验。

```
<field name = "Field Name">
    <field-validator type = "conversion">
        <message>类型转换错误!</message>
    </field-validator>
</field>
```

(2) 基于 validator 的校验。

```
<validator type = "conversion">
    <param name = "fieldName">Field Name</param>
    <message>类型转换错误!</message>
</validator>
```

10.6.2 日期校验器

日期校验器(date)用于检查提供的日期是否在指定的范围内,可应用于 field 校验和 validator 校验。如果不指定日期类型转换器,将默认使用 Date.SHORT 格式。该校验器的参数有:

- min：用于指定最小日期值,如果不指定该参数,则不检查最小日期。
- max：用于指定最大日期值,如果不指定该参数,则不检查最大日期。

例 10-10 date 校验器示例。

(1) 基于 field 的校验。

```
<field name = "bookPublishDate">
    <field-validator type = "date">
        <param name = "min">01/01/2000</param>
        <param name = "max">01/01/2010</param>
        <message>出版日期必须在 ${min}和 ${max}之间!</message>
    </field-validator>
</field>
```

(2) 基于 validator 的校验。

```
<validator type = "date">
    <param name = "fieldName">bookPublishDate</param>
    <param name = "min">01/01/2000</param>
    <param name = "max">01/01/2010</param>
    <message>出版日期必须在${min}和${max}之间!</message>
</validator>
```

10.6.3　双精度浮点数校验器

双精度浮点数校验器(double)用于检查输入的双精度浮点数是否在指定的范围内,可应用于 field 校验和 validator 校验。该校验器的参数有:

- minInclusive

用于指定双精度浮点数的最小值(包含该值)。如果不指定该参数,则不检查双精度浮点数的最小值。

- maxInclusive

用于指定双精度浮点数的最大值(包含该值)。如果不指定该参数,则不检查双精度浮点数的最大值。

- minExclusive

用于指定双精度浮点数的最小值(不包含该值)。如果不指定该参数,则不检查双精度浮点数的最小值。

- maxExclusive

用于指定双精度浮点数的最大值(不包含该值)。如果不指定该参数,则不检查双精度浮点数的最大值。

例 10-11　double 校验器示例。

(1) 基于 field 的校验。

```
<field name = "bookPrice">
    <field-validator type = "double">
        <param name = "maxExclusive">30.1</param>
        <message>单价必须小于${maxExclusive}(不包含)!</message>
    </field-validator>
</field>
```

(2) 基于 validator 的校验。

```
<validator type = "double">
    <param name = "fieldName">bookPrice</param>
    <param name = "minInclusive">10.1</param>
    <param name = "maxInclusive">30.1</param>
    <message>单价必须在${minInclusive}和${maxInclusive}(包含)之间!</message>
</validator>
```

10.6.4 电子邮件校验器

电子邮件校验器(email)用于检查输入的非空字符串是否为合法的电子邮件地址,可应用于 field 校验和 validator 校验。

用于校验字符串是否为电子邮件地址的正则表达式为:

```
\\b(^[_A-Za-z0-9-](\\.[_A-Za-z0-9-])*@([A-Za-z0-9-])+((\\.com)|(\\.net)
|(\\.org)|(\\.info)|
(\\.edu)|(\\.mil)|(\\.gov)|(\\.biz)|(\\.ws)|(\\.us)|(\\.tv)|(\\.cc)|(\\.aero)|(\\.
arpa)|(\\.coop)|(\\.int)|
(\\.jobs)|(\\.museum)|(\\.name)|(\\.pro)|(\\.travel)|(\\.nato)|(\\..{2,3})|(\\..{2,3}\\..
{2,3}))$)\\b
```

例 10-12 email 校验器示例。

(1) 基于 field 的校验。

```
<field name="userEmail">
    <field-validator type="email">
        <message>必须是合法的电子邮件地址!</message>
    </field-validator>
</field>
```

(2) 基于 validator 的校验。

```
<validator type="email">
    <param name="fieldName">userEmail</param>
    <message>必须是合法的电子邮件地址!</message>
</validator>
```

10.6.5 表达式校验器

表达式校验器(expression)基于提供的正则表达式进行校验,只可以应用于 validator 校验。如果校验失败,其错误提示信息通过<s:actionerror/>标签显示。该校验器的参数 expression 基于值栈计算的 OGNL 表达式,结果为布尔类型。如果值为 false,则显示 <message>元素设置的内容。

例 10-13 expression 校验器示例。

基于 validator 的校验。

```
<validator type="expression">
    <param name="expression">!bookName.equals("Struts 2 框架")</param>
    <message>书名不能是"Struts 2 框架"!</message>
</validator>
```

10.6.6 字段表达式校验器

字段表达式校验器(fieldexpression)使用 OGNL 表达式校验表单字段,可应用于 field

校验和 validator 校验。该校验器的参数 expression 基于值栈计算的 OGNL 表达式,结果为布尔类型。如果值为 false,则显示<message>元素设置的内容。

例 10-14 fieldexpression 校验器示例。

(1) 基于 field 的校验。

```
<field name = "myField">
    <field-validator type = "fieldexpression">
        <param name = "expression">OGNL 表达式</param>
        <message>OGNL 表达式的值为 false 时,输出该信息!</message>
    </field-validator>
</field>
```

(2) 基于 validator 的校验。

```
<validator type = "fieldexpression">
    <param name = "fieldName">myField</param>
    <param name = "expression">OGNL 表达式</param>
    <message>OGNL 表达式的值为 false 时,输出该信息!</message>
</validator>
```

10.6.7 整型校验器

整型校验器用于检查输入的整数是否在指定的范围内,可应用于 field 校验和 validator 校验。该校验器的参数有:

- min

用于指定整数的最小值。如果不指定该参数,则不检查整数的最小值。

- max

用于指定整数的最大值。如果不指定该参数,则不检查整数的最大值。

例 10-15 int 校验器示例。

(1) 基于 field 的校验。

```
<field name = "bookCount">
    <field-validator type = "int">
        <param name = "min">100</param>
        <param name = "max">200</param>
        <message>数量必须在 ${min}和 ${max}之间!</message>
    </field-validator>
</field>
```

(2) 基于 validator 的校验。

```
<validator type = "int">
    <param name = "fieldName">bookCount</param>
    <param name = "min">100</param>
    <param name = "max">200</param>
    <message>数量必须在 ${min}和 ${max}之间!</message>
</validator>
```

10.6.8 正则表达式校验器

正则表达式校验器（regex）使用正则表达式校验字段的值，可应用于 field 校验和 validator 校验。该校验器的参数有：

- expression

用于指定正则表达式。必选参数。

- caseSensitive

用于指定正则表达式匹配字符串时是否大小写敏感，默认值为 true，即大小写敏感。可选参数。

- trim

用于指定是否在匹配前删除字符串中的首尾空格，默认值为 true，即删除空格。可选参数。

例 10-16 regex 校验器示例。

（1）基于 field 的校验。

```
<field name = "phone">
    <field-validator type = "regex">
        <param name = "expression"><![CDATA[([1][3,5,8]\\d{9})]]></param>
    </field-validator>
</field>
```

（2）基于 validator 的校验。

```
<validator type = "regex">
    <param name = "fieldName">phone</param>
    <param name = "expression"><![CDATA[([1][3,5,8]\\d{9})]]></param>
</validator>
```

10.6.9 必填校验器

必填校验器（required）校验器用于检查指定的字段值是否不为 null，可应用于 field 校验和 validator 校验。

例 10-17 required 校验器示例。

（1）基于 field 的校验。

```
<field name = "bookCount">
    <field-validator type = "required">
        <message>必须输入数量!</message>
    </field-validator>
</field>
```

（2）基于 validator 的校验。

```
<validator type = "required">
```

```xml
    <param name = "fieldName">bookCount</param>
    <message>必须输入数量!</message>
</validator>
```

10.6.10 必填字符串校验器

必填字符串校验器(requiredstring)用于检查一个字符串是否不为 null,并且长度要大于 0,可应用于 field 校验和 validator 校验。该校验器的参数 trim 用于指定是否在校验前删除字符串中的首尾空格,默认值为 true,即删除空格。

例 10-18　requiredstring 校验器示例。

(1) 基于 field 的校验。

```xml
<field name = "bookName">
    <field-validator type = "requiredstring">
        <param name = "trim">true</param>
        <message>必须输入书名!</message>
    </field-validator>
</field>
```

(2) 基于 validator 的校验。

```xml
<validator type = "requiredstring">
    <param name = "fieldName">bookName</param>
    <param name = "trim">true</param>
    <message>必须输入书名!</message>
</validator>
```

10.6.11 字符串长度校验器

字符串长度校验器(stringlength)用于检查一个字符串是否为指定的长度,可应用于 field 校验和 validator 校验。该校验器的参数有:
- maxLength：用于指定字符串的最大长度。如果指定该参数,则要确保字符串最多有 maxLength 个字符。
- minLength：用于指定字符串的最小长度。如果指定该参数,则要确保字符串最少要有 minLength 个字符。
- trim：用于指定是否在计算字符串长度之前删除字符串中的首尾空格,默认值为 true,即删除空格。

例 10-19　stringlength 校验器示例。

(1) 基于 field 的校验。

```xml
<field name = "bookName">
    <field-validator type = "stringlength">
        <param name = "minLength">10</param>
```

```
            <param name = "trim"> true </param>
            <message>书名长度至少要有${minLength}个字符!</message>
        </field - validator>
</field>
```

(2) 基于 validator 的校验。

```
<validator type = "stringlength">
    <param name = "fieldName"> bookName </param>
    <param name = "maxLength"> 30 </param>
    <param name = "trim"> true </param>
    <message>书名长度最多有${maxLength}个字符!</message>
</validator>
```

10.6.12 网址校验器

网址校验器(url)用于检查给定的字段值是否为字符串,并且是否为合法的 URL,可应用于 field 校验和 validator 校验。

例 10-20 url 校验器示例。

(1) 基于 field 的校验。

```
<field name = "bookURL">
    <field - validator type = "url">
        <message>请输入合法的网址!</message>
    </field - validator>
</field>
```

(2) 基于 validator 的校验。

```
<validator type = "url">
    <param name = "fieldName"> bookURL </param>
    <message>请输入合法的网址!</message>
</validator>
```

10.6.13 visitor 校验器

visitor 校验器允许开发者使用对象自己的校验文件转发校验到 action 的对象属性。允许使用模型驱动(ModelDriven)开发模式,并在一个位置管理模型的校验。该校验器可以处理简单的对象属性,也可以处理对象的集合或者数组。visitor 校验器可应用于 field 校验和 validator 校验,参数有:

- context

指定进行校验的上下文,可选参数。

- appendPrefix

指定添加到字段的前缀,可选参数。

例 10-21 visitor 校验器示例。

(1) 基于 field 的校验。

```xml
<field name = "book">
    <field-validator type = "visitor">
        <param name = "context">myContext</param>
        <param name = "appendPrefix">true</param>
    </field-validator>
</field>
```

(2) 基于 validator 的校验。

```xml
<validator type = "visitor">
    <param name = "fieldName">book</param>
    <param name = "context">myContext</param>
    <param name = "appendPrefix">true</param>
</validator>
```

在该示例中，如果 action 的 getBook()方法返回了 book 对象，Struts 2 框架将会查找 Book-myContext-validation.xml 校验规则文件。由于 appednPrefix 属性值为 true，每一个字段名称都将加上"book."前缀，这样，即名称为"bookName"的字段的实际名称就成了 "book.bookName"。

10.6.14 conditionalvisitor 校验器

conditionalvisitor 校验器是 visitor 校验器的子类。除了具有 visitor 校验器的参数，该校验器的参数 expression 为必选参数。如果表达式的结果是 true，则调用 visitor 校验器进行校验。

例 10-22 conditionalvisitor 校验器示例。

(1) 基于 field 的校验。

```xml
<field name = "colleaguePosition">
    <field-validator type = "conditionalvisitor">
        <param name = "expression">reason == 'colleague'</param>
        <message/>
    </field-validator>
</field>
```

(2) 基于 validator 的校验。

```xml
<validator type = "conditionalvisitor">
    <param name = "fieldName">colleaguePosition</param>
    <message/>
</validator>
```

10.7 自定义校验器

使用自定义校验器主要有三个步骤：
（1）首先创建自定义校验器类。
（2）然后注册自定义校验器类。
（3）最后使用注册的自定义校验器。

Struts 2 框架提供的内置校验器已经实现了前两个步骤，因此，在 Web 应用中可以直接使用它们。Struts 2 框架是在 default.xml（位于 com.opensymphony.xwork2.validator.validator 包）文件中注册的内置校验器。

本节以验证字符串是否符合邮政编码格式为例，讲解如何开发自定义的校验器。

10.7.1 创建自定义校验器类

创建校验器类必须实现 Validator（com.opensymphony.xwork2.validator.Validator）接口，接口定义如下：

Validator.java

```
package com.opensymphony.xwork2.validator;
import com.opensymphony.xwork2.util.ValueStack;
public interface Validator<T> {
    void setDefaultMessage(String message);
    String getDefaultMessage();
    String getMessage(Object object);
    void setMessageKey(String key);
    String getMessageKey();
    void setMessageParameters(String[] messageParameters);
    String[] getMessageParameters();
    void setValidatorContext(ValidatorContext validatorContext);
    ValidatorContext getValidatorContext();
    void validate(Object object) throws ValidationException;
    void setValidatorType(String type);
    String getValidatorType();
    void setValueStack(ValueStack stack);
}
```

在该接口中有多个方法需要实现。为了简化开发，Struts 2 框架提供了 Validator 接口的两个实现类 ValidatorSupport 和 FieldValidatorSupport，其中 ValidatorSupport 实现了 Validator 接口，用于定义 validator 验证器，FieldValidatorSupport 是 ValidatorSupport 的子类，用于定义 field 校验器。开发者只要继承这两个实现类之一就可以创建自定义的校验器类。

例 10-23 创建验证字符串是否符合邮政编码格式的校验器类。

ZipCodeValidator.java

```java
package example.struts2;
import java.util.regex.Matcher;
import java.util.regex.Pattern;
import com.opensymphony.xwork2.validator.ValidationException;
import com.opensymphony.xwork2.validator.validators.FieldValidatorSupport;
public class ZipCodeValidator extends FieldValidatorSupport {
    public static final String zipCodePattern = "\\d\\d\\d\\d\\d\\d";
    public void validate(Object obj) throws ValidationException {
        String fieldName = getFieldName();
        String value = (String)getFieldValue(fieldName, obj);
        Pattern pattern = Pattern.compile(zipCodePattern);
        String compare = value.trim();
        Matcher matcher = pattern.matcher( compare );
        if (!matcher.matches()) {
            addFieldError(fieldName, obj);
        }
    }
}
```

该校验器类继承了 FieldValidatorSupport 类，并重写了 validate()方法。在该校验器类中定义了邮政编码的格式。

10.7.2　注册自定义校验器类

自定义校验器类是在 validators.xml 文件中注册的，并且必须保存在 Web 应用的 WEB-INF/classes 目录下或者 CLASSPATH 中。validators.xml 文件的结构由 http://www.opensymphony.com/xwork/xwork-validator-config-1.0.dtd 文件定义。格式如下：

xwork-validator-config-1.0.dtd

```
<?xml version = "1.0" encoding = "UTF-8"?>
<!--
    XWork Validator Config DTD.
    Used the following DOCTYPE.
    <!DOCTYPE validators PUBLIC
        "-//OpenSymphony Group//XWork Validator Config 1.0//EN"
        "http://www.opensymphony.com/xwork/xwork-validator-config-1.0.dtd">
-->
<!ELEMENT validators (validator) + >
<!ELEMENT validator (#PCDATA)>
<!ATTLIST validator
    name CDATA #REQUIRED
    class CDATA #REQUIRED
>
```

定义文件指出，注册自定义校验器的文件中要使用如下的文档声明：

```
<!DOCTYPE validators PUBLIC
    "-//OpenSymphony Group//XWork Validator Config 1.0//EN"
    "http://www.opensymphony.com/xwork/xwork-validator-config-1.0.dtd">
```

注册文件的根元素是＜validators＞，其包含一个或多个＜validator＞子元素。
＜validator＞子元素具有两个必选属性：
- name 属性：用于指定校验器的名称。
- class 属性：用于指定校验器的完整实现类。

例 10-24 注册自定义校验器类(ZipCodeValidator)。

在 src 文件夹下创建 validators.xml 文件，内容如下：

validators.xml

```xml
<?xml version="1.0" encoding="UTF-8"?>
<!DOCTYPE validators PUBLIC
    "-//OpenSymphony Group//XWork Validator Config 1.0//EN"
    "http://www.opensymphony.com/xwork/xwork-validator-config-1.0.dtd">
<validators>
    <validator name="zipcoder" class="example.struts2.ZipCodeValidator"/>
</validators>
```

10.7.3 使用自定义校验器

(1) 创建 JSP 文件。

zipCode.jsp

```jsp
<%@ page language="java" contentType="text/html; charset=UTF-8"
    pageEncoding="UTF-8"%>
<%@ taglib prefix="s" uri="/struts-tags" %>
<s:actionerror/>
<s:form action="zipCode" >
    <s:textfield name="zipCode" label="邮政编码"/>
    <s:submit value="提交"/>
</s:form>
<s:if test="zipCode!=null">
    输入的邮政编码是：<s:property value="zipCode"/>
</s:if>
```

(2) 创建 Action。

ZipCodeAction.java

```java
package example.struts2;
import com.opensymphony.xwork2.ActionSupport;
public class ZipCodeAction extends ActionSupport {
    private String zipCode;
```

```java
        public String getZipCode() {
            return zipCode;
        }
        public void setZipCode(String zipCode) {
            this.zipCode = zipCode;
        }
    }
```

(3) 创建校验规则文件。

ZipCodeAction-validation.xml

```xml
<?xml version="1.0" encoding="UTF-8"?>
<!DOCTYPE validators PUBLIC
        "-//OpenSymphony Group//XWork Validator 1.0.2//EN"
        "http://www.opensymphony.com/xwork/xwork-validator-1.0.2.dtd">
<validators>
    <field name="zipCode">
        <field-validator type="requiredstring" short-circuit="true">
            <message>必须输入邮政编码!</message>
        </field-validator>
        <field-validator type="stringlength" short-circuit="true">
            <param name="minLength">6</param>
            <param name="maxLength">6</param>
            <message>邮政编码必须是6位!</message>
        </field-validator>
        <field-validator type="zipcode">
            <message>邮政编码格式不正确!</message>
        </field-validator>
    </field>
</validators>
```

(4) 创建配置文件。

struts.xml

```xml
<?xml version="1.0" encoding="UTF-8" ?>
<!DOCTYPE struts PUBLIC
    "-//Apache Software Foundation//DTD Struts Configuration 2.0//EN"
    "http://struts.apache.org/dtds/struts-2.0.dtd">
<struts>
    <package name="default" extends="struts-default">
        <action name="zipCode" class="example.struts2.ZipCodeAction">
            <result>/zipCode.jsp</result>
            <result name="input">/zipCode.jsp</result>
        </action>
    </package>
</struts>
```

(5) 发布 Web 项目，启动 Tomcat 服务器，访问 zipCode.jsp，并进行测试。

① 不输入信息提交的结果如图 10-11 所示。

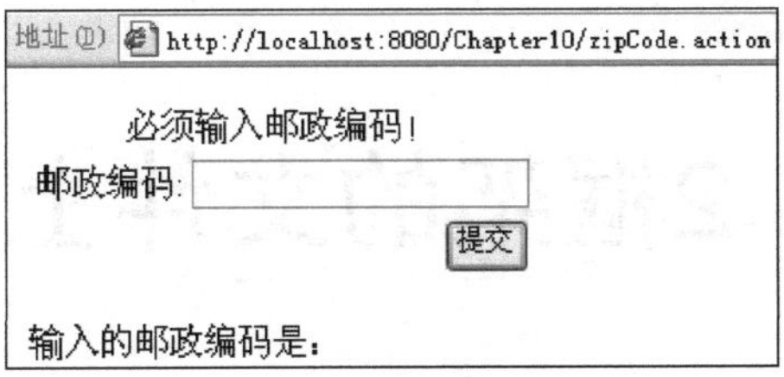

图 10-11　不输入信息提交后的效果

② 输入部分信息提交的结果如图 10-12 所示。

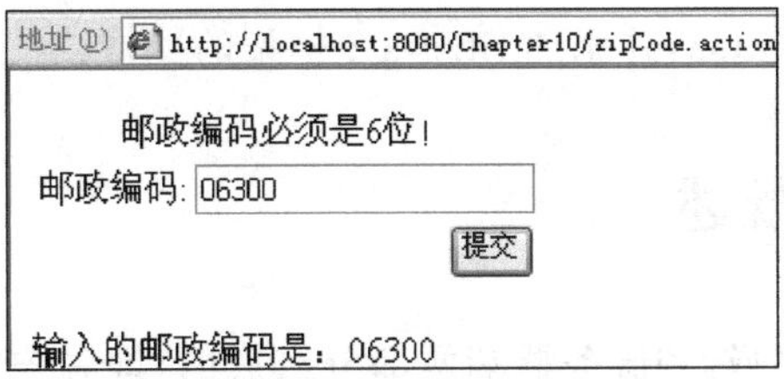

图 10-12　输入部分信息提交后的效果

③ 输入正确信息提交的结果如图 10-13 所示。

图 10-13　输入正确信息提交后的效果

习题

1. 修改例 10-3，校验规则文件名称改为 BookFieldAction-add-validation.xml。
2. 修改例 10-4，混合使用 ＜field＞ 子元素和 ＜validator＞ 子元素定义校验规则文件中的校验规则。
3. 验证校验规则的搜索顺序。
4. 修改例 10-23 中的自定义校验器类，使其继承 ValidatorSupport 类。

第11章 基于Struts 2框架的文件上传和下载

文件的上传与下载是 Web 应用中经常使用的功能，本章将介绍基于 Struts 2 框架的文件上传与下载功能的实现。为了方便本章内容的讲解，建立 Web 项目 Chapter11，在 src 文件夹下创建包 example.struts2，并添加 Struts 2 框架的支持。

11.1 文件上传概述

文件上传功能是由客户端和服务器端两部分组成的，即客户端文件选取部分和服务器端文件存储部分。其中客户端文件选取部分可以使用 HTML 中的＜input＞元素或者 Struts 2 框架中的＜s:file＞标签。服务器端文件存储部分，需要开发人员编写代码获取上传文件的内容及保存内容到服务器中的某个文件中。处理文件上传的服务器端代码通常较复杂，为了简化文件上传代码编写的工作量，一些公司开发了文件上传组件，如 Apache 文件上传组件。使用文件上传组件，可以提高开发效率。

11.1.1 文件上传组件

Apache 文件上传组件（commons-fileupload）是一个实用的第三方软件。其下载网址是：

```
http://commons.apache.org/fileupload/
```

截至作者编写本章内容，commons-fileupload 文件上传组件的最新版本是 1.2.2。从网站下载 bin 压缩包（commons-fileupload-1.2.2-bin.zip），压缩包中包含一个 Jar 文件 commons-fileupload-1.2.2.jar，该 Jar 文件是 commons-fileupload 组件的类库。

文件上传组件工作时要依赖于另外一个组件 commons-io。commons-io 组件的下载网址是：

```
http://commons.apache.org/io/
```

截至作者编写本章内容，commons-io 组件的最新版本是 2.1。下载 bin 压缩包（commons-io-2.1-bin.zip），压缩包中包含三个 Jar 文件，其中 commons-io-2.1.jar 是 commons-io 的类库，commons-io-2.1-sources.jar 是源代码，commons-io-2.1-javadoc.jar 是 API 文档。

只要将 commons-io-2.1.jar 和 commons-fileupload-1.2.2.jar 两个 Jar 文件添加到 Web 应用的 WEB-INF/lib 文件夹下，就可以轻松编写文件上传代码。

11.1.2　基于 Struts 2 框架的文件上传开发体验

Struts 2 框架对文件上传组件提供了很好的支持，在 Struts 2 框架中已经包含了 commons-io-2.0.1.jar 和 commons-fileupload-1.2.2.jar 这两个 Jar 文件，因此，基于 Struts 2 框架实现文件上传功能时，不用单独下载 commons-io 和 commons-fileupload 这两个组件。

例 11-1　基于 Struts 2 框架的文件上传。

具体操作步骤如下：

（1）编辑 web.xml 文件，文件内容同 Web 项目 Chapter10 中 web.xml 文件的内容。

（2）创建 JSP 文件，用于选取客户端文件。

uploadFile.jsp

```jsp
<%@ page language="java" contentType="text/html; charset=UTF-8"
    pageEncoding="UTF-8"%>
<%@ taglib prefix="s" uri="/struts-tags" %>
<s:form action="uploadFile" method="post" enctype="multipart/form-data">
    <s:file name="upload" label="请选择上传的文件"/>
    <s:submit value="上传"/>
</s:form>
<s:if test="upload!=null">
    上传文件信息如下：<br>
    上传文件名称：<s:property value="uploadFileName"/><br>
    上传文件类型：<s:property value="uploadContentType"/><br>
</s:if>
```

（3）创建 Action，用于服务器端上传的文件存储。

UploadFileAction.java

```java
package example.struts2;
import java.io.File;
import java.io.IOException;
import org.apache.commons.io.FileUtils;
import org.apache.struts2.ServletActionContext;
import com.opensymphony.xwork2.ActionSupport;
public class UploadFileAction extends ActionSupport {
    private File sourceFile;
    private String fileName;
    private String contentType;
    public File getUpload() {
        return sourceFile;
    }
    public void setUpload(File sourceFile) {
```

```java
            this.sourceFile = sourceFile;
        }
        public String getUploadFileName() {
            return fileName;
        }
        public void setUploadFileName(String fileName) {
            this.fileName = fileName;
        }
        public String getUploadContentType() {
            return contentType;
        }
        public void setUploadContentType(String contentType) {
            this.contentType = contentType;
        }
        public String execute() throws IOException{
            if(sourceFile != null){
                String serverRealPath = 
                    ServletActionContext.getServletContext().getRealPath("/UploadFiles");
                File targetFile = new File(serverRealPath, getUploadFileName());
                FileUtils.copyFile(sourceFile, targetFile);
            }
            else{
                addActionMessage("请选择文件!");
            }
            return SUCCESS;
        }
    }
```

（4）创建配置文件。

struts.xml

```xml
<?xml version="1.0" encoding="UTF-8" ?>
<!DOCTYPE struts PUBLIC
    "-//Apache Software Foundation//DTD Struts Configuration 2.0//EN"
    "http://struts.apache.org/dtds/struts-2.0.dtd">
<struts>
    <package name="default" extends="struts-default">
        <action name="uploadFile" class="example.struts2.UploadFileAction">
            <result>/uploadFile.jsp</result>
        </action>
    </package>
</struts>
```

（5）发布 Web 项目，启动 Tomcat 服务器，访问 uploadFile.jsp，结果如图 11-1 所示。选择文件后上传，结果如图 11-2 所示。

在文件上传时，使用了 fileUpload 拦截器，该拦截器包含在默认的拦截器栈（defaultStack）中。

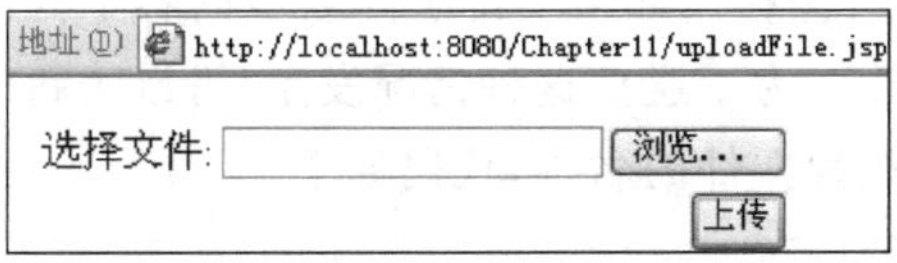

第11章　基于Struts 2框架的文件上传和下载　271

图 11-1　访问 uploadFile.jsp 的效果

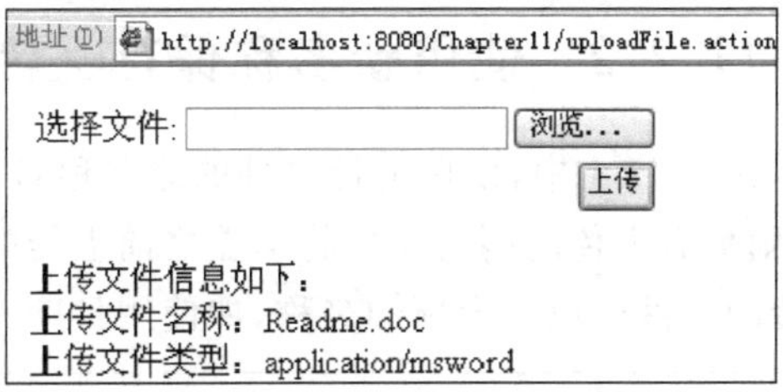

图 11-2　上传文件后的效果

11.2　上传单个文件

例 11-1 给出了基于 Struts 2 框架的文件上传功能的基本开发步骤。在本节将对文件上传功能进行丰富,将考虑以下一些问题。

11.2.1　不对保存上传文件的目录进行硬编码

在例 11-1 中,服务器端保存上传文件的目录被硬编码到了 Action 中,通过如下修改,则可以实现对保存上传文件的目录不进行硬编码。

(1) 在配置文件中使用＜param＞元素,将保存上传文件的目录传递给 Action,形如:

```
<action name = "actionName" class = "ActionClass">
    <param name = "targetDir">/UploadFiles</param>
</action>
```

(2) 在 Action 中,添加 targetDir 属性声明及该属性的 getter 和 setter 方法,并获取目标目录的绝对路径,如下:

```
…
private String targetDir;
public String getTargetDir() {
    return targetDir;
}
public void setTargetDir(String targetDir) {
    this.targetDir = targetDir;
}
public String execute() throws IOException{
    String serverRealPath =
            ServletActionContext.getServletContext().getRealPath(targetDir);
    File dir = new File(serverRealPath);
    if (!dir.exists()){
        dir.mkdir();
    }
    …
}
```

这样,就不用将保存上传文件的目录硬编码到 Action 中了。

11.2.2 使用新名称保存上传的文件

在例 11-1 中,使用上传文件的原名称对其进行保存。如果上传了同名但内容不同的文件,则最后上传的同名文件将覆盖之前上传的文件。为了避免这种情况发生,可以为每一个上传的文件生成一个新的名称,如使用日期时间作为文件的名称,代码如下:

```
…
SimpleDateFormat sf = new SimpleDateFormat("yyyyMMddHHmmssSSS");
String newFileName = sf.format(Calendar.getInstance().getTime());
int beginIndex = fileName.lastIndexOf(".");
if(beginIndex > 0){
    String fileExt = fileName.substring(beginIndex);
    newFileName = newFileName + fileExt;
}
…
```

newFileName 是使用"yyyyMMddHHmmssSSS"格式的日期时间生成的新的文件名称。

11.2.3 对上传文件的大小及类型进行限制

在上传文件的时候,通常要对上传的文件进行限制,如限制上传文件的大小及类型等。Struts 2 框架有一个内置的用于文件上传的拦截器 FileUpload,FileUpload 拦截器有三个可选参数:maximumSize、allowedTypes 和 allowedExtensions,可以实现对上传文件的限制。其中 maximumSize 参数设置拦截器允许 action 中处理的上传文件的最大字节数;allowedTypes 参数设置拦截器允许 action 中处理的用逗号分隔的上传文件的内容类型列表;allowedExtensions 设置拦截器允许 action 中处理的用逗号分隔的上传文件的扩展名列表。

因此,只要在配置文件中定义用于文件上传的 action 时,将相关的参数传递给 FileUpload 拦截器即可实现对上传文件的限制。

配置如下:

```
<package name = "packageName" extends = "struts - default">
    <action name = "actionName" class = "ActionClass">
        <result>… …</result>
        <result name = "input">… …</result>
        <interceptor - ref name = "defaultStack">
            <param name = "fileUpload.maximumSize">1024</param>
            <param name = "fileUpload.allowedTypes">application/pdf</param>
        </interceptor - ref>
    </action>
</package>
```

11.2.4 上传文件属性的配置

在 Struts 2 框架的属性配置文件中,有三个用于设置文件上传的属性:

(1) struts.multipart.parser:设置用于处理 HTTP POST 请求的解析器,该请求使用 MIME 类型 multipart/form-data 编码,默认值为 jakarta。其实,如果不使用拦截器设置上传文件的大小,在该解析器的实现类中默认上传文件的大小最大为 2MB。

(2) struts.multipart.saveDir:设置上传文件临时保存到服务器端的目录,如果指定的目录不存在,Struts 2 框架会自动创建。

(3) struts.multipart.maxSize:设置上传文件的最大字节数。设置该属性后,当上传的文件大小超过指定值时,就不能上传文件。和通过 fileUpload 拦截器设置允许上传文件大小参数(maximumSize)不同之处在于:使用 struts.multipart.maxSize 属性,当上传文件大小超过指定值时,文件并不上传到服务器,而使用 fileUpload 拦截器时,即使上传文件大小超过了指定值,文件也上传到了服务器。

配置如下:

struts.properties

```
struts.multipart.saveDir = E:/tempFiles
struts.multipart.maxSize = 1024
```

该文件保存在 Web 应用的 WEB-INF/classes 目录下。

11.2.5 对上传文件错误消息进行国际化

Struts 2 框架对文件上传错误信息提供了很好的国际化支持。如果文件上传的 action 实现了 com.opensymphony.xwork2.ValidationAware 接口或者继承了 com.opensymphony.xwork2.ActionSupport 类,那么在文件上传发生错误时,Struts 2 框架将会向 action 添加几个字段错误消息,这些错误消息是基于 struts-messages.properties 属性文件中的几个 i18n 值,该属性文件是所有的 i18n 请求都要处理的一个默认属性文件。开发者可以通过在资源文件中设置下面 i18n 值来覆盖 struts-messages.properties 属性文件中的 key 进行国际化。

(1) struts.messages.error.uploading:用于设置文件上传失败时显示的通用错误消息。

(2) struts.messages.error.file.too.large:用于设置当上传文件的大小超出了 maximumSize 指定的值时显示的错误消息。

(3) struts.messages.error.content.type.not.allowed:用于设置当上传文件是不允许的内容类型时显示的错误消息。

资源文件内容格式如下:

upload_zh.properties

```
struts.messages.error.uploading = 文件上传时出现错误:{0}
struts.messages.error.file.too.large = 文件太大:{0} "{1}" "{2}" {3}
struts.messages.error.content.type.not.allowed = 内容类型不允许:{0} "{1}" "{2}" {3}
```

11.2.6 上传单个文件示例

例 11-2 上传单个文件示例。

具体操作步骤如下：

(1) 创建 JSP 文件。

uploadOneFile.jsp

```jsp
<%@ page language="java" contentType="text/html; charset=UTF-8"
    pageEncoding="UTF-8"%>
<%@ taglib prefix="s" uri="/struts-tags" %>
<s:form action="uploadOneFile" method="post" enctype="multipart/form-data">
    <s:file name="upload" label="选择文件"/>
    <s:submit value="上传"/>
</s:form>
<s:if test="upload!=null">
    上传文件信息如下：<br>
    上传文件名称：<s:property value="uploadFileName"/><br>
    上传文件类型：<s:property value="uploadContentType"/><br>
    上传文件保存位置：<s:property value="targetDir"/><br>
    上传文件新名称：<s:property value="targetFileName"/><br>
</s:if>
```

(2) 创建 Action。

UploadOneFileAction.java

```java
package example.struts2;
import java.io.File;
import java.io.IOException;
import java.text.SimpleDateFormat;
import java.util.Calendar;
import org.apache.commons.io.FileUtils;
import org.apache.struts2.ServletActionContext;
import com.opensymphony.xwork2.ActionSupport;
public class UploadOneFileAction extends ActionSupport {
    private File sourceFile;
    private String fileName;
    private String contentType;
    private String targetDir;
    private String targetFileName;
    public File getUpload() {
        return sourceFile;
    }
    public void setUpload(File sourceFile) {
        this.sourceFile = sourceFile;
    }
    public String getUploadFileName() {
```

```java
        return fileName;
    }
    public void setUploadFileName(String fileName) {
        this.fileName = fileName;
    }
    public String getUploadContentType() {
        return contentType;
    }
    public void setUploadContentType(String contentType) {
        this.contentType = contentType;
    }
    public String getTargetDir() {
        return targetDir;
    }
    public void setTargetDir(String targetDir) {
        this.targetDir = targetDir;
    }
    public String getTargetFileName() {
        return targetFileName;
    }
    public void setTargetFileName(String targetFileName) {
        this.targetFileName = targetFileName;
    }
    public String execute() throws IOException{
        if(sourceFile != null){
            String serverRealPath = 
                ServletActionContext.getServletContext().getRealPath(targetDir);
            File dir = new File(serverRealPath);
            if (!dir.exists()){
                dir.mkdir();
            }
            targetFileName = genNewFileName(fileName);
            File targetFile = new File(serverRealPath, targetFileName);
            FileUtils.copyFile(sourceFile, targetFile);
        }
        else{
            addActionMessage("请选择文件!");
        }
        return SUCCESS;
    }
    private String genNewFileName(String fileName){
        SimpleDateFormat sf = new SimpleDateFormat("yyyyMMddHHmmssSSS");
        String newFileName = sf.format(Calendar.getInstance().getTime());
        int beginIndex = fileName.lastIndexOf(".");
        if(beginIndex > 0){
            String fileExt = fileName.substring(beginIndex);
            newFileName = newFileName + fileExt;
```

```
            }
            return newFileName;
    }
}
```

(3) 创建资源文件。

upload_zh.properties

```
struts.messages.error.uploading = 文件上传错误：{0}
struts.messages.error.file.too.large = 文件太大：{0} "{1}" "{2}" {3}
struts.messages.error.content.type.not.allowed = 内容类型不允许：{0} "{1}" "{2}" {3}
```

(4) 编辑配置文件。

struts.xml

```xml
<?xml version = "1.0" encoding = "UTF-8" ?>
<!DOCTYPE struts PUBLIC
    "-//Apache Software Foundation//DTD Struts Configuration 2.0//EN"
    "http://struts.apache.org/dtds/struts-2.0.dtd">
<struts>
    <package name = "default" extends = "struts-default">
        <action name = "uploadOneFile" class = "example.struts2.UploadOneFileAction">
            <result>/uploadOneFile.jsp</result>
            <result name = "input">/uploadOneFile.jsp</result>
            <param name = "targetDir">/Upload</param>
            <interceptor-ref name = "defaultStack">
                <param name = "fileUpload.maximumSize">40960</param>
                <param name = "fileUpload.allowedTypes">application/msword</param>
            </interceptor-ref>
        </action>
    </package>
    <constant name = "struts.custom.i18n.resources" value = "upload"/>
</struts>
```

(5) 发布 Web 项目，启动 Tomcat 服务器，进行测试。

① 如果文件上传成功，结果如图 11-3 所示。

图 11-3 文件上传成功后的效果

② 如果上传的文件超过了 40KB，结果如图 11-4 所示。

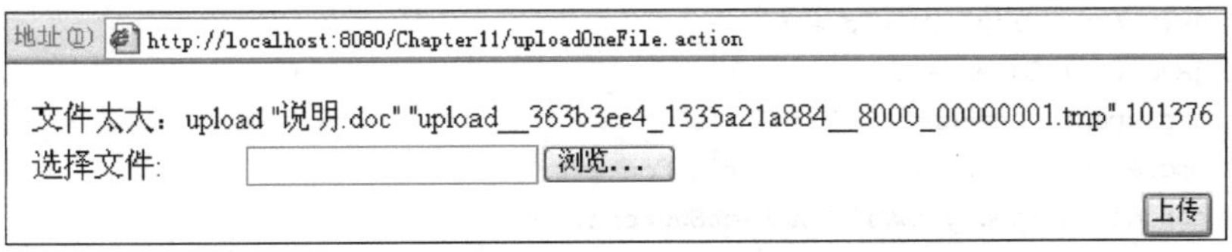

图 11-4　上传文件超过了限定值后的效果

11.3　上传多个文件

Struts 2 框架支持同时上传多个文件，有使用数组和 List 两种实现方式。

11.3.1　使用数组方式实现多文件上传

例 11-3　使用数组实现多文件上传。
具体操作步骤如下：
（1）创建 JSP 文件。

uploadFiles.jsp

```jsp
<%@ page language="java" contentType="text/html; charset=UTF-8"
                                       pageEncoding="UTF-8" %>
<%@ taglib prefix="s" uri="/struts-tags" %>
<s:form action="uploadFiles" method="post" enctype="multipart/form-data">
    <s:file name="upload" label="选择文件1"/>
    <s:file name="upload" label="选择文件2"/>
    <s:submit value="上传"/>
</s:form>

<s:if test="upload!=null">
    上传文件保存位置：<s:property value="targetDir"/><br>
    <s:iterator value="targetFileName" status="status">
        <br>上传文件<s:property value="%{#status.index+1}"/>的信息如下：<br>
        文件名称：<s:property value="%{uploadFileName[#status.index]}"/><br>
        文件类型：<s:property value="%{uploadContentType[#status.index]}"/><br>
        文件新名称：<s:property value="%{targetFileName[#status.index]}"/><br>
    </s:iterator>
</s:if>
```

使用多个同名的<s:file>标签，用于文件选取。
（2）创建 Action。

uploadFilesAction.java

```java
package example.struts2;
import java.io.File;
```

```java
import java.io.IOException;
import java.text.SimpleDateFormat;
import java.util.Calendar;
import org.apache.commons.io.FileUtils;
import org.apache.struts2.ServletActionContext;
import com.opensymphony.xwork2.ActionSupport;
public class UploadFilesAction extends ActionSupport {
    private File[] sourceFile;
    private String[] fileName;
    private String[] contentType;
    private String[] targetFileName;
    private String targetDir;
    private int filesCount;
    public File[] getUpload() {
        return sourceFile;
    }
    public void setUpload(File[] sourceFile) {
        this.sourceFile = sourceFile;
    }
    public String[] getUploadContentType() {
        return contentType;
    }
    public void setUploadContentType(String[] contentType) {
        this.contentType = contentType;
    }
    public String[] getUploadFileName() {
        return fileName;
    }
    public void setUploadFileName(String[] fileName) {
        this.fileName = fileName;
    }
    public String[] getTargetFileName() {
        return targetFileName;
    }
    public void setTargetFileNames(String[] targetFileName) {
        this.targetFileName = targetFileName;
    }
    public String getTargetDir() {
        return targetDir;
    }
    public void setTargetDir(String targetDir) {
        this.targetDir = targetDir;
    }
    public String execute() throws IOException{
        if(sourceFile != null){
            String serverRealPath =
                    ServletActionContext.getServletContext().getRealPath(targetDir);
            File dir = new File(serverRealPath);
            if (!dir.exists()){
```

```java
                    dir.mkdir();
                }
                filesCount = sourceFile.length;
                targetFileName = new String[filesCount];
                for(int i = 0; i<filesCount; i++){
                    targetFileName[i] = genNewFileName(fileName[i],i);
                    File targetFile = new File(serverRealPath, targetFileName[i]);
                    FileUtils.copyFile(sourceFile[i], targetFile);
                }
            }
            else{
                addActionMessage("请选择文件!");
            }
            return SUCCESS;
    }
    private String genNewFileName(String fileName, int curIndex){
        SimpleDateFormat sf = new SimpleDateFormat("yyyyMMddHHmmssSSS");
        String newFileName = sf.format(Calendar.getInstance().getTime());
        int beginIndex = fileName.lastIndexOf(".");
        if(beginIndex > 0){
            String fileExt = fileName.substring(beginIndex);
            newFileName = newFileName + curIndex + fileExt;
        }
        return newFileName;
    }
}
```

(3) 编辑配置文件。

struts.xml

```xml
<?xml version = "1.0" encoding = "UTF-8" ?>
<!DOCTYPE struts PUBLIC
    "-//Apache Software Foundation//DTD Struts Configuration 2.0//EN"
    "http://struts.apache.org/dtds/struts-2.0.dtd">
<struts>
    <package name = "default" extends = "struts-default">
        <action name = "uploadFiles" class = "example.struts2.UploadFilesAction">
            <result>/uploadFiles.jsp</result>
            <result name = "input">/uploadFiles.jsp</result>
            <param name = "targetDir">/Upload</param>
        </action>
    </package>
</struts>
```

(4) 发布 Web 项目,启动 Tomcat 服务器,访问 uploadFiles.jsp 进行测试。

① 选择两个文件上传,结果如图 11-5 所示。

② 选择一个文件上传,结果如图 11-6 所示。

图 11-5　上传两个文件的效果 　　　　　图 11-6　上传一个文件的效果

11.3.2　使用 List 方式实现多文件上传

例 11-4　使用 List 泛型实现多文件上传。

在例 11-3 的基础上操作，具体操作步骤如下：

(1) 创建 Action。

uploadFilesListAction.java

```java
package example.struts2;
import java.io.File;
import java.io.IOException;
import java.text.SimpleDateFormat;
import java.util.Calendar;
import java.util.List;
import org.apache.commons.io.FileUtils;
import org.apache.struts2.ServletActionContext;
import com.opensymphony.xwork2.ActionSupport;
public class UploadFilesListAction extends ActionSupport {
    private List<File> sourceFile;
    private List<String> fileName;
    private List<String> contentType;
    private String[] targetFileName;
    private String targetDir;
    private int filesCount;
    public List<File> getUpload() {
        return sourceFile;
    }
    public void setUpload(List<File> sourceFile) {
        this.sourceFile = sourceFile;
    }
    public List<String> getUploadContentType() {
        return contentType;
```

```java
    }
    public void setUploadContentType(List<String> contentType) {
        this.contentType = contentType;
    }
    public List<String> getUploadFileName() {
        return fileName;
    }
    public void setUploadFileName(List<String> fileName) {
        this.fileName = fileName;
    }
    public String[] getTargetFileName() {
        return targetFileName;
    }
    public void setTargetFileNames(String[] targetFileName) {
        this.targetFileName = targetFileName;
    }
    public String getTargetDir() {
        return targetDir;
    }
    public void setTargetDir(String targetDir) {
        this.targetDir = targetDir;
    }
    public String execute() throws IOException{
        if(sourceFile != null){
            String serverRealPath =
                    ServletActionContext.getServletContext().getRealPath(targetDir);
            File dir = new File(serverRealPath);
            if (!dir.exists()){
                dir.mkdir();
            }
            filesCount = sourceFile.size();
            targetFileName = new String[filesCount];
            for (int i = 0; i<filesCount; i++){
                targetFileName[i] = genNewFileName(fileName.get(i),i);
                File targetFile = new File(serverRealPath, targetFileName[i]);
                FileUtils.copyFile(sourceFile.get(i), targetFile);
            }
        }
        else{
            addActionMessage("请选择文件!");
        }
            return SUCCESS;
    }
    private String genNewFileName(String fileName, int curIndex){
        SimpleDateFormat sf = new SimpleDateFormat("yyyyMMddHHmmssSSS");
        String newFileName = sf.format(Calendar.getInstance().getTime());
        int beginIndex = fileName.lastIndexOf(".");
        if(beginIndex > 0){
            String fileExt = fileName.substring(beginIndex);
```

```
            newFileName = newFileName + curIndex + fileExt;
        }
        return newFileName;
    }
}
```

(2) 编辑配置文件。

struts.xml

```xml
<?xml version = "1.0" encoding = "UTF-8" ?>
<!DOCTYPE struts PUBLIC
    "-//Apache Software Foundation//DTD Struts Configuration 2.0//EN"
    "http://struts.apache.org/dtds/struts-2.0.dtd">

<struts>
    <package name = "default" extends = "struts-default">
        <action name = "uploadFiles" class = "example.struts2.UploadFilesListAction">
            <result>/uploadFiles.jsp</result>
            <result name = "input">/uploadFiles.jsp</result>
            <param name = "targetDir">/Upload</param>
        </action>
    </package>
</struts>
```

(3) 发布 Web 项目,启动 Tomcat 服务器,访问 uploadFiles.jsp 进行测试。

例 11-5 不使用 List 泛型实现多文件上传。

在例 11-3 的基础上进行操作,具体操作步骤如下:

(1) 创建类型转换文件。

uploadFilesList1Action-conversion.properties

```
Element_upload = java.io.File
```

(2) 创建 Action。

uploadFilesList1Action.java

```java
package example.struts2;
import java.io.File;
import java.io.IOException;
import java.text.SimpleDateFormat;
import java.util.Calendar;
import java.util.List;
import org.apache.commons.io.FileUtils;
import org.apache.struts2.ServletActionContext;
import com.opensymphony.xwork2.ActionSupport;
public class UploadFilesList1Action extends ActionSupport {
    private List sourceFile;
```

```java
    private List fileName;
    private List contentType;
    private String[] targetFileName;
    private String targetDir;
    private int filesCount;
    public List getUpload() {
        return sourceFile;
    }
    public void setUpload(List<File> sourceFile) {
        this.sourceFile = sourceFile;
    }
    public List getUploadContentType() {
        return contentType;
    }
    public void setUploadContentType(List contentType) {
        this.contentType = contentType;
    }
    public List getUploadFileName() {
        return fileName;
    }
    public void setUploadFileName(List fileName) {
        this.fileName = fileName;
    }
    public String[] getTargetFileName() {
        return targetFileName;
    }
    public void setTargetFileNames(String[] targetFileName) {
        this.targetFileName = targetFileName;
    }
    public String getTargetDir() {
        return targetDir;
    }
    public void setTargetDir(String targetDir) {
        this.targetDir = targetDir;
    }
    public String execute() throws IOException{
        if(sourceFile != null){
            String serverRealPath =
                    ServletActionContext.getServletContext().getRealPath(targetDir);
            File dir = new File(serverRealPath);
            if (!dir.exists()){
                dir.mkdir();
            }
            filesCount = sourceFile.size();
            targetFileName = new String[filesCount];
            for (int i = 0; i < filesCount; i++){
                targetFileName[i] = genNewFileName(fileName.get(i).toString(),i);
                File targetFile = new File(serverRealPath, targetFileName[i]);
                FileUtils.copyFile((File)sourceFile.get(i), targetFile);
```

```
                }
            }
            else{
                addActionMessage("请选择文件!");
            }
            return SUCCESS;
    }
    private String genNewFileName(String fileName, int curIndex){
        SimpleDateFormat sf = new SimpleDateFormat("yyyyMMddHHmmssSSS");
        String newFileName = sf.format(Calendar.getInstance().getTime());
        int beginIndex = fileName.lastIndexOf(".");
        if(beginIndex > 0){
            String fileExt = fileName.substring(beginIndex);
            newFileName = newFileName + curIndex + fileExt;
        }
        return newFileName;
    }
}
```

（3）编辑配置文件。

struts.xml

```
<?xml version = "1.0" encoding = "UTF-8" ?>
<!DOCTYPE struts PUBLIC
    "-//Apache Software Foundation//DTD Struts Configuration 2.0//EN"
    "http://struts.apache.org/dtds/struts-2.0.dtd">
<struts>
    <package name = "default" extends = "struts-default">
        <action name = "uploadFiles" class = "example.struts2.UploadFilesList1Action">
            <result>/uploadFiles.jsp</result>
            <result name = "input">/uploadFiles.jsp</result>
            <param name = "targetDir">/Upload</param>
        </action>
    </package>
</struts>
```

（4）发布 Web 项目，启动 Tomcat 服务器，访问 uploadFiles.jsp 进行测试。

11.4 文件下载概述

文件下载和文件上传相比，前者要容易些。可以通过 FTP 协议和 HTTP 协议实现。使用 FTP 协议，通常是设置一台 FTP 服务器提供文件下载服务，使用 HTTP 协议，通常是采用链接的形式。这两种方式都是针对已经存在的文件提供的下载方法。

如果下载的文件是根据用户的请求动态产生的，或者用户不想将下载文件的链接显示给用户，或者网站想统计某个文件的下载次数等，这就需要对下载的文件进行控制。这可以通过处理 HttpServletReponse 对象来完成，这种方式对开发人员来说相对比较复杂，为了简化文件下载的操作,Struts 2 框架提供了对文件下载的支持。

基于 Struts 2 框架的文件下载,是通过使用 stream 结果类型实现的。stream 结果类型支持的参数如下:

(1) contentType:用于指定发送给 Web 浏览器的流的 MIME 类型,默认值是 text/plain。MIME 类型如表 11-1 所示。

表 11-1 MIME 类型

扩 展 名	内 容 类 型
gif	image/gif
jpg	image/jpg,image/jpeg,image/pjpeg
bmp	image/bmp
png	image/png
swf	application/x-shockwave-flash
doc	application/msword
txt	text/plain
xls	application/vnd. ms-excel
ppt	application/vnd. ms-powerpoint
pdf	application/pdf
exe	application/octet-stream

(2) contentLength:用于指定流的长度(浏览器显示一个进度条),以字节为单位。

(3) contentDisposition:用于设置响应的 Content-Disposition 头,指定下载文件的名称。该参数有两个可选项:

- inline;filename="下载文件的名称";
- attachment;filename="下载文件的名称"。

其中,inline 表示在当前页面打开下载的文件,attachment 表示打开一个"文件下载"对话框,将文件作为附件下载,默认值是 inline。

(4) inputName:用于指定 action 链中 InputStream 属性的名称,默认值是 inputStream。

(5) bufferSize:用于设置下载文件时缓冲区的大小,默认值是 1024 字节。

(6) allowCaching:用于设置是否允许客户端缓存下载的内容,默认值是 true。如果设置为 false,将设置响应头 'Pragma' 和 'Cache-Control' 为 'no-cache',即不缓存。

(7) contentCharSet:设置字符集。如果设置为一个字符串,则在请求头的内容类型(content-type)中包含 ';charset=value';如果设置为一个表达式,则使用表达式的计算结果。如果没有设置该参数,请求头中不包含字符集(charset)项。

例 11-6 下载一幅图片,默认以"test.jpg"文件名保存下载的文件。

struts.xml 文件的配置如下:

struts.xml

```
<?xml version = "1.0" encoding = "UTF-8" ?>
<!DOCTYPE struts PUBLIC
    "-//Apache Software Foundation//DTD Struts Configuration 2.0//EN"
```

```xml
    "http://struts.apache.org/dtds/struts-2.0.dtd">
<struts>
    <package name="default" extends="struts-default">
        <action name="..." class="...">
            <result name="success" type="stream">
                <param name="contentType">image/jpeg</param>
                <param name="inputName">inputStream</param>
                <param name="contentDisposition">attachment;filename="test.jpg"</param>
                <param name="bufferSize">1024</param>
            </result>
        </action>
    </package>
</struts>
```

如果将 contentDisposition 参数的值中的 attachment 改为 inline，则不打开下载对话框，而是在浏览器中显示该文件。

11.5 基于 Struts 2 框架的文件下载

例 11-7 通过链接的方式提供下载的文件。

具体操作步骤如下：

（1）创建 Action。

struts.xml

```java
package example.struts2;
import java.io.InputStream;
import org.apache.struts2.ServletActionContext;
import com.opensymphony.xwork2.Action;
public class DownloadFileAction implements Action {
    private String downloadPath;
    private String contentType;
    private String fileName;
    public String getContentType() {
        return contentType;
    }
    public void setContentType(String contentType) {
        this.contentType = contentType;
    }
    public String getDownloadPath() {
        return downloadPath;
    }
    public void setDownloadPath(String downloadPath) {
        this.downloadPath = downloadPath;
    }
    public String getFileName() {
```

```java
        return fileName;
    }
    public void setFileName(String fileName) {
        this.fileName = fileName;
    }
    public String execute() throws Exception {
        downloadPath = ServletActionContext.getRequest().getParameter("download");
        int position = downloadPath.lastIndexOf("/");
        if (position > 0){
            fileName = downloadPath.substring(position + 1);
        }
        else{
            fileName = downloadPath;
        }
        contentType = "application/msword";
        return SUCCESS;
    }
    public InputStream getInputStream() throws Exception{
        return ServletActionContext.getServletContext().
                                   getResourceAsStream(downloadPath);
    }
}
```

(2) 编辑 struts.xml 文件。

struts.xml

```xml
<?xml version = "1.0" encoding = "UTF-8" ?>
<!DOCTYPE struts PUBLIC
    "-//Apache Software Foundation//DTD Struts Configuration 2.0//EN"
    "http://struts.apache.org/dtds/struts-2.0.dtd">

<struts>
    <package name = "default" extends = "struts-default">
        <action name = "downloadFile" class = "example.struts2.DownloadFileAction">
            <result name = "success" type = "stream">
                <param name = "contentType">${contentType}</param>
                <param name = "inputName">inputStream</param>
                <param name = "contentDisposition">
                    attachment;filename = ${fileName}
                </param>
            </result>
        </action>
    </package>
</struts>
```

(3) 发布 Web 项目，启动 Tomcat 服务器，访问如下的 URL，结果如图 11-7 所示。

```
http://localhost:8080/Chapter11/downloadFile?download = /Upload/Readme.doc
```

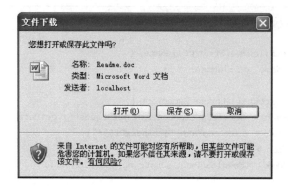

图 11-7　文件下载对话框

11.6　任意内容类型的文件下载

在例 11-7 中，将 contentType 的值进行了硬编码，这样下载的文件默认是 Word 类型的文件。例如，如果访问如下的 URL，则结果如图 11-8 所示。

```
http://localhost:8080/Chapter11/downloadFile?download = /Upload/test.pdf
```

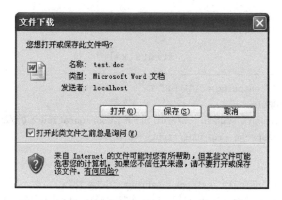

图 11-8　文件下载对话框

由图 11-8 可以看出，下载的文件默认为 Word 文件，而不是 PDF 文件。

可以根据文件扩展名动态设置 contentType 的值，而不是将 contentType 的值进行了硬编码。

例 11-8　根据文件扩展名动态设置 contentType 的值。

具体操作步骤如下：

（1）创建属性文件。

DownloadFileContentTypeAction.properties

```
struts.contentType.gif = image/gif
struts.contentType.jpg = image/jpeg
struts.contentType.bmp = image/bmp
```

```
struts.contentType.png = image/png
struts.contentType.swf = application/x-shockwave-flash
struts.contentType.doc = application/msword
struts.contentType.txt = text/plain
struts.contentType.xls = application/vnd.ms-excel
struts.contentType.ppt = application/vnd.ms-powerpoint
struts.contentType.pdf = application/pdf
struts.contentType.exe = application/octet-stream
```

该属性文件必须和 Action 同名,且保存在同一个包中。

(2) 创建 Action。

DownloadFileContentTypeAction.java

```java
package example.struts2;
import java.io.InputStream;
import org.apache.struts2.ServletActionContext;
import com.opensymphony.xwork2.ActionSupport;
public class DownloadFileContentTypeAction extends ActionSupport {
    private String downloadPath;
    private String contentType;
    private String fileName;
    public String getContentType() {
        return contentType;
    }
    public void setContentType(String contentType) {
        this.contentType = contentType;
    }
    public String getDownloadPath() {
        return downloadPath;
    }
    public void setDownloadPath(String downloadPath) {
        this.downloadPath = downloadPath;
    }
    public String getFileName() {
        return fileName;
    }
    public void setFileName(String fileName) {
        this.fileName = fileName;
    }
    public String execute() throws Exception {
        downloadPath = ServletActionContext.getRequest().getParameter("download");
        int position = downloadPath.lastIndexOf("/");
        if (position > 0){
            fileName = downloadPath.substring(position + 1);
        }
        else{
```

```
            fileName = downloadPath;
        }
        int extPos = fileName.lastIndexOf(".");
        String contentTypeKey;
        if (extPos > 0){
            contentTypeKey = "struts.contentType" + fileName.substring(extPos);
        }
        else{
            contentTypeKey = "struts.contentType.txt";
        }
        contentType = getText(contentTypeKey);
        return SUCCESS;
    }
    public InputStream getInputStream() throws Exception{
        return ServletActionContext.getServletContext().
                                    getResourceAsStream(downloadPath);
    }
}
```

（3）编辑 struts.xml 文件，修改 name 属性值为 downloadFile 的 action 定义中 class 属性值为如下内容：

```
<action name = "downloadFile" class = " example.struts2.DownloadFileContentTypeAction">
```

（4）发布 Web 项目，启动 Tomcat 服务器，访问如下的 URL，结果如图 11-9 所示。

```
http://localhost:8080/Chapter11/downloadFile?download = /Upload/test.pdf
```

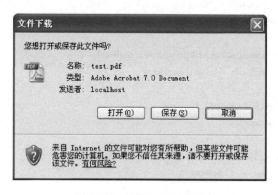

图 11-9　文件下载对话框

11.7　统计文件下载的次数

例 11-9　统计文件下载次数示例。

通常使用一个同名的文本文件记录文件下载的次数。具体操作步骤如下：

(1) 创建 Action。

DownloadFileCountAction.java

```java
package example.struts2;
import java.io.File;
import java.io.InputStream;
import org.apache.commons.io.FileUtils;
import org.apache.struts2.ServletActionContext;
import com.opensymphony.xwork2.ActionSupport;
public class DownloadFileCountAction extends ActionSupport  {
    private String downloadPath;
    private String contentType;
    private String fileName;
    public String getContentType() {
        return contentType;
    }
    public void setContentType(String contentType) {
        this.contentType = contentType;
    }
    public String getDownloadPath() {
        return downloadPath;
    }
    public void setDownloadPath(String downloadPath) {
        this.downloadPath = downloadPath;
    }
    public String getFileName() {
        return fileName;
    }
    public void setFileName(String fileName) {
        this.fileName = fileName;
    }
    public String execute() throws Exception {
        downloadPath = ServletActionContext.getRequest().getParameter("download");
        int position = downloadPath.lastIndexOf("/");
        String downloadFilePath;
        if (position > 0){
            fileName = downloadPath.substring(position+1);
            downloadFilePath = downloadPath.substring(0,position+1);
        }
        else{
            fileName = downloadPath;
            downloadFilePath = "";
        }
        int extPos = fileName.lastIndexOf(".");
        String contentTypeKey;
        if (extPos > 0){
            contentTypeKey = "struts.contentType" + fileName.substring(extPos);
        }
        else{
```

```java
                contentTypeKey = "struts.contentType.txt";
            }
            contentType = getText(contentTypeKey);
            String counterFile = genNewFileName(downloadFilePath,fileName);
            setDownloadCount(counterFile);
            return SUCCESS;
        }
        public InputStream getInputStream() throws Exception{
            return ServletActionContext.getServletContext().
                                getResourceAsStream(downloadPath);
        }
        private String genNewFileName(String filePath,String fileName){
            String serverRealPath = ServletActionContext.getServletContext().
                                getRealPath(filePath) + "/DownloadCounter/";
            File dir = new File(serverRealPath);
            if (!dir.exists()){
                dir.mkdir();
            }
            String downloadCountFile = serverRealPath + fileName + ".txt";
            return downloadCountFile;
        }
        private void setDownloadCount(String fullFileName) throws Exception{
            int curCount = 1;
            File file = new File(fullFileName);
            if(file.exists()){
                curCount =
                    Integer.parseInt(FileUtils.readFileToString(new File(fullFileName))) + 1;
            }
            FileUtils.writeStringToFile(file, Integer.toString(curCount));
        }
    }
```

（2）编辑 struts.xml 文件。

struts.xml

```xml
<?xml version = "1.0" encoding = "UTF-8" ?>
<!DOCTYPE struts PUBLIC
    "-//Apache Software Foundation//DTD Struts Configuration 2.0//EN"
    "http://struts.apache.org/dtds/struts-2.0.dtd">
<struts>
    <package name = "default" extends = "struts-default">
        <action name = "downloadFileCount"
                class = "example.struts2.DownloadFileCountAction">
            <result name = "success" type = "stream">
                <param name = "contentType">${contentType}</param>
                <param name = "inputName">inputStream</param>
                <param name = "contentDisposition">
                    attachment;filename = ${fileName}
```

```
            </param>
        </result>
    </action>
  </package>
</struts>
```

（3）发布 Web 项目，启动 Tomcat 服务器，访问如下的 URL，结果如图 11-10 所示。

```
http://localhost:8080/Chapter11/downloadFileCount?download = /Upload/Readme.doc
```

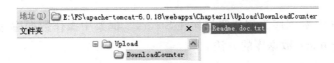

图 11-10　记录文件下载次数的文件

习题

1．修改例 11-2，使用属性配置文件设置上传文件的最大字节数。如果多处配置上传文件大小，哪个配置将会起作用？

2．修改例 11-8，使其具有用户登录之后才能下载的功能。

第 12 章 Struts 2框架中使用数据库

本章以 MySQL 数据库为例介绍如何在 Struts 2 框架中操作数据库,并给出操作 SQL Server 数据库和 Oracle 数据库的方法。

12.1 连接数据库

连接数据库可以采用 JDBC 和 Tomcat 数据源的方式,本节介绍连接三种数据库的方式。

12.1.1 连接 MySQL 数据库

MySQL 数据库的数据库驱动程序为 mysql-connector-java-5.0.5-bin.jar。其下载网址是:

```
http://www.mysql.com/downloads/connector/j/
```

下载 mysql-connector-java-5.0.5.zip 压缩文件,解压后就可以得到数据库驱动程序。将 MySQL 的数据库驱动程序复制到 Tomcat 服务器的 lib 文件夹中,或者保存到 Web 应用的 WEB-INF\classes 文件夹中。

1. 使用 JDBC 方式连接数据库

首先要加载 MySQL 数据库的数据库驱动程序,然后和指定的数据库建立连接。
(1) 加载数据库驱动程序,代码如下:

```
try{
    Class.forName("com.mysql.jdbc.Driver");
}
catch(ClassNotFoundException e){
    out.print("" + e);
}
```

(2) 连接指定的数据库,代码如下:

```
String url = "jdbc:mysql://localhost:3306/mydata";
String user = "test";
```

```
String password = "test";
try{
    Connection conn = DriverManager.getConnection(url,user,password);
}
catch(SQLException e){
    out.print("" + e);
}
```

其中 mydata 是所要连接数据库的名称。

2. 使用 Tomcat 数据源连接数据库

（1）在 context.xml 文件中配置数据源，该文件保存在 WEB 应用的 META-INF 目录下。内容如下：

context.xml

```
<?xml version = '1.0' encoding = 'utf-8'?>
<Resource name = "jdbc/Struts2DB" auth = "Container"
          type = "javax.sql.DataSource" username = "test" password = "test"
          driverClassName = "com.mysql.jdbc.Driver"
          url = "jdbc:mysql://localhost:3306/mydata?characterEncoding = UTF-8"
          maxActive = "8" maxIdle = "4"/>
```

（2）数据源配置完成后，需要在 web.xml 文件中引用配置好的数据源。代码如下：

```
<resource-ref>
    <description>DB Connection</description>
    <res-ref-name>jdbc/Struts2DB</res-ref-name>
    <res-type>javax.sql.DataSource</res-type>
    <res-auth>Container</res-auth>
</resource-ref>
```

代码中的<res-ref-name>、<res-type>和<res-auth>子元素的值分别和 context.xml 文件中<Resource>元素的 name、type 和 auth 属性值相同。

12.1.2 连接 Oracle 数据库

Oracle11g 数据库的数据库驱动程序为 classes12.jar。在 Oracle11g 的安装目录中可以找到。classes12.jar 在安装目录中的位置如下：

```
product\11.1.0\db_1\oui\jlib
```

将 Oracle11g 的数据库驱动程序复制到 Tomcat 服务器的 lib 文件夹中，或者保存到 Web 应用的 WEB-INF\classes 文件夹中。

1. 使用 JDBC 方式连接数据库

首先要加载 MySQL 数据库的数据库驱动程序，然后和指定的数据库建立连接。

(1) 加载数据库驱动程序，代码如下：

```java
try{
    Class.forName("oracle.jdbc.driver.OracleDriver");
}
catch(ClassNotFoundException e){
    out.print("" + e);
}
```

(2) 连接指定的数据库，代码如下：

```java
String url = "jdbc:oracle:thin:@localhost:1521:mydata";
String user = "test";
String password = "test";
try{
    Connection conn = DriverManager.getConnection(url,user,password);
}
catch(SQLException e){
    out.print("" + e);
}
```

其中 mydata 是所要连接数据库的名称。

2. 使用 Tomcat 数据源连接数据库

(1) 在 context.xml 文件中配置数据源，该文件保存在 WEB 应用的 META-INF 目录下。内容如下：

context.xml

```xml
<?xml version = '1.0' encoding = 'utf-8'?>
<Context>
    <Resource name = "jdbc/Struts2DB" auth = "Container"
            type = "javax.sql.DataSource" username = "test" password = "test"
            driverClassName = "oracle.jdbc.driver.OracleDriver"
            url = "jdbc:oracle:thin:@127.0.0.1:1521:mydata"
            maxActive = "50" maxIdle = "30" maxWait = "10000"/>
</Context>
```

(2) 数据源配置完成后，需要在 web.xml 文件中引用配置好的数据源。代码如下：

```xml
<resource-ref>
    <description>DB Connection</description>
    <res-ref-name>jdbc/Struts2DB</res-ref-name>
    <res-type>javax.sql.DataSource</res-type>
    <res-auth>Container</res-auth>
</resource-ref>
```

代码中的<res-ref-name>、<res-type>和<res-auth>子元素的值分别和 context.xml 文件中<Resource>元素的 name、type 和 auth 属性值相同。

12.1.3 连接 SQL Server 数据库

SQL Server 2000 数据库的数据库驱动程序为 sqljdbc.jar。其下载网址是：

```
http://www.mysql.com/downloads/connector/j/
```

下载 sqljdbc_1.2.2828.100_chs.exe 文件并安装，在安装目录的 chs 目录中可以找到 SQL Server 数据库驱动程序。将 SQL Server 的数据库驱动程序复制到 Tomcat 服务器的 lib 文件夹中，或者保存到 Web 应用的 WEB-INF\classes 文件夹中。

1. 使用 JDBC 方式连接数据库

首先要加载 SQL Server 数据库的数据库驱动程序，然后和指定的数据库建立连接。
（1）加载数据库驱动程序，代码如下：

```
try{
    Class.forName("com.microsoft.sqlserver.jdbc.SQLServerDriver");
}
catch(ClassNotFoundException e){
    out.print("" + e);
}
```

（2）连接指定的数据库，代码如下：

```
String url = "jdbc:sqlserver://localhost:1433;DatabaseName=mydata";
String user = "test";
String password = "test";
try{
    Connection conn = DriverManager.getConnection(url,user,password);
}
catch(SQLException e){
    out.print("" + e);
}
```

其中 mydata 是所要连接数据库的名称。

2. 使用 Tomcat 数据源连接数据库

（1）在 context.xml 文件中配置数据源，该文件保存在 WEB 应用的 META-INF 目录下。内容如下：

context.xml

```
<?xml version = '1.0' encoding = 'utf-8'?>
<Context>
    <Resource name = "jdbc/Struts2DB" auth = "Container"
        type = "javax.sql.DataSource" username = "test" password = "test"
```

```
                driverClassName = "com.microsoft.sqlserver.jdbc.SQLServerDriver"
                url = " jdbc:sqlserver://localhost:1433;DatabaseName = mydata "
                maxActive = "50" maxIdle = "30" maxWait = "10000"/>
</Context>
```

(2) 数据源配置完成后,需要在 web.xml 文件中引用配置好的数据源。代码如下:

```
<resource-ref>
    <description>DB Connection</description>
    <res-ref-name>jdbc/Struts2DB</res-ref-name>
    <res-type>javax.sql.DataSource</res-type>
    <res-auth>Container</res-auth>
</resource-ref>
```

代码中的<res-ref-name>、<res-type>和<res-auth>子元素的值分别和 context.xml 文件中<Resource>元素的 name、type 和 auth 属性值相同。

12.2 MySQL 数据库的下载与安装

1. 下载

MySQL 数据库是一个开源项目,其官方下载网址如下:

```
http://www.mysql.com/downloads/mysql/
```

选择 Windows 平台下的压缩文件下载,截至作者书写本章内容时,MySQL 的最新版本为 mysql-5.5.17-win32.zip。选择某一版本,按网站的提示下载。本书使用 MySQL 数据库的版本是 mysql-5.0.45-win32.zip,将下载文件解压后,在解压目录中有一个 setup.exe 安装文件。

2. 安装

(1) 双击 setup.exe 文件,开始安装。
(2) 在出现的安装界面中,单击 next 按钮。
(3) 在选择安装方式界面中,有三种安装方式可供选择,分别是 Typical、Complete 和 Custom。默认的安装方式是 Typical。其中 Typical 和 Complete 两种安装方式不允许改变安装路径,只能安装在 C 盘,因此,如果想要改变 MySQL 的安装路径,需要选择 Custom 安装方式。
(4) 在出现的确认安装结束界面,询问是否配置 MySQL,这里选择否,使用默认配置,单击"确定"按钮完成安装。安装之后的目录结构如图 12-1 所示。

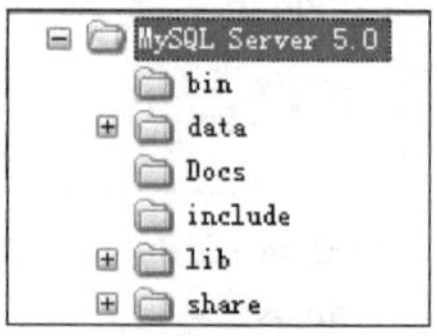

3. 启动 MySQL 数据库服务器

必须先启动 MySQL 数据库服务器,才能使用。启动

图 12-1 MySQL 安装完成后的目录结构

MySQL 数据库服务器的命令是 mysqld-nt.exe，在安装目录的 bin 子目录中。可以通过如下两种方式启动 MySQL 数据库服务器。

（1）双击 bin 子目录中的 mysqld-nt.exe 文件，将启动 MySQL 数据库服务器。启动成功后的窗口如图 12-2 所示。

图 12-2　启动成功的 MySQL 数据库服务器

（2）打开命令提示符窗口，进入到 mysqld-nt.exe 文件所在的 bin 子目录，并输入 mysqld-nt 命令。启动成功后的窗口如图 12-3 所示。

图 12-3　启动成功的 MySQL 数据库服务器

4．使用 MySQL 数据库服务器

启动 MySQL 数据库服务器后，就可以对其进行操作。通过位于 bin 子目录中的 mysql.exe 启动操作界面。

（1）双击 bin 子目录中的 mysql.exe 文件，将以 root 用户启动 MySQL 数据库的操作界面，如图 12-4 所示。

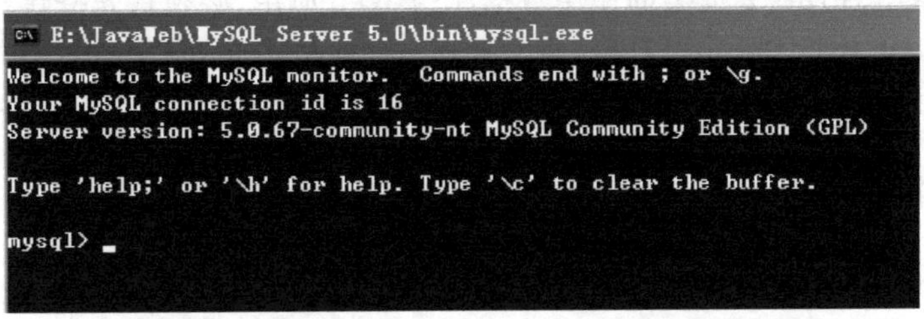

图 12-4　双击 mysql.exe 文件的界面

(2) 打开命令提示符窗口,进入到 mysql.exe 文件所在的 bin 子目录,准备启动 MySQL 数据库的操作界面。命令格式如下:

```
mysql [-h hostName] [-u userName -p]
```

其中 hostName 是 MySQL 数据库服务器所在机器的名称,如果省略,则默认为本机。userName 是登录 MySQL 数据库的用户名。
- 如果省略所有参数,只输入 mysql 命令,启动成功后的窗口如图 12-4 所示。
- 如果使用-u 和-p 参数,则窗口如图 12-5 所示。

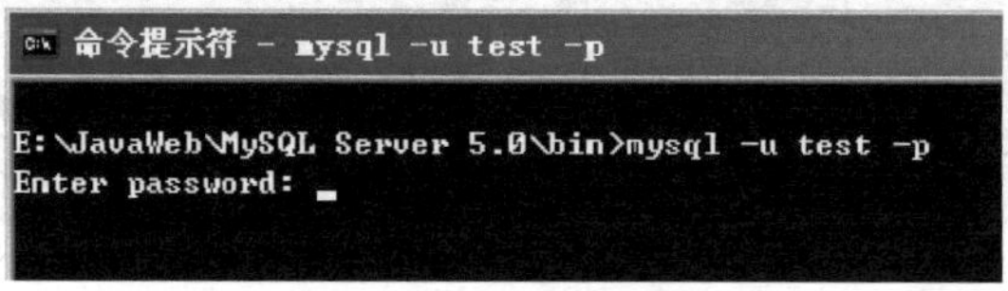

图 12-5 等待输入密码的界面

输入密码后的结果如图 12-6 所示。

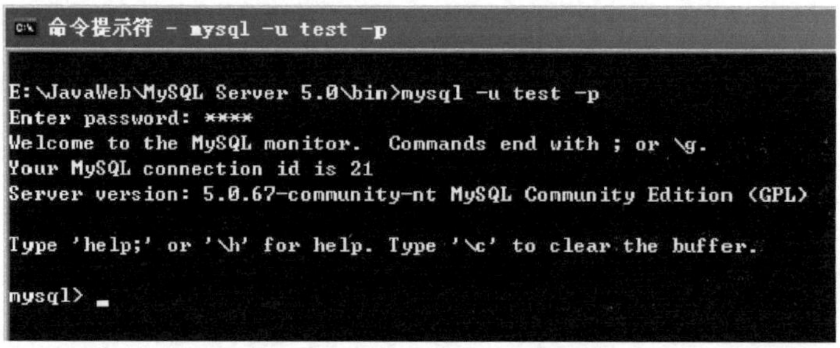

图 12-6 输入正确密码后的界面

5. 操作

打开 MySQL 数据库的操作界面后,就可以使用 SQL 语句进行操作,如创建、删除数据库,对表进行增、删、改、查等。在 SQL 语句后必须使用分号(";")结束。

12.3 连接测试

为了开发基于 Struts 2 框架的使用数据库的 Web 应用,必须首先搭建开发环境。
首先创建 Web 项目 Chapter12,在 src 文件夹下创建包 example.struts2,并添加 Struts 2 框架的支持。然后进行如下操作。

1. 创建数据库和表

(1) 创建数据库,语句如下:

```
create database mydata character set utf8;
```

（2）在 mydata 数据库中创建表，语句如下：

```sql
use mydata;
CREATE TABLE stu(
    id int(11) NOT NULL auto_increment,
    name varchar(10) NOT NULL,
    password varchar(10) NOT NULL,
    PRIMARY KEY (id) ,
    UNIQUE KEY stuId (id)
) ENGINE = InnoDB DEFAULT CHARSET = utf8;
```

（3）插入数据到表中，语句如下：

```sql
INSERT INTO stu(name, password) VALUES ('aa', 'aa');
INSERT INTO stu(name, password) VALUES ('bb', 'bb');
```

2．测试

（1）使用 JDBC 方式访问数据库

① 编辑 web.xml 文件，内容同 Chapter11 中 web.xml 文件的内容。

② 创建 JSP 文件。

testMySQLJDBC.jsp

```jsp
<%@ page language = "java" contentType = "text/html; charset = UTF-8"
    pageEncoding = "UTF-8" %>
<%@ page import = "java.sql.*" %>
<%
    try{
        Class.forName("com.mysql.jdbc.Driver");
    }
    catch(ClassNotFoundException e){
        out.print("" + e);
    }
    String url = "jdbc:mysql://localhost:3306/mydata";
    String user = "test";
    String password = "test";
    try{
        Connection conn = DriverManager.getConnection(url,user,password);
        Statement stmt = conn.createStatement();
        String sql = "select * from stu";
        ResultSet rs = stmt.executeQuery(sql);
        while(rs.next()){
            out.println("UserName:" + rs.getString("name") + "<br/>");
            out.println("UserPassword:" + rs.getString("password") + "<br/>");
            out.println("--------------------------------<br/>");
        }
        out.print("恭喜你,数据库操作(JDBC)成功!");
```

```
            rs.close();
            stmt.close();
            conn.close();
        }
        catch(SQLException ee){
            out.print("" + ee);
        }
    %>
```

③ 发布 Web 项目，启动 Tomcat 服务器，并启动 MySQL 数据库服务器，访问 testMySQLJDBC.jsp，结果如图 12-7 所示。

图 12-7　访问 testMySQLJDBC.jsp 的结果

(2) 使用 Tomcat 数据源方式访问数据库

① 创建 context.xml。

context.xml

```
<?xml version = '1.0' encoding = 'utf-8'?>
<Resource name = "jdbc/Struts2DB" auth = "Container"
          type = "javax.sql.DataSource" username = "test" password = "test"
          driverClassName = "com.mysql.jdbc.Driver"
          url = "jdbc:mysql://localhost:3306/mydata"
          maxActive = "8" maxIdle = "4"/>
```

② 编辑 web.xml。

web.xml

```
<?xml version = "1.0" encoding = "UTF-8"?>
<web-app version = "2.4"
    xmlns = "http://java.sun.com/xml/ns/j2ee"
    xmlns:xsi = "http://www.w3.org/2001/XMLSchema-instance"
    xsi:schemaLocation = "http://java.sun.com/xml/ns/j2ee
    http://java.sun.com/xml/ns/j2ee/web-app_2_4.xsd">

    <resource-ref>
        <description>DB Connection</description>
        <res-ref-name>jdbc/Struts2DB</res-ref-name>
        <res-type>javax.sql.DataSource</res-type>
```

```xml
      <res-auth>Container</res-auth>
    </resource-ref>
    <filter>
      <filter-name>Struts 2</filter-name>
      <filter-class>
          org.apache.struts2.dispatcher.ng.filter.StrutsPrepareAndExecuteFilter
      </filter-class>
    </filter>
    <filter-mapping>
        <filter-name>Struts 2</filter-name>
        <url-pattern>/*</url-pattern>
    </filter-mapping>
</web-app>
```

③ 创建 JSP 文件。

testMySQLDataSource.jsp

```jsp
<%@ page language="java" contentType="text/html; charset=UTF-8"
    pageEncoding="UTF-8"%>
<%@ page import="java.sql.*,javax.sql.DataSource,javax.naming.*" %>
<%
    try{
        Context initCtx = new InitialContext();
        DataSource ds = (DataSource)initCtx.lookup("java:comp/env/jdbc/Struts2DB");
        Connection conn = ds.getConnection();
        Statement stmt = conn.createStatement();
        ResultSet rs = stmt.executeQuery("select * from stu");
        while(rs.next()){
            out.println("UserName:" + rs.getString("name") + "<br/>");
            out.println("UserPassword:" + rs.getString("password") + "<br/>");
            out.println("--------------------------------<br/>");
        }
        out.print("恭喜你,数据库操作(DataSource)成功!");
        rs.close();
        stmt.close();
        conn.close();
    }
    catch(NameNotFoundException e){
        out.print("" + e);
    }
    catch(SQLException ee){
        out.print("" + ee);
    }
%>
```

④ 将 MySQL 数据库驱动 mysql-connector-java-5.0.5-bin.jar 复制到 Tomcat 服务器安装目录的 lib 子目录下,否则将报错,如图 12-8 所示。

地址(D) http://localhost:8080/Chapter12/testMySQLDataSource.jsp

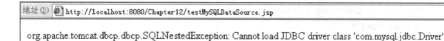

org.apache.tomcat.dbcp.dbcp.SQLNestedException: Cannot load JDBC driver class 'com.mysql.jdbc.Driver'

图 12-8 数据库驱动没在指定位置的结果

⑤ 发布 Web 项目,启动 Tomcat 服务器,访问 testMySQLDataSource.jsp,结果如图 12-9 所示。

地址(D) http://localhost:8080/Chapter12/testMySQLDataSource.jsp

UserName:aa
UserPassword:aa

UserName:bb
UserPassword:bb

恭喜你,数据库操作(DataSource)成功!

图 12-9 访问 testMySQLDataSource.jsp 的结果

12.4 使用数据库的示例

本节以注册、登录以及显示用户列表为例进行介绍。

在 12.3 节的基础上进行操作。

12.4.1 创建国际化资源文件

用于 Web 项目的国际化,这里只给出中文信息的资源文件。

ApplicationResources_zh_CN.properties

```
label.text.userName = 用户名称
label.text.password = 用户密码
label.text.confirmPassword = 确认密码
label.text.register = 注册
label.text.login = 登录
label.text.ordinal = 序号
label.text.id = 用户 ID
label.text.loginSuccess = 登录成功!
label.text.registerSuccess = 注册成功!
struts.messages.invalid.token = 重复提交!
userNameRequired = 必须输入 ${getText("label.text.userName")}
passwordRequired = 必须输入 ${getText("label.text.password")}
confirmPasswordRequired = 必须输入 ${getText("label.text.confirmPassword")}
confirmPasswordError = ${getText("label.text.password")}和
                       ${getText("label.text.confirmPassword")}必须一致!
```

12.4.2 创建数据库操作的辅助类

该辅助类用于将对数据库的操作代码进行封装。

DbUtils.java

```java
package example.struts2;
import java.sql.Connection;
import java.sql.PreparedStatement;
import java.sql.ResultSet;
import java.sql.ResultSetMetaData;
import java.sql.SQLException;
import java.sql.Statement;
import java.util.ArrayList;
import java.util.HashMap;
import java.util.List;
import java.util.Map;
import javax.naming.InitialContext;
import javax.sql.DataSource;
public class DbUtils {
    private static Connection getConnection() throws Exception {
        InitialContext cxt = new InitialContext();
        if ( cxt == null ) {
            throw new Exception("no context!");
        }
        DataSource ds = (DataSource) cxt.lookup("java:comp/env/jdbc/Struts2DB");
        if ( ds == null ) {
            throw new Exception("Data source not found!");
        }
        return ds.getConnection();
    }
    private static PreparedStatement getPreparedStatement(Connection conn, String sql) throws Exception {
        if(conn == null || sql == null || sql.trim().equals("")) {
            return null;
        }
        PreparedStatement pstmt = conn.prepareStatement(sql.trim());
        return pstmt;
    }
    private static void setPreparedStatementParam(PreparedStatement pstmt, List<Object> paramList) throws Exception {
        if(pstmt == null || paramList == null || paramList.isEmpty()) {
            return;
        }
        for (int i = 0; i < paramList.size(); i++) {
            if(paramList.get(i) instanceof Integer) {
                int paramValue = ((Integer)paramList.get(i)).intValue();
                pstmt.setInt(i + 1, paramValue);
            } else if(paramList.get(i) instanceof String) {
```

```java
                    pstmt.setString(i + 1, (String)paramList.get(i));
                }
            }
            return;
        }
        private static ResultSet getResultSet(PreparedStatement pstmt) throws Exception {
            if(pstmt == null) {
                return null;
            }
            ResultSet rs = pstmt.executeQuery();
            return rs;
        }
        public static int save(String userName, String password) throws Exception {
            int result = 0;
            String sql = "insert into stu(name, password) values(?, ?)";
            List<Object> paramList = new ArrayList<Object>();
            paramList.add(userName);
            paramList.add(password);
            Connection conn = null;
            PreparedStatement pstmt = null;
            try {
                conn = DbUtils.getConnection();
                pstmt = DbUtils.getPreparedStatement(conn, sql);
                setPreparedStatementParam(pstmt, paramList);
                if(pstmt == null) {
                    return -1;
                }
                result = pstmt.executeUpdate();
            } catch (Exception e) {
                throw new Exception(e);
            } finally {
                closeStatement(pstmt);
                closeConn(conn);
            }
            return result;
        }
        public static int find(String userName, String password) throws Exception {
            int result = 0;
            String sql = "select * from stu where name = ? and password = ?";
            List<Object> paramList = new ArrayList<Object>();
            paramList.add(userName);
            paramList.add(password);
            Connection conn = null;
            PreparedStatement pstmt = null;
            ResultSet rs = null;
            try {
                conn = DbUtils.getConnection();
                pstmt = DbUtils.getPreparedStatement(conn, sql);
                setPreparedStatementParam(pstmt, paramList);
```

```java
                rs = getResultSet(pstmt);
                if(rs.next()) {
                    return 1;
                }
        } catch (Exception e) {
            throw new Exception(e);
        } finally {
            closeResultSet(rs);
            closeStatement(pstmt);
            closeConn(conn);
        }
        return result;
    }
    public static boolean findUser(String userName) throws Exception {
        boolean result = false;
        String sql = "select * from stu where name = ?";
        List<Object> paramList = new ArrayList<Object>();
        paramList.add(userName);
        Connection conn = null;
        PreparedStatement pstmt = null;
        ResultSet rs = null;
        try {
            conn = DbUtils.getConnection();
            pstmt = DbUtils.getPreparedStatement(conn, sql);
            setPreparedStatementParam(pstmt, paramList);
            rs = getResultSet(pstmt);
            if(rs.next()) {
                result = true;
            }
        } catch (Exception e) {
            throw new Exception(e);
        } finally {
            closeResultSet(rs);
            closeStatement(pstmt);
            closeConn(conn);
        }
        return result;
    }
    public static List<Map<String, String>> findAll() throws Exception {
        String sql = "select * from stu";
        List<Map<String, String>> queryList = null;
        if(sql == null || sql.trim().equals("")) {
            return null;
        }
        Connection conn = null;
        PreparedStatement pstmt = null;
        ResultSet rs = null;
        try {
            conn = DbUtils.getConnection();
```

```java
                pstmt = DbUtils.getPreparedStatement(conn, sql);
                rs = getResultSet(pstmt);
                queryList = DbUtils.getQueryList(rs);
        } catch (Exception e) {
                throw new Exception(e);
        } finally {
                closeResultSet(rs);
                closeStatement(pstmt);
                closeConn(conn);
        }
        return queryList;
    }
    private static List<Map<String, String>> getQueryList(ResultSet rs) throws Exception {
        if(rs == null) {
                return null;
        }
        ResultSetMetaData rsMetaData = rs.getMetaData();
        int columnCount = rsMetaData.getColumnCount();
        List<Map<String, String>> dataList = new ArrayList<Map<String, String>>();
        while (rs.next()) {
                Map<String, String> dataMap = new HashMap<String, String>();
                for (int i = 0; i < columnCount; i++) {
                        dataMap.put(rsMetaData.getColumnName(i + 1), rs.getString(i + 1));
                }
                dataList.add(dataMap);
        }
        return dataList;
    }
    private static void closeConn(Connection conn) throws Exception {
        if(conn == null) {
                return;
        }
        try {
                conn.close();
        } catch (SQLException e) {
                throw new Exception(e);
        }
    }
    private static void closeStatement(Statement stmt) throws Exception {
        if(stmt == null) {
                return;
        }
        try {
                stmt.close();
        } catch (SQLException e) {
                throw new Exception(e);
        }
    }
    private static void closeResultSet(ResultSet rs) throws Exception {
```

```
            if(rs == null) {
                return;
            }
            try {
                rs.close();
            } catch (SQLException e) {
                throw new Exception(e);
            }
        }
    }
```

DbUtils 类封装了数据库的连接、获取结果集、关闭连接、关闭结果集、保存用户数据、查询用户数据等操作。

12.4.3 创建 Action 类

实现用户注册、登录和用户显示等操作。

1. 用户注册 Action

RegisterAction.java

```java
package example.struts2;
import java.util.ArrayList;
import java.util.List;
import com.opensymphony.xwork2.ActionSupport;
public class RegisterAction extends ActionSupport {
    private String userName;
    private String password;
    private String confirmPassword;
    public String getPassword() {
        return password;
    }
    public void setPassword(String password) {
        this.password = password;
    }
    public String getUserName() {
        return userName;
    }
    public void setUserName(String userName) {
        this.userName = userName;
    }
    public String getConfirmPassword() {
        return confirmPassword;
    }
    public void setConfirmPassword(String confirmPassword) {
        this.confirmPassword = confirmPassword;
    }
    public String register() throws Exception {
```

```java
            int result = DbUtils.save(userName, password);
            if(result == 1) {
                return SUCCESS;
            }
            return INPUT;
        }
        public String input() {
            return INPUT;
        }
    }
```

2. 用户登录 Action

LoginAction.java

```java
    package example.struts2;
    import java.util.ArrayList;
    import java.util.List;
    import com.opensymphony.xwork2.ActionSupport;
    public class LoginAction extends ActionSupport {
        private String userName;
        private String password;
        public String getPassword() {
            return password;
        }
        public void setPassword(String password) {
            this.password = password;
        }
        public String getUserName() {
            return userName;
        }
        public void setUserName(String userName) {
            this.userName = userName;
        }
        public String login() throws Exception {
            int result = DbUtils.find(userName, password);
            if(result == 1) {
                return SUCCESS;
            }
            return INPUT;
        }
    }
```

3. 检查用户名是否重复 Action

UserCheckAction.java

```java
    package example.struts2;
    import javax.servlet.http.HttpServletRequest;
```

```java
import javax.servlet.http.HttpServletResponse;
import org.apache.struts2.ServletActionContext;
import org.apache.struts2.interceptor.ServletRequestAware;
import com.opensymphony.xwork2.ActionSupport;
public class UserCheckAction extends ActionSupport implements ServletRequestAware {
    private String userName;
    protected HttpServletRequest servletRequest = null;
    public String getUserName() {
        return userName;
    }
    public void setUserName(String userName) {
        this.userName = userName;
    }
    public String userCheck() throws Exception {
        String userName = servletRequest.getParameter("userName");
        boolean isNameValid = DbUtils.findUser(userName);
        HttpServletResponse response = ServletActionContext.getResponse();
        response.getWriter().print(isNameValid);
        return null;
    }
    public void setServletRequest(HttpServletRequest request) {
        this.servletRequest = request;
    }
}
```

4. 用户列表显示 Action

UserListAction.java

```java
package example.struts2;
import java.util.List;
import java.util.Map;
import javax.servlet.http.HttpServletRequest;
import org.apache.struts2.interceptor.ServletRequestAware;
import com.opensymphony.xwork2.ActionSupport;
public class UserListAction extends ActionSupport implements ServletRequestAware {
    protected HttpServletRequest servletRequest = null;
    public String list() throws Exception {
        List<Map<String, String>> userList = null;
        try {
            userList = DbUtils.findAll();
            System.out.println("findAll");
        } catch (RuntimeException e) {
        }
        servletRequest.setAttribute("userList", userList);
        return "userlist";
    }
    public void setServletRequest(HttpServletRequest request) {
```

```java
            this.servletRequest = request;
        }
    }
```

12.4.4　创建输入校验的配置文件

用户注册和用户登录时,需要进行输入校验。

1. 用户注册输入校验

RegisterAction-validation.xml

```xml
<?xml version="1.0" encoding="UTF-8"?>
<!DOCTYPE validators PUBLIC
        "-//OpenSymphony Group//XWork Validator 1.0.2//EN"
        "http://www.opensymphony.com/xwork/xwork-validator-1.0.2.dtd">
<validators>
    <field name="userName">
        <field-validator type="requiredstring">
            <message key="userNameRequired" />
        </field-validator>
    </field>
    <field name="password">
        <field-validator type="requiredstring">
            <message key="passwordRequired" />
        </field-validator>
    </field>
    <field name="confirmPassword">
        <field-validator type="requiredstring">
            <message key="confirmPasswordRequired" />
        </field-validator>
        <field-validator type="fieldexpression">
            <param name="expression">password.equals(confirmPassword)</param>
            <message key="confirmPasswordError" />
        </field-validator>
    </field>
</validators>
```

2. 用户登录输入校验

LoginAction-validation.xml

```xml
<?xml version="1.0" encoding="UTF-8"?>
<!DOCTYPE validators PUBLIC
        "-//OpenSymphony Group//XWork Validator 1.0.2//EN"
        "http://www.opensymphony.com/xwork/xwork-validator-1.0.2.dtd">
<validators>
```

```xml
<field name="userName">
    <field-validator type="requiredstring">
        <message key="userNameRequired"/>
    </field-validator>
</field>
<field name="password">
    <field-validator type="requiredstring">
        <message key="passwordRequired"/>
    </field-validator>
</field>
</validators>
```

12.4.5 编辑配置文件

struts.xml

```xml
<?xml version="1.0" encoding="UTF-8" ?>
<!DOCTYPE struts PUBLIC
    "-//Apache Software Foundation//DTD Struts Configuration 2.0//EN"
    "http://struts.apache.org/dtds/struts-2.0.dtd">
<struts>
    <package name="default" extends="struts-default">
        <interceptors>
            <interceptor-stack name="tokenStack">
                <interceptor-ref name="token">
                    <param name="excludeMethods">input</param>
                </interceptor-ref>
                <interceptor-ref name="defaultStack"/>
            </interceptor-stack>
        </interceptors>
        <action name="register_*" method="{1}"
                                    class="example.struts2.RegisterAction">
            <interceptor-ref name="tokenStack"/>
            <result name="invalid.token">/register.jsp</result>
            <result name="input">/register.jsp</result>
            <result>/registerSuccess.jsp</result>
        </action>
        <action name="login" method="login" class="example.struts2.LoginAction">
            <result name="input">/login.jsp</result>
            <result>/loginSuccess.jsp</result>
        </action>
        <action name="userList" method="list" class="example.struts2.UserListAction">
            <result name="userlist">/userList.jsp</result>
        </action>
        <action name="userCheck" method="userCheck"
                                    class="example.struts2.UserCheckAction"/>
    </package>
    <constant name="struts.custom.i18n.resources" value="ApplicationResources"/>
</struts>
```

在配置文件中定义了 tokenStack 拦截器栈,用于用户注册时,防止注册数据的重复提交。

12.4.6 创建 JSP 文件

1. 创建首页

index.jsp

```jsp
<%@ page language="java" contentType="text/html; charset=UTF-8" pageEncoding="UTF-8"%>
<%@ taglib uri="/struts-tags" prefix="s" %>
<a href='<s:url value="/login.jsp" />'>登录</a>  
<a href='<s:url action="register" method="input" />'>注册</a>  
<a href='<s:url action="userList" method="list" />'>用户列表</a>
```

2. 创建用户注册页面

register.jsp

```jsp
<%@ page language="java" contentType="text/html; charset=UTF-8" pageEncoding="UTF-8"%>
<%@ taglib uri="/struts-tags" prefix="s" %>
<html>
<head>
<script language="javascript">
    var xmlHttp;
    function createXMLHttpRequest(){
        if(window.ActiveXObject){
            xmlHttp = new ActiveXObject("Microsoft.XMLHTTP");
        }else if(window.XMLHttpRequest){
            xmlHttp = new XMLHttpRequest();
        }
    }
    function checkUserName(){
        var userName = document.getElementById("userName").value;
        createXMLHttpRequest();
        var url = "userCheck.action?userName=" + userName;
        xmlHttp.open("POST",url,true);
        xmlHttp.onreadystatechange = ajaxStatus;
        xmlHttp.send(null);
    }
    function ajaxStatus(){
        var responseContext = "";
        if(xmlHttp.readyState == 4){
            if(xmlHttp.status == 200){
                responseContext = xmlHttp.responseText;
                if (responseContext == "true"){
                    document.getElementById("userNameSpan").innerHTML =
                                        "该用户名已经被占用";
```

```
                        document.getElementById("userName").focus();
                    }
                    else{
                        document.getElementById("userNameSpan").innerHTML =
                                            "该用户名可以使用";
                    }
                }
            }
        }
</script>
</head>
<s:actionerror/>
<s:form action = "register_register" method = "post">
    <s:token />
    <s:textfield name = "userName" id = "userName" key = "label.text.userName"
                                    onblur = "checkUserName()"/><br/>
    <span id = "userNameSpan"></span><br />
    <s:password name = "password" key = "label.text.password" /><br/>
    <s:password name = "confirmPassword" key = "label.text.confirmPassword" /><br/>
    <s:submit name = "submit" key = "label.text.register" />
</s:form>
```

3. 创建用户注册成功显示页面

registerSuccess.jsp

```
<%@ page language = "java" contentType = "text/html; charset = UTF-8" pageEncoding = "UTF-8" %>
<%@ taglib uri = "/struts-tags" prefix = "s" %>
<s:text name = "label.text.registerSuccess"/><br/>
<s:text name = "label.text.userName"/>
<b><s:property value = "userName"></s:property></b><br/>
<s:text name = "label.text.password"/><b><s:property value = "password"></s:property></b>
```

4. 创建用户登录页面

login.jsp

```
<%@ page language = "java" contentType = "text/html; charset = UTF-8" pageEncoding = "UTF-8" %>
<%@ taglib uri = "/struts-tags" prefix = "s" %>
<s:form action = "login" method = "post">
    <s:textfield name = "userName" key = "label.text.userName"/><br/>
    <s:password name = "password" key = "label.text.password" /><br/>
    <s:submit name = "submit" key = "label.text.login" />
</s:form>
```

5. 创建用户登录成功显示页面

loginSuccess.jsp

```jsp
<%@ page language="java" contentType="text/html;charset=UTF-8" pageEncoding="UTF-8"%>
<%@ taglib uri="/struts-tags" prefix="s" %>
<b><s:property value="userName"></s:property></b>
<s:text name="label.text.loginSuccess"/>
```

6. 创建用户列表显示页面

userList.jsp

```jsp
<%@ page language="java" contentType="text/html;charset=UTF-8" pageEncoding="UTF-8"%>
<%@ taglib uri="/struts-tags" prefix="s" %>
<table border="1">
    <tr>
        <td><s:text name="label.text.ordinal"/></td>
        <td><s:text name="label.text.id"/></td>
        <td><s:text name="label.text.userName"/></td>
        <td><s:text name="label.text.password"/></td>
    </tr>
    <s:iterator value="#request.userList" status="status">
    <tr>
        <td width="40"><s:property value="#status.index+1"></s:property></td>
        <td width="80"><s:property value="id"></s:property></td>
        <td width="80"><s:property value="name"></s:property></td>
        <td width="80"><s:property value="password"></s:property></td>
    </tr>
    </s:iterator>
</table>
```

12.4.7 测试

发布 Web 项目，启动 Tomcat 服务器以及 MySQL 服务器，访问 index.jsp，结果如图 12-10 所示。

（1）单击"注册"链接，结果如图 12-11 所示。

图 12-10　访问 index.jsp 的结果

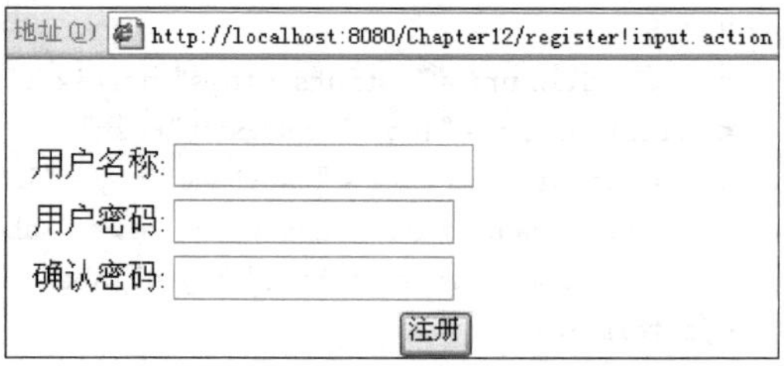

图 12-11　注册页面

① 输入信息,单击"注册"按钮,结果如图 12-13 所示。
② 输入部分信息,单击"注册"按钮,结果如图 12-13 所示。

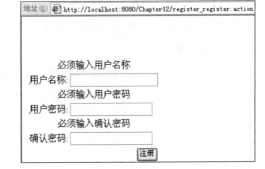

图 12-12 注册成功页面　　　　图 12-13 输入部分信息注册的页面

(2) 单击"登录"链接,结果如图 12-14 所示。
① 登录成功页面如图 12-15 所示。

图 12-14 登录页面　　　　图 12-15 登录成功页面

② 登录失败页面如图 12-16 所示。
③ 输入部分信息,单击"登录"按钮的页面如图 12-17 所示。

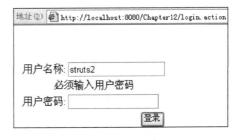

图 12-16 登录失败页面　　　　图 12-17 输入部分信息登录的页面

(3) 单击"用户列表"链接,结果如图 12-18 所示。

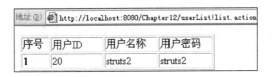

图 12-18 用户列表显示页面

习题

1. 将 12.4.3 节中的实现用户注册、登录和用户显示操作的 Action 用一个 Action(如 UserAction)实现。
2. 修改 12.4 节示例,使用 SQL Server 数据库替换 MySQL 数据库。

第13章 在线图片管理

本章介绍一个基于 Struts 2 框架的在线图片管理系统的设计与实现,该系统实现了用户注册、登录和显示用户列表以及允许登录用户上传图片、编辑图片和显示图片等操作的功能。

13.1 概述

13.1.1 功能简介

在线图片管理的功能由以下几部分组成。

1. 用户管理

通过网站首页的"注册"链接进入注册页面,实现用户的注册。用户注册后可以通过网站首页的"登录"链接进入登录页面进行登录。

登录后的用户可以通过网站首页的"注销"链接从网站注销。

另外,用户也可以通过网站首页的"用户列表"链接查看注册的用户信息。

2. 图片管理

未登录用户可以通过网站首页的"图片列表"查看所有图片。

当用户登录到网站后,可以对自己上传的图片进行操作,如图片列表显示、按图片标题进行查找、上传图片、编辑和删除照片。

网站首页如图 13-1 所示。

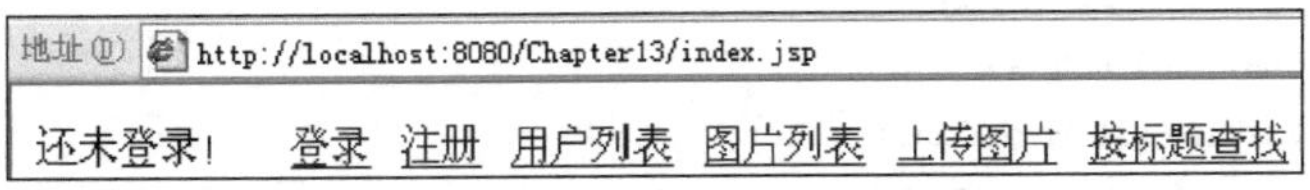

图 13-1 在线图片管理网站首页

登录成功之后的页面如图 13-2 所示。

图 13-2 登录成功之后的页面

13.1.2　总体设计

系统使用 Struts 2 框架技术和 MySQL 数据库进行开发，持久化解决方案使用 Tomcat 数据源的方式实现。

系统业务处理遵循实际项目中的分层，即使用 Action 处理用户的请求，业务逻辑操作在 Service 层实现，基本的持久化操作在 DAO 层实现。Service 层和 DAO 层也都遵循实际项目开发中的实现方式，使用面向接口编程，在接口中定义业务方法，在实现类中实现方法的业务逻辑处理。项目中的命名遵循下面的规则，接口以字母"I"开头，例如，IUserService，实现类以"Impl"结尾，例如，UserSercieImpl。通过使用接口，可以让系统中的业务逻辑组件面向接口编程，让系统的业务逻辑组件依赖于接口，而不是依赖于具体的实现类，从而提供代码间良好的解耦。

13.2　准备工作

在实现在线图片管理功能之前，首先做好如下的准备工作。

13.2.1　创建数据库和表

1. 创建数据库 mydata

请参考第 12.3 节中创建数据库的语句。

2. 创建数据表

该系统需要两个数据表 stu 和 album。其中创建 stu 数据表的语句请参考第 12.3 节中创建数据表的语句。

创建 album 数据表的语句如下：

```sql
use mydata;
CREATE TABLE album (
  id int(11) NOT NULL auto_increment,
  imgTitle varchar(30) NOT NULL,
  fileName varchar(30) NOT NULL,
  description varchar(100) default NULL,
  updateTime timestamp NOT NULL default '0000-00-00 00:00:00' on update CURRENT_TIMESTAMP,
  userId int(11) NOT NULL,
  PRIMARY KEY (id),
  KEY stuId(userId),
  CONSTRAINT stu_album FOREIGN KEY (userId) REFERENCES stu(id) ON DELETE CASCADE) ENGINE = InnoDB DEFAULT CHARSET = utf8;
```

13.2.2　使用 Log4j 输出信息

系统使用 Log4j 输出信息，在实际开发中，使用 Log4j 输出信息代替使用 System.out 输出信息，可以配置输出信息的级别，也可以提高系统的性能。使用 Log4j，则需要添加 Log4j 的 Jar 包（log4j-1.2.14.jar），并配置 log4j.properties 文件。

log4j.properties

```
log4j.rootLogger = INFO, stdout, logfile
# Console Appender
log4j.appender.stdout = org.apache.log4j.ConsoleAppender
log4j.appender.stdout.layout = org.apache.log4j.PatternLayout
log4j.appender.stdout.layout.ConversionPattern = %d %-5p [%r] [%t] %c{2} - %m%n
# File Appender
log4j.appender.logfile = org.apache.log4j.FileAppender
log4j.appender.logfile.File = logs/struts2.log
log4j.appender.logfile.layout = org.apache.log4j.PatternLayout
log4j.appender.logfile.layout.ConversionPattern = %d %p [%c] - %m%n
```

该配置文件不但在控制台输出信息，而且还将信息保存在 struts2.log 文件中。

13.2.3　国际化

为了实现在线图片管理 Web 应用的国际化，需要配置国际化资源文件。本章为了简化，只配置了显示中文信息的国际化资源文件 ApplicationResources_zh_CN.properties。

ApplicationResources_zh_CN.properties

```
label.text.userName = 用户名称
label.text.password = 用户密码
label.text.confirmPassword = 确认密码
label.text.register = 注册
label.text.login = 登录
label.text.ordinal = 序号
label.text.userId = 用户ID
label.text.loginSuccess = 登录成功！
label.text.registerSuccess = 注册成功！
label.text.albumTitle = 图片标题
label.text.albumDesc = 描述信息
label.text.browse = 选择图片
label.text.image = 图片
label.text.updateTime = 更新时间
label.text.save = 保存
label.text.update = 更新
label.text.operation = 操作
label.text.search = 查找
struts.messages.invalid.token = 重复提交！
userNameRequired = 必须输入 ${getText("label.text.userName")}
```

```
passwordRequired = 必须输入 ${getText("label.text.password")}
confirmPasswordRequired = 必须输入 ${getText("label.text.confirmPassword")}
confirmPasswordError = ${getText("label.text.password")}和
                       ${getText("label.text.confirmPassword")}必须一致!
errors.userNameOrPassword = 用户名或密码错误!
```

该文件需要使用 native2ascii 工具将其中的中文转换为 Unicode 编码序列。

13.2.4　异步交互

当用户注册时,用户名称不能重复,这里使用了异步提交方式进行检查注册的用户名称是否重复。异步提交方式使用 jQuery 的 Ajax 实现,为了使用 jQuery,需要复制 jquery.js 文件到 Web 应用中。

13.2.5　数据库配置

该系统使用 Tomcat 数据源连接数据库,其配置方式请参考第 12.1.1 节中 web.xml 和 context.xml 文件的配置方式。

13.2.6　Web 应用的目录结构

(1) 创建 Web 项目 Chapter13,并添加 Struts 2 框架支持,MySQL 数据库的驱动程序以及 Log4j 的 Jar 包。配置 web.xml,内容如下:

web.xml

```xml
<?xml version="1.0" encoding="UTF-8"?>
<web-app version="2.4"
    xmlns="http://java.sun.com/xml/ns/j2ee"
    xmlns:xsi="http://www.w3.org/2001/XMLSchema-instance"
    xsi:schemaLocation="http://java.sun.com/xml/ns/j2ee
    http://java.sun.com/xml/ns/j2ee/web-app_2_4.xsd">
    <filter>
        <filter-name>Struts 2</filter-name>
        <filter-class>
            org.apache.struts2.dispatcher.ng.filter.StrutsPrepareAndExecuteFilter
        </filter-class>
    </filter>
    <filter-mapping>
        <filter-name>Struts 2</filter-name>
        <url-pattern>/*</url-pattern>
    </filter-mapping>
</web-app>
```

(2) 系统的目录结构如图 13-3 所示。

(3) src 目录结构如图 13-4 所示。

(4) WebRoot 目录结构如图 13-5 所示。

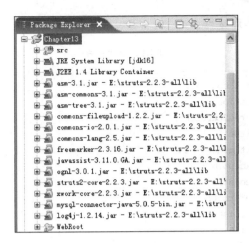

图 13-3 系统的目录结构

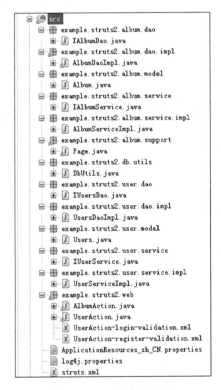

图 13-4 src 的目录结构

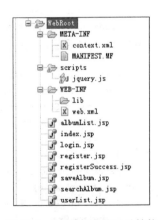

图 13-5 WebRoot 的目录结构

13.3 辅助类

为了实现代码复用,该系统提供了封装数据库操作和数据显示分页的辅助类。

13.3.1 封装数据库操作的辅助类

数据库操作辅助类封装的操作包括获得数据库连接,将从数据库中查询的结果集转换成 Map 对象,关闭数据库连接等。

DbUtils.java

```java
package example.struts2.db.utils;
import java.sql.Connection;
import java.sql.PreparedStatement;
import java.sql.ResultSet;
import java.sql.ResultSetMetaData;
import java.sql.SQLException;
import java.sql.Statement;
import java.text.DateFormat;
import java.util.ArrayList;
import java.util.Date;
import java.util.HashMap;
import java.util.List;
import java.util.Map;
import java.util.logging.Logger;
import javax.naming.InitialContext;
import javax.sql.DataSource;
import com.sun.rowset.CachedRowSetImpl;
public class DbUtils {
    private static Logger logger = Logger.getLogger("DbUtils");
    public static int execute(String sql, List<Object> paramList) throws Exception {
        if(sql == null || sql.trim().equals("")) {
            logger.info("parameter is valid!");
        }
        Connection conn = null;
        PreparedStatement pstmt = null;
        int result = 0;
        try {
            conn = DbUtils.getConnection();
            pstmt = DbUtils.getPreparedStatement(conn, sql);
            setPreparedStatementParam(pstmt, paramList);
            if(pstmt == null) {
                return -1;
            }
            result = pstmt.executeUpdate();
        } catch (Exception e) {
            logger.info(e.getMessage());
            throw new Exception(e);
        } finally {
            closeStatement(pstmt);
            closeConn(conn);
        }
        return result;
    }
    public static CachedRowSetImpl query(String sql, List<Object> paramList) throws Exception {
        if(sql == null || sql.trim().equals("")) {
```

```java
            logger.info("parameter is valid!");
        }
        Connection conn = null;
        PreparedStatement pstmt = null;
        ResultSet rs;
        CachedRowSetImpl cachedRowSet;
        try {
            conn = DbUtils.getConnection();
            pstmt = DbUtils.getPreparedStatement(conn, sql);
            setPreparedStatementParam(pstmt, paramList);
            rs = getResultSet(pstmt);
            cachedRowSet = new CachedRowSetImpl();
            cachedRowSet.populate(rs);
            cachedRowSet.last();
        } catch (Exception e) {
            logger.info(e.getMessage());
            throw new Exception(e);
        } finally {
            closeStatement(pstmt);
            closeConn(conn);
        }
        return cachedRowSet;
    }
    public static List<Map<String, String>> getQueryList(String sql, List<Object> paramList) throws Exception {
        if(sql == null || sql.trim().equals("")) {
            logger.info("parameter is valid!");
            return null;
        }
        Connection conn = null;
        PreparedStatement pstmt = null;
        ResultSet rs = null;
        List<Map<String, String>> queryList = null;
        try {
            conn = DbUtils.getConnection();
            pstmt = DbUtils.getPreparedStatement(conn, sql);
            setPreparedStatementParam(pstmt, paramList);
            if(pstmt == null) {
                return null;
            }
            rs = DbUtils.getResultSet(pstmt);
            queryList = DbUtils.getQueryList(rs);
        } catch (RuntimeException e) {
            logger.info(e.getMessage());
            throw new Exception(e);
        } finally {
            closeResultSet(rs);
            closeStatement(pstmt);
            closeConn(conn);
```

```java
            }
            return queryList;
    }
    private static void setPreparedStatementParam(PreparedStatement pstmt, List < Object >
paramList) throws Exception {
            if(pstmt == null || paramList == null || paramList.isEmpty()) {
                return;
            }
            DateFormat df = DateFormat.getDateTimeInstance();
            for (int i = 0; i < paramList.size(); i++) {
                if(paramList.get(i) instanceof Integer) {
                    int paramValue = ((Integer)paramList.get(i)).intValue();
                    pstmt.setInt(i + 1, paramValue);
                } else if(paramList.get(i) instanceof Float) {
                    float paramValue = ((Float)paramList.get(i)).floatValue();
                    pstmt.setFloat(i + 1, paramValue);
                } else if(paramList.get(i) instanceof Double) {
                    double paramValue = ((Double)paramList.get(i)).doubleValue();
                    pstmt.setDouble(i + 1, paramValue);
                } else if(paramList.get(i) instanceof Date) {
                    pstmt.setString(i + 1, df.format((Date)paramList.get(i)));
                } else if(paramList.get(i) instanceof Long) {
                    long paramValue = ((Long)paramList.get(i)).longValue();
                    pstmt.setLong(i + 1, paramValue);
                } else if(paramList.get(i) instanceof String) {
                    pstmt.setString(i + 1, (String)paramList.get(i));
                }
            }
            return;
    }
    private static Connection getConnection() throws Exception {
            InitialContext cxt = new InitialContext();
            if ( cxt == null ) {
                throw new Exception("no context!");
            }
            DataSource ds = (DataSource) cxt.lookup( "java:/comp/env/jdbc/Struts2DB" );
            if ( ds == null ) {
                throw new Exception("Data source not found!");
            }
            return ds.getConnection();
    }
     private static PreparedStatement getPreparedStatement(Connection conn, String sql)
throws Exception {
            if(conn == null || sql == null || sql.trim().equals("")) {
                return null;
            }
            PreparedStatement pstmt = conn.prepareStatement(sql.trim());
            return pstmt;
    }
```

```java
    private static ResultSet getResultSet(PreparedStatement pstmt) throws Exception {
        if(pstmt == null) {
            return null;
        }
        ResultSet rs = pstmt.executeQuery();
        return rs;
    }
    private static List<Map<String, String>> getQueryList(ResultSet rs) throws Exception {
        if(rs == null) {
            return null;
        }
        ResultSetMetaData rsMetaData = rs.getMetaData();
        int columnCount = rsMetaData.getColumnCount();
        List<Map<String, String>> dataList = new ArrayList<Map<String, String>>();
        while (rs.next()) {
            Map<String, String> dataMap = new HashMap<String, String>();
            for (int i = 0; i < columnCount; i++) {
                dataMap.put(rsMetaData.getColumnName(i + 1), rs.getString(i + 1));
            }
            dataList.add(dataMap); //将 Map 保存在 List<Map>中
        }
        return dataList;
    }
    private static void closeConn(Connection conn) {
        if(conn == null) {
            return;
        }
        try {
            conn.close();
        } catch (SQLException e) {
            logger.info(e.getMessage());
        }
    }
    private static void closeStatement(Statement stmt) {
        if(stmt == null) {
            return;
        }
        try {
            stmt.close();
        } catch (SQLException e) {
            logger.info(e.getMessage());
        }
    }
    private static void closeResultSet(ResultSet rs) {
        if(rs == null) {
            return;
        }
        try {
            rs.close();
```

```
            } catch (SQLException e) {
                logger.info(e.getMessage());
            }
        }
    }
```

13.3.2 数据分页的辅助类

在显示图片列表中,需要将图片分页显示,该系统定义了一个用于信息分页的辅助类 Page。

Page.java

```java
package example.struts2.album.support;
import java.io.Serializable;
import java.util.ArrayList;
import java.util.List;
public class Page implements Serializable {
    private static final int DEFAULT_PAGE_SIZE = 20; // 每页数据容量的默认值
    private int pageSize;              // 每页数据容量
    private int start;                 // 当前页第一条数据在 List 中的位置,默认为 0
    private List data;                 // 当前页中存放的记录
    private long totalCount;           // 总记录数
    public Page() {
        this(0, 0, DEFAULT_PAGE_SIZE, new ArrayList());
    }
    public Page(int start, long totalSize, int pageSize, List data) {
        if (pageSize <= 0 || start < 0 || totalSize < 0) {
            throw new IllegalArgumentException("Illegal Arguments to Initiate Page Object");
        }
        this.pageSize = pageSize;
        this.start = start;
        this.totalCount = totalSize;
        this.data = data;
    }
    public long getTotalCount() {
        return this.totalCount;
    }
    public long getTotalPageCount() {
        return (totalCount + pageSize - 1) / pageSize;
    }
    public int getPageSize() {
        return pageSize;
    }
    public void setResult(List data) {
        this.data = data;
    }
    public List getResult() {
```

```
            return data;
        }
        public int getCurrentPageNo() {
            return start / pageSize + 1;
        }
        public boolean hasNextPage() {
            return this.getCurrentPageNo() < this.getTotalPageCount();
        }
        public boolean hasPreviousPage() {
            return this.getCurrentPageNo() > 1;
        }
        public boolean isEmpty() {
            return data == null || data.isEmpty();
        }
        public int getStartIndex() {
            return (getCurrentPageNo() - 1) * pageSize;
        }
        public int getEndIndex() {
            int endIndex = getCurrentPageNo() * pageSize - 1;
            return endIndex >= totalCount ? (int)totalCount - 1 : endIndex;
        }
        protected static int getStartOfPage(int pageNo) {
            return getStartOfPage(pageNo, DEFAULT_PAGE_SIZE);
        }
        public static int getStartOfPage(int pageNo, int pageSize) {
            return (pageNo - 1) * pageSize;
        }
}
```

13.4 实现数据模型

为了实现代码复用和将对数据库的操作转换成对 Java 对象的操作，该系统定义了两个 Java 类（Users.java 和 Album.java）与数据库中的对应的表对应。

13.4.1 用户数据模型

用户类中的属性和数据库中的 stu 表中的字段对应。

Users.java

```
package example.struts2.user.model;
public class Users {
    private int id;
    private String name;
    private String password;
    public int getId() {
        return id;
```

```java
    }
    public void setId(int id) {
        this.id = id;
    }
    public String getName() {
        return name;
    }
    public void setName(String name) {
        this.name = name;
    }
    public String getPassword() {
        return password;
    }
    public void setPassword(String password) {
        this.password = password;
    }
}
```

13.4.2 图片数据模型

图片类中的属性和数据库中的 Album 表中的字段对应。

Album.java

```java
package example.struts2.album.model;
import java.util.Date;
import example.struts2.user.model.Users;
public class Album {
    private int id;
    private String imgTitle;
    private String fileName;
    private String description;
    private Date updateTime;
    private Users user;
    public String getDescription() {
        return description;
    }
    public void setDescription(String description) {
        this.description = description;
    }
    public String getFileName() {
        return fileName;
    }
    public void setFileName(String fileName) {
        this.fileName = fileName;
    }
```

```java
    public int getId() {
        return id;
    }
    public void setId(int id) {
        this.id = id;
    }
    public String getImgTitle() {
        return imgTitle;
    }
    public void setImgTitle(String imgTitle) {
        this.imgTitle = imgTitle;
    }
    public Date getUpdateTime() {
        return updateTime;
    }
    public void setUpdateTime(Date updateTime) {
        this.updateTime = updateTime;
    }
    public Users getUser() {
        return user;
    }
    public void setUser(Users user) {
        this.user = user;
    }
}
```

13.5 实现 DAO 层

DAO 层负责与持久化对象的交互，封装了数据的增加、删除、查询、修改等操作。

13.5.1 定义 DAO 层接口

1．管理用户数据的接口

IUsersDao.java

```java
package example.struts2.user.dao;
import java.util.List;
import example.struts2.user.model.Users;
public interface IUsersDao {
    int save(Users user) throws Exception;
    List<Users> findAll() throws Exception;
    Users getByName(String userName) throws Exception;
    Users getByNameAndPwd(String userName, String userPwd) throws Exception;
}
```

2. 管理图片数据的接口

IAlbumDao.java

```java
package example.struts2.album.dao;
import example.struts2.album.model.Album;
import example.struts2.album.support.Page;
public interface IAlbumDao {
    int save(Album album) throws Exception;                     //新增图片
    int update(Album album) throws Exception;                   //更新图片
    void delete(int albumId) throws Exception;                  //根据 ID 删除图片
    Page findAll(int pageNo, int pageSize) throws Exception;    //获得所有图片
    Page findAllByUser(int userId, int pageNo, int pageSize) throws Exception;
    Album get(int albumId) throws Exception;
    Page getByTitle(int userId, String albumTitle, int pageNo, int pageSize) throws Exception;
}
```

13.5.2　实现 DAO 层接口

DAO 接口中只声明了方法，必须提供其实现类来完成实际的业务操作。实现类中调用了数据库操作辅助类，来实现数据库业务操作。

1. 实现管理用户数据的接口

UsersDaoImpl.java

```java
package example.struts2.user.dao.impl;
import java.util.ArrayList;
import java.util.Iterator;
import java.util.List;
import java.util.Map;
import java.util.Map.Entry;
import example.struts2.db.utils.DbUtils;
import example.struts2.user.dao.IUsersDao;
import example.struts2.user.model.Users;
public class UsersDaoImpl implements IUsersDao {
    public int save(Users user) throws Exception {
        String sql = "insert into stu(name, password) values(?, ?)";
        List<Object> paramList = new ArrayList<Object>();
        paramList.add(user.getName());
        paramList.add(user.getPassword());
        int result = DbUtils.execute(sql, paramList);
        if(result == 1) {
            user = getByName(user.getName());
        }
        return result;
    }
    public List<Users> findAll() throws Exception {
```

```java
            String sql = "select * from stu";
            List<Users> userList = getUserList(sql, null);
            return userList;
    }
    public Users getByName(String userName) throws Exception {
            String sql = "select * from stu where name = ?";
            List<Object> paramList = new ArrayList<Object>();
            paramList.add(userName.trim());
            List<Users> userList = getUserList(sql, paramList);
            if(userList.isEmpty()) {
                    return null;
            }
            return userList.get(0);
    }
    public Users getByNameAndPwd(String userName, String userPwd)
                    throws Exception {
            String sql = "select * from stu where name = ? and password = ?";
            List<Object> paramList = new ArrayList<Object>();
            paramList.add(userName.trim());
            paramList.add(userPwd.trim());
            List<Users> userList = getUserList(sql, paramList);
            if(!userList.isEmpty()){
                    return userList.get(0);
            }
            return null;
    }
    private List<Users> getUserList(String sql,List<Object> paramList) throws Exception {
            List<Users> userList = new ArrayList<Users>();
            List<Map<String, String>> userMapList = DbUtils.getQueryList(sql, paramList);
            if(userMapList == null || userMapList.isEmpty()) {
                    return userList;
            }
            for (Map<String, String> userMap : userMapList) {
                    Iterator<Entry<String, String>> userEntryIt = userMap.entrySet().iterator();
                    Users user = new Users();
                    while (userEntryIt.hasNext()) {
                            Entry<String, String> userEntry = userEntryIt.next();
                            if(userEntry.getKey().equals("id")) {
                                    user.setId(Integer.parseInt(userEntry.getValue()));
                            } else if(userEntry.getKey().equals("name")) {
                                    user.setName(userEntry.getValue());
                            } else if(userEntry.getKey().equals("password")) {
                                    user.setPassword(userEntry.getValue());
                            }
                    }
                    userList.add(user);
            }
            return userList;
    }
}
```

2. 实现管理图片数据的接口

AlbumDaoImpl.java

```java
package example.struts2.album.dao.impl;
import java.sql.ResultSetMetaData;
import java.sql.SQLException;
import java.text.SimpleDateFormat;
import java.util.ArrayList;
import java.util.Date;
import java.util.HashMap;
import java.util.Iterator;
import java.util.List;
import java.util.Map;
import java.util.Map.Entry;
import com.sun.rowset.CachedRowSetImpl;
import example.struts2.album.dao.IAlbumDao;
import example.struts2.album.model.Album;
import example.struts2.album.support.Page;
import example.struts2.db.utils.DbUtils;
import example.struts2.user.model.Users;
public class AlbumDaoImpl implements IAlbumDao {
    public int save(Album album) throws Exception {
        String sql = " insert into album ( imgTitle, fileName, description, updateTime, userId) values(?, ?, ?, ?, ?)";
        List<Object> paramList = new ArrayList<Object>();
        paramList.add(album.getImgTitle());
        paramList.add(album.getFileName());
        paramList.add(album.getDescription());
        paramList.add(new Date());
        paramList.add(album.getUser().getId());
        int result = DbUtils.execute(sql, paramList);
        return result;
    }
    public int update(Album album) throws Exception {
        String sql = "update album set imgTitle = ?, description = ?, updateTime = ? where id = ?";
        List<Object> paramList = new ArrayList<Object>();
        paramList.add(album.getImgTitle());
        paramList.add(album.getDescription());
        paramList.add(new Date());
        paramList.add(album.getId());
        int result = DbUtils.execute(sql, paramList);
        return result;
    }
    public void delete(int albumId) throws Exception {
        String sql = "delete from album where id = ?";
        List<Object> paramList = new ArrayList<Object>();
        paramList.add(albumId);
        DbUtils.execute(sql, paramList);
```

```java
    }
    public Page findAll(int pageNo, int pageSize) throws Exception {
        String sql = "select * from album order by updateTime desc";
        return pagedQuery(sql, pageNo, pageSize, null);
    }
    public Page findAllByUser(int userId, int pageNo, int pageSize)    throws Exception {
        String sql = "select * from album where userId = ? order by updateTime desc";
        List<Object> paramList = new ArrayList<Object>();
        paramList.add(userId);
        return pagedQuery(sql, pageNo, pageSize, paramList);
    }
    public Album get(int albumId) throws Exception {
        String sql = "select al.*,name,password from album as al, stu where al.id = ? and al.userId = stu.id";
        List<Object> paramList = new ArrayList<Object>();
        paramList.add(albumId);
        List<Album> albumList = getAlbumList(sql, paramList);
        if(albumList.isEmpty()) {
            return null;
        }
        return albumList.get(0);
    }
    public Page getByTitle(int userId, String albumTitle, int pageNo, int pageSize) throws Exception {
         String sql = "select * from album where userId = ? and imgTitle = ? order by updateTime desc";
        List<Object> paramList = new ArrayList<Object>();
        paramList.add(userId);
        paramList.add(albumTitle);
        return pagedQuery(sql, pageNo, pageSize, paramList);
    }
    public Page pagedQuery(String sql, int pageNo, int pageSize, List<Object> paramList) throws Exception{
        CachedRowSetImpl cachedRowSet = DbUtils.query(sql, paramList);
        long totalCount = cachedRowSet.getRow();
        if (totalCount < 1) {
            return new Page();
        }
        int startIndex = Page.getStartOfPage(pageNo, pageSize);
        List list = toList(cachedRowSet, pageNo, pageSize);
        return new Page(startIndex, totalCount, pageSize, list);
    }
    private static List toList(CachedRowSetImpl cachedRowSet, int pageNo, int pageSize) {
        try {
            ResultSetMetaData md = cachedRowSet.getMetaData();
            int columnCount = md.getColumnCount();
            List list = new ArrayList();
            cachedRowSet.absolute((pageNo - 1) * pageSize + 1);
            for(int i = 1; i <= pageSize; i++) {
```

```java
                    Map rowData = new HashMap();
                    for (int j = 1; j <= columnCount; j++) {
                        rowData.put(md.getColumnName(j), cachedRowSet.getObject(j));
                    }
                    list.add(rowData);
                    if(!cachedRowSet.next()){
                        break;
                    }
                }
                return list;
            }
            catch (SQLException e) {
                e.printStackTrace();
                return null;
            }
        }
        private List<Album> getAlbumList(String sql, List<Object> paramList) throws Exception{
            List<Album> albumList = new ArrayList<Album>();
            List<Map<String, String>> albumMapList = DbUtils.getQueryList(sql, paramList);
            if(albumMapList == null || albumMapList.isEmpty()) {
                return albumList;
            }
            for (Map<String, String> albumMap : albumMapList) {
                Iterator<Entry<String, String>> albumEntryIt = albumMap.entrySet().iterator();
                Album album = new Album();
                Users user = new Users();
                while (albumEntryIt.hasNext()) {
                    Entry<String, String> albumEntry = albumEntryIt.next();
                    if(albumEntry.getKey().equals("id")) {
                        album.setId(Integer.parseInt(albumEntry.getValue()));
                    } else if(albumEntry.getKey().equals("imgTitle")) {
                        album.setImgTitle(albumEntry.getValue());
                    } else if(albumEntry.getKey().equals("fileName")) {
                        album.setFileName(albumEntry.getValue());
                    } else if(albumEntry.getKey().equals("description")) {
                        album.setDescription(albumEntry.getValue());
                    }
                    else if(albumEntry.getKey().equals("updateTime")) {
                        SimpleDateFormat sdf = new SimpleDateFormat("yyyy-MM-dd HH:mm:ss");
                        album.setUpdateTime(sdf.parse(albumEntry.getValue()));
                    }
                    else if(albumEntry.getKey().equals("userId")) {
                        user.setId(Integer.parseInt(albumEntry.getValue()));
                    } else if(albumEntry.getKey().equals("name")) {
                        user.setName(albumEntry.getValue());
                    } else if(albumEntry.getKey().equals("password")) {
                        user.setName(albumEntry.getValue());
                    }
```

```
            }
            album.setUser(user);
            albumList.add(album);
        }
        return albumList;
    }
}
```

13.6 实现业务逻辑层

系统的业务逻辑（Service）组件应该包含的方法取决于系统的业务需求，用户的每个业务请求通常都应该提供一个与之对应的业务逻辑方法。

13.6.1 定义 Service 层接口

1. 用户业务逻辑的接口

本系统用户管理部分需要处理的请求包括：用户注册、用户登录、判断新注册的用户名是否可以使用、用户列表显示。业务逻辑接口定义如下：

IUserService.java

```
package example.struts2.user.service;
import java.util.List;
import example.struts2.user.model.Users;
public interface IUserService {
    int addUser(Users user) throws Exception;
    Users getUserByNameAndPassword(String userName, String userPwd) throws Exception;
    boolean isNameValid(String userName) throws Exception; //用户名是否可用
    List<Users> findAll() throws Exception; //获得所有用户
}
```

2. 图片业务逻辑的接口

本系统图片管理部分需要处理的请求包括：新增图片、修改图片、删除图片、查询和列表图片。业务逻辑接口定义如下：

IAlbumService.java

```
package example.struts2.album.service;
import example.struts2.album.model.Album;
import example.struts2.album.support.Page;
public interface IAlbumService {
    int save(Album album) throws Exception;
    int update(Album album) throws Exception;
    void delete(int albumId) throws Exception;
```

```java
    Page findAll(int pageNo, int pageSize) throws Exception;
    Page findAllByUser(int userId, int pageNo, int pageSize) throws Exception;
    Album get(int albumId) throws Exception;
    Page getByTitle(int userId, String albumTitle, int pageNo, int pageSize) throws Exception;
}
```

13.6.2 实现 Service 层接口

定义了业务逻辑层接口,必须提供实现类才能完成实际的业务逻辑操作。业务逻辑层的业务逻辑处理依赖底层的 DAO 组件提供的操作来实现业务逻辑,这样就可以专注于处理业务逻辑,而无需关心底层的实现细节。

业务逻辑层依赖于 DAO 组件,但这是一种松耦合的依赖,是接口层次的依赖。

1. 实现用户业务逻辑的接口

用户管理的业务逻辑层的实现类为 UserServiceImpl.java,代码如下:

UserServiceImpl.java

```java
package example.struts2.user.service.impl;
import java.util.List;
import example.struts2.user.dao.IUsersDao;
import example.struts2.user.dao.impl.UsersDaoImpl;
import example.struts2.user.model.Users;
import example.struts2.user.service.IUserService;
public class UserServiceImpl implements IUserService {
    private static IUsersDao userDao = new UsersDaoImpl();
    public int addUser(Users user) throws Exception {
        if(user == null || user.getName() == null || user.getPassword() == null || user.getName().equals("") || user.getPassword().equals("")){
            return 0;
        }
        return userDao.save(user);
    }
    public List<Users> findAll() throws Exception {
        return userDao.findAll();
    }
    public Users getUserByNameAndPassword(String userName, String userPwd) throws Exception {
        return userDao.getByNameAndPwd(userName, userPwd);
    }
    public boolean isNameValid(String userName) throws Exception {
        Users user = userDao.getByName(userName);
        if(user == null){
            return true;
        }
        else{
            return false;
```

```
            }
        }
    }
```

2. 实现图片业务逻辑的接口

图片管理的业务逻辑层的实现类为AlbumServiceImpl.java,代码如下:

AlbumServiceImpl.java

```java
package example.struts2.album.service.impl;
import example.struts2.album.dao.IAlbumDao;
import example.struts2.album.dao.impl.AlbumDaoImpl;
import example.struts2.album.model.Album;
import example.struts2.album.service.IAlbumService;
import example.struts2.album.support.Page;
public class AlbumServiceImpl implements IAlbumService {
    private static IAlbumDao albumDao = new AlbumDaoImpl();
    public int save(Album album) throws Exception {
        if(album == null || album.getImgTitle() == null || album.getImgTitle().equals("")){
            return 0;
        }
        return albumDao.save(album);
    }
    public int update(Album album) throws Exception {
        if(album == null || album.getImgTitle() == null || album.getImgTitle().equals("")){
            return 0;
        }
        return albumDao.update(album);
    }
    public void delete(int albumId) throws Exception {
        albumDao.delete(albumId);
    }
    public Page findAll(int pageNo, int pageSize) throws Exception {
        return albumDao.findAll(pageNo, pageSize);
    }
    public Page findAllByUser(int userId, int pageNo, int pageSize)     throws Exception {
        return albumDao.findAllByUser(userId, pageNo, pageSize);
    }
    public Album get(int albumId) throws Exception {
        return albumDao.get(albumId);
    }
    public Page getByTitle(int userId, String albumTitle, int pageNo, int pageSize) throws Exception {
        return albumDao.getByTitle(userId, albumTitle, pageNo, pageSize);
    }
}
```

13.7 实现控制器 Action

13.7.1 管理用户的控制器

本系统中,使用同一个 Action 类处理所有的用户请求,即 UserAction,代码如下:

UserAction.java

```java
package example.struts2.web;
import java.util.List;
import javax.servlet.http.HttpServletRequest;
import javax.servlet.http.HttpServletResponse;
import org.apache.struts2.ServletActionContext;
import org.apache.struts2.interceptor.ServletRequestAware;
import com.opensymphony.xwork2.ActionSupport;
import example.struts2.user.model.Users;
import example.struts2.user.service.IUserService;
import example.struts2.user.service.impl.UserServiceImpl;
public class UserAction extends ActionSupport implements ServletRequestAware {
    private String userName;
    private String password;
    private String confirmPassword;
    private Users user;
    private static IUserService userService = new UserServiceImpl();
    protected HttpServletRequest servletRequest = null;
    public String getPassword() {
        return password;
    }
    public void setPassword(String password) {
        this.password = password;
    }
    public String getUserName() {
        return userName;
    }
    public void setUserName(String userName) {
        this.userName = userName;
    }
    public String getConfirmPassword() {
        return confirmPassword;
    }
    public void setConfirmPassword(String confirmPassword) {
        this.confirmPassword = confirmPassword;
    }
    public Users getUser() {
        return user;
    }
    public void setUser(Users user) {
```

```java
        this.user = user;
    }
    public String register() throws Exception {
        Users userReg = new Users();
        userReg.setName(userName);
        userReg.setPassword(password);
        try{
            userService.addUser(userReg);
        }
        catch(RuntimeException e){
            servletRequest.setAttribute("fail", "failure, please re-regist!");
            return INPUT;
        }
        setUser(userReg);
        return SUCCESS;
    }
    public String login() throws Exception {
        Users userLog = null;
        try{
            userLog = userService.getUserByNameAndPassword(userName, password);
        }
        catch(Exception e){
            System.out.println(e);
        }
        if(userLog == null) {
            addActionError(getText("errors.userNameOrPassword"));
            return INPUT;
        }
        setUser(userLog);
        servletRequest.getSession().setAttribute("user", userLog);
        return SUCCESS;
    }
    public String logout() throws Exception {
        servletRequest.getSession().setAttribute("user", null);
        return SUCCESS;
    }
    public String list() throws Exception {
        List<Users> userList = null;
        try {
            userList = userService.findAll();
        }
        catch (RuntimeException e) {
            System.out.println(e);
        }
        servletRequest.setAttribute("userList", userList);
        return "userlist";
    }
    public String checkUser() throws Exception {
        String userName = servletRequest.getParameter("userName");
```

```java
            boolean isNameValid = userService.isNameValid(userName);
            HttpServletResponse response = ServletActionContext.getResponse();
            response.setHeader("Cache-Control", "no-store");
            response.setHeader("Pragma", "no-cache");
            response.setDateHeader("Expires", 0);
            response.setContentType("text/html");
            response.getWriter().write("{\"isNameValid\":" + isNameValid + "}");
            return null;
    }
    public String input() {
            return INPUT;
    }
    public void setServletRequest(HttpServletRequest request) {
            this.servletRequest = request;
    }
}
```

该 Action 类包含用户注册、登录、注销、列表、检查用户名是否重复等方法。

13.7.2 管理图片的控制器

本系统中,图片管理使用同一个 Action 类处理,即 AlbumAction,代码如下:

AlbumAction.java

```java
package example.struts2.web;
import java.io.File;
import java.text.SimpleDateFormat;
import java.util.Calendar;
import java.util.Date;
import java.util.List;
import javax.servlet.http.HttpServletRequest;
import org.apache.commons.io.FileUtils;
import org.apache.struts2.ServletActionContext;
import org.apache.struts2.interceptor.ServletRequestAware;
import com.opensymphony.xwork2.ActionSupport;
import example.struts2.album.model.Album;
import example.struts2.album.service.IAlbumService;
import example.struts2.album.service.impl.AlbumServiceImpl;
import example.struts2.album.support.Page;
import example.struts2.user.model.Users;
public class AlbumAction extends ActionSupport implements ServletRequestAware {
        private static IAlbumService albumService = new AlbumServiceImpl();
        protected HttpServletRequest servletRequest = null;
        private int id;
        private String imgTitle;
        private String description;
        private int userId;
        private File albumImage;
```

```java
    private String albumImageContentType;
    private String albumImageFileName;
    private String targetDir;
    private String targetFileName;
    private int curPage = 1;
    private int pageSize = 3;
    private int totalRecords;
    public String getTargetDir() {
        return targetDir;
    }
    public void setTargetDir(String targetDir) {
        this.targetDir = targetDir;
    }
    public File getAlbumImage() {
        return albumImage;
    }
    public void setAlbumImage(File albumImage) {
        this.albumImage = albumImage;
    }
    public String getAlbumImageContentType() {
        return albumImageContentType;
    }
    public void setAlbumImageContentType(String albumImageContentType) {
        this.albumImageContentType = albumImageContentType;
    }
    public String getAlbumImageFileName() {
        return albumImageFileName;
    }
    public void setAlbumImageFileName(String albumImageFileName) {
        this.albumImageFileName = albumImageFileName;
    }
    public int getCurPage() {
        return curPage;
    }
    public void setCurPage(int curPage) {
        this.curPage = curPage;
    }
    public String getDescription() {
        return description;
    }
    public void setDescription(String description) {
        this.description = description;
    }
    public int getId() {
        return id;
    }
    public void setId(int id) {
        this.id = id;
    }
```

```java
        public String getImgTitle() {
            return imgTitle;
        }
        public void setImgTitle(String imgTitle) {
            this.imgTitle = imgTitle;
        }
        public int getPageSize() {
            return pageSize;
        }
        public void setPageSize(int pageSize) {
            this.pageSize = pageSize;
        }
        public int getTotalRecords() {
            return totalRecords;
        }
        public void setTotalRecords(int totalRecords) {
            this.totalRecords = totalRecords;
        }
        public int getUserId() {
            return userId;
        }
        public void setUserId(int userId) {
            this.userId = userId;
        }
        public String getTargetFileName() {
            return targetFileName;
        }
        public void setTargetFileName(String targetFileName) {
            this.targetFileName = targetFileName;
        }
        public void setServletRequest(HttpServletRequest servletRequest) {
            this.servletRequest = servletRequest;
        }
        public String save() throws Exception {
            Users user = (Users)servletRequest.getSession().getAttribute("user");
            if(user == null) {
                return "login";
            }
            Album album = null;
            if(id == 0) {
                album = new Album();
            } else {
                album = albumService.get(id);
            }
            album.setImgTitle(imgTitle);
            album.setDescription(description);
            album.setUpdateTime(new Date());
            album.setUser(user);
            if(album == null || album.getId() == 0) {
```

```java
            if(albumImage == null) {
                servletRequest.setAttribute("album", album);
                return INPUT;
            }
        }
        if(id == 0) { //保存
            if(albumImageFileName != null) {
                targetFileName = genNewFileName(albumImageFileName);
            }
        }
        else{ //更新
            targetFileName = album.getFileName();
        }
        album.setFileName(targetFileName);
        try {
            if(id == 0){
                albumService.save(album);
            }
            else{
                albumService.update(album);
            }
            if(albumImageFileName != null) {
                String destPath = ServletActionContext.getServletContext()
                                    .getRealPath("/" + targetDir);
                File destDir = new File(destPath);
                if(!destDir.exists()) {
                    destDir.mkdirs();
                }
                if(albumImage != null) {
                    File targetFile = new File(destPath + "/" + targetFileName);
                    FileUtils.copyFile(albumImage, targetFile);
                }
            }
        } catch (Exception e) {
            e.printStackTrace();
        }
        return "saveAlbum";
    }
    private String genNewFileName(String fileName){
        int num = (int)Math.round((Math.random() * 100));
        SimpleDateFormat sf = new SimpleDateFormat("yyyyMMddHHmmssSSS");
        String newFileName = sf.format(Calendar.getInstance().getTime()) + num;
        int beginIndex = fileName.lastIndexOf(".");
        if(beginIndex > 0){
            String fileExt = fileName.substring(beginIndex);
            newFileName = newFileName + fileExt;
        }
        return newFileName;
    }
```

```java
public String list() throws Exception {  //显示图片
    Users user = (Users)servletRequest.getSession().getAttribute("user");
    List albumList = null;
    Page page = null;
    try {
        if(user == null) {
            page = albumService.findAll(curPage, pageSize);
        }
        else{
            if((imgTitle == null) || (imgTitle.equals(""))){
                page = albumService.findAllByUser(user.getId(),curPage,pageSize);
            }
            else{
                page = albumService.getByTitle(user.getId(), imgTitle,
                                                curPage, pageSize);
            }
        }
        albumList = page.getResult();
        totalRecords = (int)page.getTotalCount();
        servletRequest.setAttribute("page", page);
    }
    catch (RuntimeException e) {
        System.out.println(e);
    }
    servletRequest.setAttribute("albumList", albumList);
    return "albumList";
}
public String edit() throws Exception {  //编辑图片
    Users user = (Users)servletRequest.getSession().getAttribute("user");
    Album album = null;
    if(user == null) {
        return "login";
    }
    else{
        if(id != 0){
            album = albumService.get(id);
        }
        if(!album.getUser().getName().equals(user.getName())){
            album = null;
        }
    }
    servletRequest.setAttribute("album", album);
    return "saveAlbum";
}
public String delete() throws Exception {  //删除图片
    Users user = (Users)servletRequest.getSession().getAttribute("user");
    Album album = null;
    if(user == null) {
        return "login";
```

```java
            }
            else{
                if(id != 0){
                    album = albumService.get(id);
                }
                if(album.getUser().getId() == user.getId()){
                    albumService.delete(id);
                }
                String destPath = ServletActionContext.getServletContext().getRealPath("/"
                                                                + targetDir);
                File targetFile = new File(destPath + "/" + album.getFileName());
                targetFile.delete();
            }
            return list();
    }
}
```

AlbumAction 类中定义了保存与更新图片、显示图片、编辑图片、删除图片的方法。该类实现了 ServletRequestAware 接口，这样可以使用 Servlet API。

13.8 编写配置文件

13.8.1 struts.xml

struts.xml 文件中配置了国际化资源文件、拦截器和 Action，内容如下：

struts.xml

```xml
<?xml version="1.0" encoding="UTF-8"?>
<!DOCTYPE struts PUBLIC
    "-//Apache Software Foundation//DTD Struts Configuration 2.0//EN"
    "http://struts.apache.org/dtds/struts-2.0.dtd">
<struts>
    <package name="default" extends="struts-default">
        <interceptors>
            <interceptor-stack name="tokenStack">
                <interceptor-ref name="token">
                    <param name="excludeMethods">input</param>
                </interceptor-ref>
                <interceptor-ref name="defaultStack"/>
            </interceptor-stack>
        </interceptors>
        <action name="register" method="{1}" class="example.struts2.web.UserAction">
            <interceptor-ref name="tokenStack"/>
            <result name="invalid.token">/register.jsp</result>
            <result name="input">/register.jsp</result>
            <result>/registerSuccess.jsp</result>
```

```xml
            </action>
            <action name="login" method="login" class="example.struts2.web.UserAction">
                <result name="input">/login.jsp</result>
                <result>/index.jsp</result>
            </action>
            <action name="logout" method="logout"
                                    class="example.struts2.web.UserAction">
                <result>/index.jsp</result>
            </action>
            <action name="user" method="{1}" class="example.struts2.web.UserAction">
                <result name="userlist">/userList.jsp</result>
            </action>
            <action name="*Album" method="{1}"
                                    class="example.struts2.web.AlbumAction">
                <result name="saveAlbum">/saveAlbum.jsp</result>
                <result name="input">/saveAlbum.jsp</result>
                <result name="albumList">/albumList.jsp</result>
                <result name="login">login.jsp</result>
                <param name="targetDir">uploadImages</param>
                <interceptor-ref name="defaultStack">
                    <param name="fileUpload.maximumSize">409600</param>
                    <param name="fileUpload.allowedTypes">
                        image/jpg,image/jpeg,image/pjpeg
                    </param>
                </interceptor-ref>
            </action>
        </package>
        <constant name="struts.custom.i18n.resources" value="ApplicationResources"/>
</struts>
```

对于用户注册实现使用了 TOKEN，这样可以实现防止表单重复提交的功能。

13.8.2 输入校验文件

本系统中提供了两个基于 Struts 2 框架的输入校验配置文件，其中用于注册的输入校验配置文件为 UserAction-register-validation.xml，用于登录的输入校验配置文件为 UserAction-login-validation.xml。两个输入校验配置文件的代码分别如下：

1. 用于注册的输入校验配置文件

UserAction-register-validation.xml

```xml
<?xml version="1.0" encoding="UTF-8"?>
<!DOCTYPE validators PUBLIC
        "-//OpenSymphony Group//XWork Validator 1.0.2//EN"
        "http://www.opensymphony.com/xwork/xwork-validator-1.0.2.dtd">
<validators>
    <field name="userName">
```

```xml
            <field-validator type="requiredstring">
                <message key="userNameRequired" />
            </field-validator>
        </field>
        <field name="password">
            <field-validator type="requiredstring">
                <message key="passwordRequired" />
            </field-validator>
        </field>
        <field name="confirmPassword">
            <field-validator type="requiredstring">
                <message key="confirmPasswordRequired" />
            </field-validator>
            <field-validator type="fieldexpression">
                <param name="expression">password.equals(confirmPassword)</param>
                <message key="confirmPasswordError" />
            </field-validator>
        </field>
</validators>
```

2. 用于登录的输入校验配置文件

UserAction-login-validation.xml

```xml
<?xml version="1.0" encoding="UTF-8"?>
<!DOCTYPE validators PUBLIC
        "-//OpenSymphony Group//XWork Validator 1.0.2//EN"
        "http://www.opensymphony.com/xwork/xwork-validator-1.0.2.dtd">
<validators>
    <field name="userName">
        <field-validator type="requiredstring">
            <message key="userNameRequired" />
        </field-validator>
    </field>
    <field name="password">
        <field-validator type="requiredstring">
            <message key="passwordRequired" />
        </field-validator>
    </field>
</validators>
```

13.9 编写 JSP 文件

13.9.1 操作入口界面

本系统提供了一个所有操作的入口页面 index.jsp，代码如下：

index.jsp

```jsp
<%@ page language="java" contentType="text/html;charset=UTF-8" pageEncoding="UTF-8"%>
<%@ taglib uri="/struts-tags" prefix="s" %>
<s:if test="#session.user==null">还未登录!</s:if>
<s:else>
    <a href='<s:url action="logout" method="logout"/>'>注销</a>!
</s:else>  
<a href='<s:url value="/login.jsp"/>'>登录</a>  
<a href='<s:url action="register" method="input"/>'>注册</a>  
<a href='<s:url action="user" method="list"/>'>用户列表</a>  
<a href='<s:url action="listAlbum"/>'>图片列表</a>  
<a href='<s:url value="/saveAlbum.jsp"/>'>上传图片</a>  
<a href='<s:url value="/searchAlbum.jsp"/>'>按标题查找</a>
```

13.9.2 用户注册界面

(1) 实现用户注册的相关信息输入,代码如下:

register.jsp

```jsp
<%@ page language="java" contentType="text/html;charset=UTF-8" pageEncoding="UTF-8"%>
<%@ taglib uri="/struts-tags" prefix="s" %>
<html>
<head>
    <script type="text/javascript" src="scripts/jquery.js" type="text/javascript"></script>
    <script type="text/javascript">
        function checkUserName() {
            var userNameInput = document.getElementById("userName");
            var userName = userNameInput.value;
            if(jQuery.trim(userName) == "") {
                document.getElementById("userNameSpan").innerHTML = "";
                return false;
            }
            var isNameValid = true;
            jQuery.ajax({
                type: "POST",
                url: '<s:url action="user" method="checkUser.action"/>',
                data: "userName=" + userName,
                dataType: "json",
                cache: false,
                async: false,
                success: function (data, textStatus) {
                    isNameValid = data['isNameValid'];
                    if(isNameValid) {
                        document.getElementById("userNameSpan").innerHTML = 
                                                "该用户名可以使用";
                    }
```

```
                    else {
                        document.getElementById("userNameSpan").innerHTML =
                                                "该用户名已经被占用";
                        userNameInput.focus();
                    }
                }
            });
            return isNameValid;
        }
    </script>
</head>
<body>
<a href="index.jsp">网站首页</a>
<s:actionerror/>
<s:form action="register!register" method="post">
    <s:token />
    <s:textfield name="userName" id="userName" key="label.text.userName" onblur=
"checkUserName()"/><br/>
    <span id="userNameSpan"></span><br />
    <s:password name="password" key="label.text.password" /><br/>
    <s:password name="confirmPassword" key="label.text.confirmPassword" /><br/>
```

```
    <s:submit name="submit" value="%{getText('label.text.register')}" />
</s:form>
</body>
</html>
```

注册功能实现了检查用户是否可以使用的功能,在鼠标焦点离开用户名输入框时,会执行检验操作,并根据结果给出相应的提示信息。检查用户名是否可用是使用 Ajax 来实现的。

(2) 用户注册成功后进入 registerSuccess.jsp 页面,代码如下:

registerSuccess.jsp

```
<%@ page language="java" contentType="text/html; charset=UTF-8" pageEncoding="UTF-8" %>
<%@ taglib uri="/struts-tags" prefix="s" %>
<s:text name="label.text.registerSuccess"/><br/>
<s:text name="label.text.userName"/><b><s:property value="userName"></s:property>
</b><br/>
<s:text name="label.text.password"/><b><s:property value="password"></s:property>
</b><br/>
<a href="index.jsp">网站首页</a>
```

13.9.3 用户登录界面

用户登录界面为 login.jsp,代码如下:

login.jsp

```jsp
<%@ page language="java" contentType="text/html; charset=UTF-8" pageEncoding="UTF-8"%>
<%@ taglib uri="/struts-tags" prefix="s" %>
<s:actionerror/>
<s:form action="login" method="post">
    <s:textfield name="userName" key="label.text.userName"/><br/>
    <s:password name="password" key="label.text.password" /><br/>
    <s:submit name="submit" value="%{getText('label.text.login')}" />
</s:form>
```

13.9.4 用户列表界面

用户列表界面为 userList.jsp，代码如下：

userList.jsp

```jsp
<%@ page language="java" contentType="text/html; charset=UTF-8" pageEncoding="UTF-8"%>
<%@ taglib uri="/struts-tags" prefix="s" %>
<table border="1">
    <tr>
        <td><s:text name="label.text.ordinal"/></td>
        <td><s:text name="label.text.userId"/></td>
        <td><s:text name="label.text.userName"/></td>
        <td><s:text name="label.text.password"/></td>
    </tr>
    <s:iterator value="#request.userList" status="status">
    <tr>
        <td width="40"><s:property value="#status.index+1"></s:property></td>
        <td width="80"><s:property value="id"></s:property></td>
        <td width="80"><s:property value="name"></s:property></td>
        <td width="80"><s:property value="password"></s:property></td>
    </tr>
    </s:iterator>
</table>
<a href="index.jsp">网站首页</a>
```

13.9.5 图片上传与编辑界面

图片上传与编辑界面 saveAlbum.jsp，代码如下：

saveAlbum.jsp

```jsp
<%@ page language="java" contentType="text/html; charset=UTF-8" pageEncoding="UTF-8"%>
<%@ taglib uri="/struts-tags" prefix="s" %>
<script type="text/javascript">
    function check() {
        var imgTitle = document.getElementById("imgTitle").value;
        var description = document.getElementById("description").value;
```

```
            var albumImage = document.getElementById("albumImage").value;
            if(albumImage == ""){
            alert("albumImage");
            }
            if(imgTitle == "" || description == "" || albumImage == "" ||
                        imgTitle.replace(/^\s+|\s+$/g,"").length == 0 ||
                        description.replace(/^\s+|\s+$/g,"").length == 0){
                alert("必须输入信息!");
                return false;
            }
        }
</script>
<a href = "index.jsp">网站首页</a>
<s:actionerror/>
<s:form method = "POST" enctype = "multipart/form-data" action = "saveAlbum" onsubmit = "return check()">
    <s:textfield name = "imgTitle" id = "imgTitle" key = "label.text.albumTitle" value = "%{#request.album.imgTitle}"/><br/>
    <s:textfield name = "description" id = "description" key = "label.text.albumDesc" value = "%{#request.album.description}"/><br/>
    <s:file contenteditable = "false" name = "albumImage" id = "albumImage" key = "label.text.browse"/>
    <s:if test = "!#request.album.id.equals('')">
        <s:hidden name = "id" value = "%{#request.album.id}"></s:hidden>
        <s:submit type = "input" value = "%{getText('label.text.update')}"/>
    </s:if>
    <s:else>
        <s:submit type = "input" value = "%{getText('label.text.save')}"/>
    </s:else>
</s:form>
```

<s:file>标签的 contenteditable 属性设置为 false,目的是为了实现只能通过单击"浏览"按钮实现上传文件的选择。

13.9.6 图片列表界面

图片列表界面 albumList.jsp,代码如下:

albumList.jsp

```
<%@ page language = "java" contentType = "text/html; charset = UTF-8" pageEncoding = "UTF-8" %>
<%@ taglib uri = "/struts-tags" prefix = "s" %>
<table border = "1">
    <tr>
        <td><s:text name = "label.text.ordinal"/></td>
        <td><s:text name = "label.text.albumTitle"/></td>
        <td><s:text name = "label.text.albumDesc"/></td>
        <td><s:text name = "label.text.image"/></td>
        <td><s:text name = "label.text.updateTime"/></td>
```

```
            <td><s:text name = "label.text.userId"/></td>
            <s:if test = "#session.user! = null">
                <td><s:text name = "label.text.operation"/></td>
            </s:if>
        </tr>
        <s:iterator value = "#request.albumList" status = "status">
        <tr>
            <td width = "40"><s:property value = "#status.index + 1"></s:property></td>
            <td width = "80"><s:property value = "imgTitle"></s:property></td>
            <td width = "80"><s:property value = "description"></s:property></td>
            <td width = "80"><a href = '<s:property value = "targetDir"/>/<s:property value = "fileName"/>' target = _blank><img src = '<s:property value = "targetDir"/>/<s:property value = "fileName"/>' width = "100" height = "100"/></a></td>
            <td width = "160"><s:property value = "updateTime"></s:property></td>
            <td width = "80"><s:property value = "userId"></s:property></td>
            <s:if test = "#session.user! = null"><td width = "80"><a href = "editAlbum?id = <s:property value = 'id'/>">编辑</a>、<a href = "deleteAlbum?id = <s:property value = 'id'/>">删除</a></td></s:if>
        </tr>
        </s:iterator>
</table>
<s:iterator value = "#request.page">
<table border = "1">
    <tr>
        <td><s:if test = "hasPreviousPage()"><a href = "listAlbum?imgTitle = <s:property value = "imgTitle" />">首页</a></s:if><s:else>首页</s:else></td>
        <td><s:if test = "hasPreviousPage()"><a href = "listAlbum?imgTitle = <s:property value = "imgTitle" /> &curPage = <s:property value = "currentPageNo - 1" />">上页</a></s:if><s:else>上页</s:else></td>
        <td><s:if test = "hasNextPage()"><a href = "listAlbum?imgTitle = <s:property value = "imgTitle" /> &curPage = <s:property value = "currentPageNo + 1" />">下页</a></s:if><s:else>下页</s:else></td>
        <td><s:if test = "hasNextPage()"><a href = 'listAlbum?imgTitle = <s:property value = "imgTitle" /> &curPage = <s:property value = "totalPageCount" />'>尾页</a></s:if><s:else>尾页</s:else></td>
        <td>当前页 <s:property value = "currentPageNo" /> / <s:property value = "TotalPageCount" /></td>
        <td>每页记录数: <s:property value = "pageSize" /></td>
        <td><a href = "index.jsp">网站首页</a></td>
    </tr>
</table>
</s:iterator>
```

13.9.7 图片查找界面

图片查找界面 searchAlbum.jsp,代码如下:

searchAlbum.jsp

```
<%@ page language = "java" contentType = "text/html; charset = UTF - 8" pageEncoding = "UTF -
8" %>
<%@ taglib uri = "/struts - tags" prefix = "s" %>
<a href = "index.jsp">网站首页</a>
<s:form method = "POST" action = "listAlbum">
    <s:textfield name = "imgTitle" id = "imgTitle" key = "label.text.albumTitle" /><br />
    <s:submit type = "input" value = "%{getText('label.text.search')}"/>
</s:form>
```

13.10 测试

发布 Web 项目 Chapter13,启动 Tomcat 服务器和 MySQL 服务器,访问 index.jsp,结果如图 13-1 所示,登录成功后结果如图 13-2 所示。

13.10.1 上传图片

登录成功后,单击"上传图片"链接,结果如图 13-6 所示。

输入信息,单击"保存"按钮后仍返回如图 13-6 所示界面,这样可以上传多张图片。

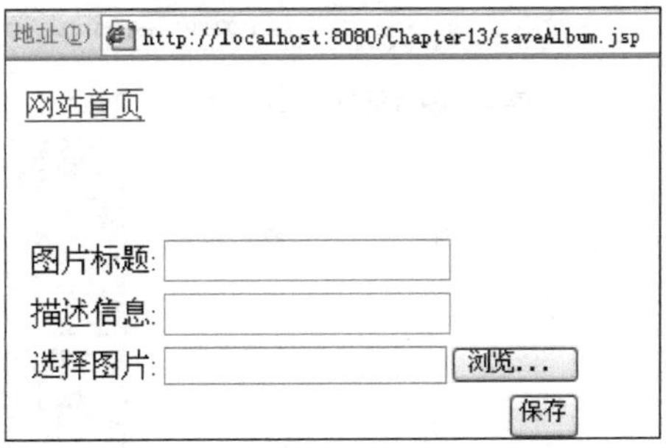

图 13-6 上传图片页面

13.10.2 显示图片

回到网站首页,单击"图片列表"链接,结果如图 13-7 所示。

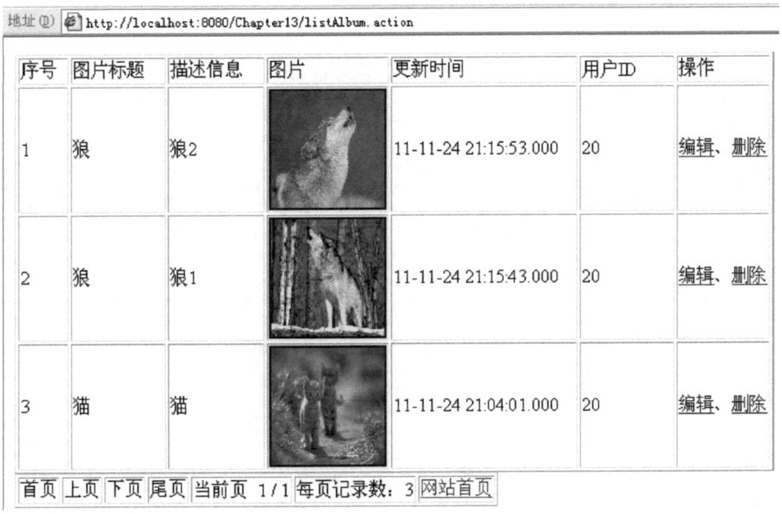

图 13-7 显示图片页面

13.10.3 查找图片

回到网站首页,单击"按标题查找"链接,结果如图 13-8 所示。

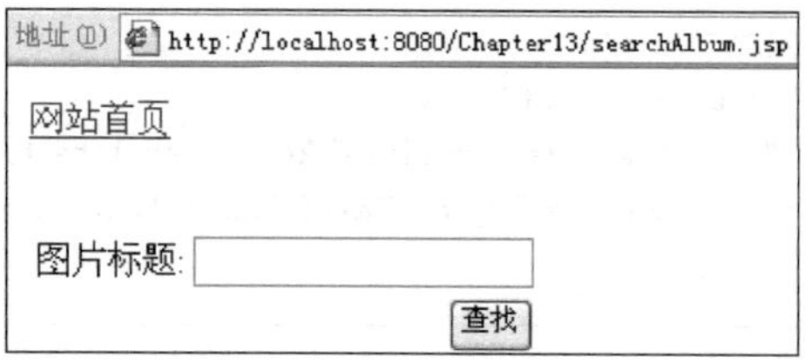

图 13-8 查找图片页面

输入"狼",结果如图 13-9 所示。

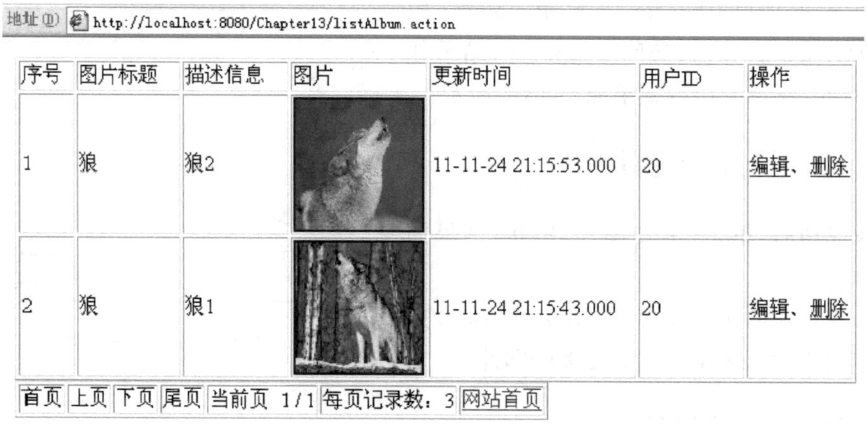

图 13-9 查找图片的结果

参 考 文 献

[1] http://struts.apache.org/2.2.3/index.html.
[2] 闫术卓,杨强.Struts 2技术详解:基于WebWork核心的MVC开发与实践.北京:电子工业出版社,2008.
[3] 孙鑫.Struts 2深入详解.北京:电子工业出版社,2008.
[4] 白广元.Java Web整合开发完全自学手册.北京:机械工业出版社,2009.
[5] 杨涛,王建桥,杨晓云译.深入浅出Struts 2.北京:人民邮电出版社,2009.
[6] 王建国,王建英.Struts+Spring+Hibernate框架及应用开发.北京:清华大学出版社,2011.
[7] http://commons.apache.org/fileupload/.

图书资源支持

感谢您一直以来对清华版图书的支持和爱护。为了配合本书的使用,本书提供配套的素材,有需求的用户请到清华大学出版社主页(http://www.tup.com.cn)上查询和下载,也可以拨打电话或发送电子邮件咨询。

如果您在使用本书的过程中遇到了什么问题,或者有相关图书出版计划,也请您发邮件告诉我们,以便我们更好地为您服务。

我们的联系方式:

地　　址:北京海淀区双清路学研大厦A座707

邮　　编:100084

电　　话:010-62770175-4604

资源下载:http://www.tup.com.cn

电子邮件:weijj@tup.tsinghua.edu.cn

QQ:883604(请写明您的单位和姓名)

用微信扫一扫右边的二维码,即可关注清华大学出版社公众号"书圈"。

扫一扫
资源下载、样书申请
新书推荐、技术交流